# MAKING ART SAFELY

...

## Alternative Methods and Materials in Drawing, Painting, Printmaking, Graphic Design, and Photography

### MERLE SPANDORFER
### DEBORAH CURTISS
### JACK SNYDER, M.D.

VNR VAN NOSTRAND REINHOLD
—————— New York

*To my husband, Lester Spandorfer; to my parents, Bernice Bank Altschul and Simon Louis Bank; and to Cathy, John, Amy, and David.* **M.S.**

*In loving memory of Lucy Glick, Frances M. Cox, and all artists who may have learned the hard way.* **D.C.**

*To Michele, Jacqueline, and Danielle.* **J.S.**

Copyright © 1993, 1996 by Van Nostrand Reinhold

I(T)P™  A division of International Thomson Publishing, Inc.
The ITP logo is a trademark under license.

Library of Congress Catalog Card Number 92-4841 (Hardcover)
ISBN 0-442-23489-9 (Hardcover)
ISBN 0-442-02131-3 (Paperback)

Printed in Mexico

For more information, contact:

Van Nostrand Reinhold
115 Fifth Avenue
New York, NY 10003

Chapman & Hall GmbH
Pappelallee 3
69469 Weinheim
Germany

Chapman & Hall
2-6 Boundary Row
London
SE1 8HN
United Kingdom

International Thomson Publishing Asia
221 Henderson Road #05-10
Henderson Building
Singapore 0315

Thomas Nelson Australia
102 Dodds Street
South Melbourne, 3205
Victoria, Australia

International Thomson Publishing Japan
Hirakawacho Kyowa Building, 3F
2-2-1 Hirakawacho
Chiyoda-ku, 102 Tokyo
Japan

Nelson Canada
1120 Birchmount Road
Scarborough, Ontario
Canada M1K 5G4

International Thomson Editores
Campos Eliseos 385, Piso 7
Col. Polanco
11560 Mexico D.F. Mexico

2  3  4  5  6  7  8  9  10   RRD/DUN   01  00  99  98  97

Library of Congress Cataloging-in-Publication Data

Spandorfer, Merle, 1934–
    Making art safely: alternatives in drawing, painting, printmaking, graphic design, and photography / Merle Spandorfer, Deborah Curtiss, Jack W. Snyder.
        p.      cm.
    Includes bibliographical references and index.
    ISBN 0-442-23489-9
    1. Artists—Health and hygiene.   2. Artists' materials—Safety measures.
I. Curtiss, Deborah, 1937–. II. Snyder, Jack W. III. Title.
RC963.6.A78S62   1992                                    92-4841
363.11'976—dc20                                          CIP

# Contents

# 3. The Ideal Studio

# 4. The Healthy Artist

# 5. Drawing

# 6. Painting

# 7. Printmaking

# 8. Graphic Design/ Illustration

# Preface

Our intent in writing *Making Art Safely* is threefold:

◆ to call attention to the growing evidence of dangers in art materials

◆ to present alternative, safer materials and procedures for the two-dimensional art media

◆ to recommend precautions to avoid injury from toxic substances that are present in some art materials.

We, as two artists and a toxicologist, have written this book for both professional and recreational artists, art students, and art educators. With information that is daunting in both scope and implications, we have attempted to be as "artist friendly" as possible—to use language in the way that artists, rather than scientists, communicate. Additionally, whenever appropriate we augment data with artwork and comments provided by artists from around the world.

We hope *Making Art Safely* will have many uses and applications and will be accessible to artists and students working independently as well as in art schools and in the workplace. It should be pertinent for art educators responsible for training students at all levels. For example, they may adapt information about protective attire to establish dress and hygiene codes for the art room or studio. All readers are encouraged to apply information and techniques to their own work, workplaces and educational contexts.

This is not, however, a handbook or manual on how to do drawing, painting, printmaking, graphic design or photography. We assume that readers and users of this book have familiarity with, or are learning, the techniques of these media. While we wish we could have addressed all art media and techniques, there is so much information relevant to safety in the two-dimensional arts that we felt it necessary to limit our scope. Artists who work with ceramics, fibers, glass, metals, plastic, stone and wood, however, will also find much useful information in this book.

Here's to your Health and Creativity!

# *Introduction*

We have divided *Making Art Safely* into ten chapters. Chapters 1 through 4 address issues of art hazards and the precautions that are relevant to working with all art media. Chapters 5 through 9 are identified by the specific media that are addressed. A concluding chapter looks ahead to the implications of making art safely, and is recommended reading at any time.

Chapter 1, "Facing Facts," gives historical examples of the havoc art hazards have wreaked, describes how we are vulnerable, and gives an overview of the primary dangers in specific art materials and how they threaten injury. The hazardous substances artists are likely to encounter are identified.

Chapter 2, "Staying in the Know," provides resource information and identifies art materials that are nontoxic. We describe the consumer's (and artist's) Right to Know, present the Hazard Communication Standard (HAZCOM) and Labeling Act of 1988, and discuss how to read labels and use Material Safety Data Sheets. The chapter concludes with a section on the importance of health and safety programs in schools and workplaces.

Chapter 3, "The Ideal Studio," recommends studio upkeep, safe studio practices, appropriate and adequate ventilation, safe storage and disposal of hazardous materials, and fire prevention for both individual and institutional studios. This includes descriptions of the Occupational Safety and Health Administration enforceable standards for ventilation, first aid, and eye wash stations in studios that have employees, and the Hazard Communication Standard of 1983, which mandates that employers with ten or more employees provide hazard education programs to inform workers and students about the materials they encounter in the workplace and studio. This chapter includes guidelines which allow artists and students who work in commercial or educational studios to evaluate the safety of their work environments, and compliance with these standards.

Chapter 4, "The Healthy Artist," discusses individual susceptibility, and the basic precautions that all students, artists, designers and art educators should take in their active creative repertoires. This includes personal hygiene, protective attire for use with hazardous materials, as well as emergency first-aid. We do not wish to alarm artmakers unnecessarily, but because hazards in art materials and procedures are real, and many are hidden, we describe possible illnesses and when to seek medical attention.

Chapters 5 through 9, respectively, identify hazards associated with drawing, painting, printmaking, graphic design and photographic materials and processes, with primary focus on the positive aspects of safe alternatives and procedures. Each chapter includes visual examples made by various artists who vividly demonstrate our basic premise: that safety in artmaking need not compromise quality. Guidelines for handling hazardous materials with optimum safety are included in each of these chapters.

The conclusion, "Looking Ahead," contains thoughts that have occurred to us in the process of compiling the information herein and about the implications of safe artmaking for the future.

The Bibliography lists references we consulted in the process of writing *Making Art Safely.* Appendix I, "Art Material Chemicals," lists toxicity, health effects, uses, precautions and substitutions of chemicals used in making art. Appendix II is a directory of information and help resources for health and safety in the arts. Appendix III provides the names and addresses of suppliers of safety equipment and attire. Appendix IV lists acronyms used in *Making Art Safely* and in the world of occupational health and safety. Finally, there is an index, and following this a section on emergency first-aid procedures.

In summary, our purpose is to equip and empower artists and art educators with information about the development of nonoffending materials and safer ways of handling hazardous materials and procedures. We hope that *Making Art Safely* will enable artists to continue to produce their art in ways that will reduce risks to their health, save unnecessary expense by avoiding hazardous materials, and, at the same time, offer alternatives that will expand their creative repertoires.

## How This Book Came to Be Written

**A**s an artist and teacher of painting, printmaking and photography, I have worked for many years with an extensive range of art materials. This has included painting with oils, watercolors, and acrylics; printmaking including silkscreen, collagraphs, etching, and lithography; and working with both silver and nonsilver photographic methods. Rarely was I concerned that my art materials might be hazardous,

for the spiritual and philosophical content of the work was uppermost in my mind. Whatever techniques I used were governed by the image I wished to create, and I was willing to employ any means necessary to achieve that goal. Safe materials and procedures had no priority.

As a teacher, however, I felt a responsibility to educate my students about hazards and safeguards in using art materials. Due to my initiative, the Temple University, Tyler School of Art, silkscreen studio was converted from oil-based to the less toxic water-base system. Meanwhile, in my own studio, I worked without proper ventilation or safeguards. Why change? I was healthy and satisfied with my work.

Then one day a malignant tumor was discovered in my breast that required surgery and weeks of radiation therapy. Due to early discovery of the cancer, the prognosis was for a full recovery. Although I was shocked and weakened by the experience, I continued to teach and even had a solo exhibit during this period.

At the end of the radiation treatment, I was stricken again, this time with pneumonia in both lungs. I was rushed to the hospital in critical condition. The doctors could not determine what caused the pneumonia or why I was hit so severely.

While the cause of the cancer and pneumonia was not known, it was felt that exposure to hazardous art materials, without proper safeguards, may have been a factor in my susceptibility to illness. Of particular suspicion was ammonium bichromate, a chemical formed of crystals and a known poison, which I had used regularly and often in non-silver photographic processes. A biopsy had, in fact, revealed the presence of unexplained crystal deposits in my lungs.

Five years have passed since the original diagnosis. My health is excellent, and I treasure it. During the first year of convalescence, I learned that eight artist friends were diagnosed as having cancer; two have succumbed to the disease. My illness, combined with these subsequent events, caused me to indelibly associate health risks with making art.

During my illness I had time to reflect. Could I have avoided all this? Why didn't I practice what I taught? Why are so many of my friends fighting cancer? Gradually, the questioning evolved into something more positive: there must be a way to alert artists to the hazards they are exposed to and to develop procedures that are safe.

As I resumed my artwork and teaching, it was with a different spirit. I dedicated myself to adopting safety measures and to discovering alternative, safe techniques that would allow me to achieve the quality and creativity I felt was central to making art. My students and I were rewarded with results that far exceeded our expectations, and I feel personally that I am doing my best work ever. As my discoveries and excitement grow, I search for ways to reach and educate a broader audience of artists to the hazards that I ignored, to share working experiences, and show examples of how excellent art can be made safely.

**Merle Spandorfer**

Late in 1987, Merle Spandorfer called me to discuss an idea. I'd known Merle casually for a number of years as another artist active in the art world, and had used some examples of her prints and printmaking in my book, *Introduction to Visual Literacy.* She wished to write a book that would empower and educate artists about the hazards of art materials.

Since my first year in art school in the late 1950s, I experienced sensitivity to a number of art materials and procedures. Vapors and odors of printmaking seemed to affect my breathing, to give me a vague sense of unease. The distant odor of turpentine seemed romantic and exciting, but close up, solvents caused my eyes and nasal passages to burn. As a result, I experimented with mediums and emulsions that were less irritating and used laundry soap (and elbow grease) instead of turpentine to clean my brushes and hands.

In my second year we pioneered acrylic paints before they were available in tubes. This required ordering powdered pigments in five-pound bags and acrylic mediums in 55-gallon drums. Ammonia vapors from distributing the mediums were so unpleasant that, thinking it was only my own hypersensitivity, I asked fellow students to fill my jugs and then left the studio. Likewise, with all offending materials since, I have taken a wary path and worked with, what are to me, the least irritating and, I hope, least dangerous art materials.

In the meantime, as a teacher of drawing and painting, including oils, at Philadelphia College of Art (now The University of the Arts), I felt responsible to my students to be informed about the growing concern over potential hazards in the materials they used. I frequently consulted Michael McCann's *Artist Beware* and *Art Hazard News* from the time they were first published and recommended to my students to do the same.

Because Merle knew my book and did not feel comfortable about writing herself, she thought of me as a potential collaborator. Despite my early reluctance, by the time the proposal was written, my name was on it as a committed co-author. It was agreed that she would coordinate the project, gather the information and illustrations, and I would concentrate on writing.

Before completing *Making Art Safely,* I had joined with a group of artists in buying a 1919 factory building for rehabilitation for an artist cooperative of living and studio spaces. From the outset our plans included an agreement to eliminate as many potentially hazardous materials as possible prior to moving into the building, to install pertinent effective exhaust ventilation in each studio, and to use the safest materials and procedures possible in our work henceforth.

Merle and I knew we would need a consulting toxicologist, and through her son, then a medical student at Thomas Jefferson University, Merle found Jack.

**Deborah Curtiss**

When Merle Spandorfer told me about a book she was preparing on non-toxic art and design, I was immediately interested. I had just completed several clinic sessions evaluating two artists with problems related to their work with art materials. These were rather simple situations; both were resolved by educating their employers about the inappropriate uses of solvents in poorly ventilated work areas. As a physician, toxicologist, educator, and art appreciator, this book is an opportunity for me to communicate a type of knowledge that I feel is important to artists and the quality of their lives.

As our society continuously nurtures, develops, and maintains creative artmaking, this effort increasingly requires enlightened approaches to healthy work habits and environments for artists. The work of McCann, Rossol, Waller, and others has clearly enhanced the ability of art professionals to recognize and respond appropriately to potential health hazards in their workplaces. More of these efforts are forthcoming, and all contribute to a much needed awareness among artists for protecting themselves. A growing network of health professionals, including chemists, toxicologists, occupational medicine specialists, industrial hygienists, and poison control centers, can now provide information relevant to chemicals and materials used by artists, who must learn to use these resources effectively.

Although fragmentary knowledge of human health effects will always characterize some aspects of artmaking, recent developments indicate that many products are quite safe for both short- and long-term use by artists. It is important to bear in mind, however, that much is yet to be learned about health hazards in art. Moreover, complete removal of all risks to health and safety is unrealistic, and no document about art safety will ever be definitive. These uncertainties suggest that prudent or "conservative" approaches to making art are appropriate.

It is encouraging, however, to observe that new information is constantly becoming available that is potentially useful to how artists work and create, and we know enough to act on this developing information. As artists aggressively seek information to become more effective participants in the "information-gathering society" of which they are such a vital and integral part, they can strike an appropriate balance between concern for human health and safety, and unencumbered artistic creation.

It is with deep personal satisfaction that I contribute my time and effort to this needed book.

**Jack Snyder**

As should be evident from our accounts, we represent different perspectives on the issues of making art safely, a complementarity that has been enriching for us as we have worked together to create this book. We sincerely hope that *Making Art Safely* will be similarly enlightening and inspiring for a multitude of readers and artists. While we have made every effort to be accurate and current with regard to the safest approaches to making art, the nature of the topic prevents us from being definitive. Further:

## WARNING

The authors and publisher take no responsibility for any harm that may be caused by the use or misuse of information contained in this book or by the use or misuse of any materials, techniques or processes suggested herein. Moreover, it is not the purpose of this book to provide medical diagnosis, suggest individualized treatment, or provide regulatory or legal counsel.

Here's to your health!

# Acknowledgments

*Making Art Safely* would not exist without the generous contribution of many individuals. Special appreciation is extended to the hundreds of safety-conscious artists who generously contributed visual examples of their work accompanied by personal statements describing their techniques and studio safeguards. These contributions demonstrate that quality of artwork can be maintained while making art safely! Encouragement and enthusiasm was received from many, including Ted Rickard, Director of Health and Safety at The Ontario College of Art, Milton Glaser, and Nathan Oliveira. Due to topic and space constraints it was not possible to include all submissions. We have avoided the use of quotation marks with the artists' statements since in many instances some amount of editing was undertaken, although every effort was made to retain the spirit and voice of the contributor.

Technical guidance and encouragement was provided by Dr. Michael McCann, Executive Director of the Center for Safety in the Arts, Monona Rossol, President of Arts, Crafts, and Theater Safety, and Deborah M. Fanning, Executive Vice President of the Art and Craft Materials Institute. Early support was provided by a technical assistance grant from the Pennsylvania Council on the Arts, David Stephens, Director, administered through The Cheltenham Center for the Arts; thanks are due to Marie Alexander for personal and administrative support throughout the project.

We are greatly indebted to Cathy Spandorfer, who provided and edited much of the material on graphic design. Stephanie Knopp and Alison Carson also made important contributions to this chapter. John Spandorfer, M.D., made a key contribution: he introduced the artist authors to Jack Snyder, M.D. Individuals who offered expert advice on specific sections included Anthony Batchelor and Timothy Sheesley on printmaking; and Bruce Katsiff, David Porter, and Thomas White on photography. Important support was provided by Dr. Amy Jordan, Jerrold V. Moss, Lewis F. Gould, Jr., Marsha Moss, Lily Yeh, Luther W. Brady, M.D., and Dov Leis. Thanks are also due to those unnamed individuals from the art, toxicology, and occupational health communities who critically reviewed portions of the manuscript. Our thanks also go to Geri Biscardi of

Thomas Jefferson University, who provided word processing assistance.

Numerous individuals at galleries and museums kindly supported our undertaking. Worthy of special mention are Ann Percy, Curator of Drawings at the Philadelphia Museum of Art, and Sande Webster, Sande Webster Gallery, Philadelphia. Support was also provided by The Metropolitan Museum of Art, O.K. Harris Gallery, New York, Crown Point Press, San Francisco and New York, and Benjamin and Deborah Mangel, Mangel Gallery, Philadelphia.

None of us worked in a vacuum, and personal, loving appreciation is due to Hugh Sutherland, Margaret Morley, Leslie Smith, Teresa Benzwie, Gretchen and Jeremy Klotz, and Oddy Curtiss; Michele, Jacqueline and Danielle Snyder; and Todd H. Freedman, Bernice Altschul and Lillian Spandorfer.

Special thanks go to Dr. Lester Spandorfer for his unflagging encouragement, technical assistance on the computer technology and electromagnetic field–related material, and critical readings of the manuscript during its various stages.

Finally we extend our gratitude to Lilly Kaufman, Amanda Miller, Ken Allen, Monika Keano and all at Van Nostrand Reinhold who have enthusiastically and patiently supported us in bringing *Making Art Safely* into being and taking it onward into the world.

A rtists of the twenty-first century will work differently after learning about the potential hazards in making art. Dangers in art materials and procedures are not new, but have existed for centuries. Recently, however, growing holistic consciousness about health, fitness, and the

# Facing Facts

environment in which we live and work has contributed to increased awareness about all aspects of our lives. During the 1980s especially, toxic substances in our environment and in our food became significant personal, public, and political concerns. While we do not and apparently cannot live in a risk-free world, there is much we can do to minimize risks associated with art materials and procedures.

◆ ◆ ◆

A number of books and articles have been written about health risks in art materials, and the reaction of artists has often been an ostrichlike "I don't want to hear about it." No one wants to hear bad news, especially when it threatens to stop or interrupt the way we do things. When as artists we achieve a technique that serves our vision, we tend to invest ourselves in it and want to feel free to explore it fully. We may even feel that attention to potential hazards will hinder both our development as artists and the appreciation by others of our artwork's content and expression.

Another factor in artists' refusal to acknowledge hazards may be that they are unaware of creative, nonencumbering substitutes for the techniques and media with which they are familiar. In *Making Art Safely* we present art materials and processes—with which practicing artists Spandorfer or Curtiss are familiar—that are safe, or at least safer than traditional methods. Materials and procedures are presented with due regard for high-quality artwork and individual creativity. Just as dieters do not lose weight when they feel they are depriving themselves, we know that artists will be reluctant to change their ways of working if it threatens to limit standards of quality and expression.

A third aspect of resistance to change is habit, those mini-rituals we establish in our work. Resistance may also spring from what some artists describe as a mystical relationship with art materials, and the physical or emotional effects they may have upon us.

As we learn that art has an effect on health as well as on the pocketbook, both individuals and institutions are more likely to change. A number of art schools know that limiting the use of solvent-based inks and paints and replacing them with water-soluble products can save thousands of dollars per year. Ventilation requirements are reduced, and waste disposal procedures are simpler and cheaper. The savings can be significant. By changing to water-based inks, the screen printing studio at the Art Academy of Cincinnati saved $11,000 in safety equipment costs (Bachelor and Reed 1984). Similarly, by switching to water-soluble inks, a gift wrap company in Tennessee cut its industrial waste expenses by $35,000 per year (Caplan 1990). What artists, studios, and institutions will save in improved employee and student health cannot easily be measured, but we predict reduced sick leave, fewer health payments, fewer workers' compensation claims, and improved performance.

Concern for the health and well-being of artists has increased due to expansion of art into all levels of education. In addition, art remains a popular av-ocation among persons of all ages and walks of life. According to a poll by Louis Harris and Associates, one of four Americans routinely encounters art materials or processes. Whether artist or hobbyist, educator or student, too few know the potential dangers that exist in many of the supplies we enthusiastically employ. It is essential that we learn more about and develop respect for the materials and procedures we use in the creation of art.

Some individuals may have experienced an immediate reaction to an art material, such as skin rash, nausea, or dizziness that could be readily treated and subsequently managed or avoided. Of growing concern are ill effects that are insidious and slow to develop. These effects may require a number of years to manifest themselves. Delayed reactions, known as long-term health effects, carry serious implications for the health and well-being of all artists, at all ages, and in all media.

Potentially hazardous products are those that have one or more of the following properties:

◆ Toxic: can cause physical injury via inhalation, ingestion, or absorption through the skin

◆ Caustic or corrosive: may burn living tissue on contact

◆ Irritant: can cause skin, mucous membrane, and eye irritation, inflammation, or soreness

◆ Flammable: can be ignited or set on fire at any temperature

◆ Explosive: may explode when exposed to heat, pressure, or shock

Hazards in art materials have threatened artists for ages. Potential problems are documented with increasing frequency due to advances in science and medicine. Nevertheless, many art-making procedures are so firmly established that most of them, no matter how hazardous, remain in frequent use. This is of special concern because many years are needed to identify and regulate toxic substances. Meanwhile, individual artists, art educators, and employers of artists must aggressively seek information to protect themselves as well as their families, students, and employees.

A growing number of laws and regulations in the United States, Canada, and elsewhere pertain to the manufacture of art materials and dissemination of art hazard information in the workplace and classroom. The U.S. Department of Labor's Occupational Safety and Health Administration (OSHA) enforces "right-to-know" laws, which require employers and teachers to provide a safe environment and to notify employees of any hazards in the workplace or classroom. This is addressed further in chapter 3.

Regulation of harmful substances lags far behind their use because the potential for hazard must first be recognized, evaluated, and confirmed before legal action is commenced. Once regulations are promulgated, enforcement is likely to be delayed while staffing, investigation, and verification are undertaken and charges are brought in the due process of law. As a result, it can be years from the time a hazard is suspected until protective restrictions are actually applied.

No work on the safety of art materials can be definitive for a number of reasons. The combination of new materials and growing knowledge creates an atmosphere of flux and uncertainty. To date, few studies of health hazards in art materials have been made, for artists seldom work together in numbers large enough to warrant such research. We are also at risk because many artists work long hours in places not covered by state and federal occupational safety standards. Our personal studios may not be equipped for prudent use of hazardous materials.

We must assume individual responsibility for our health and well-being through effective, creative action. Make time to assess work habits and to identify and reorganize materials in the studio. Procedures for evaluation of studio and personal care are provided in chapters 3 and 4.

## Dangers in the Arts Are Not a Twentieth-Century Phenomenon

**F**or centuries, ignorance of health hazards in the arts has had serious consequences. Forensic historians suggest that many of our old masters may have experienced illness and even death because of the materials they used. In 1700 Bernadino Ramazzini, a Venetian physician, published *Diseases of Workers*, where he observed that "painters are attacked by various ailments such as palsy of the limbs, cachexia [emaciation and general ill health], blackened teeth, unhealthy complexions, melancholia and loss of the sense of smell. It very seldom happens that painters look florid or healthy" (Karlen 1984).

Though he lived to age 82, Francisco Goya (1746-1828) suffered recurring episodes of excruciating illness. At 46 he suffered "an agonizing and mysterious illness [that] almost killed him, left him stone deaf, and transformed his personality and art" (Kar-

len 1984). Some believe Goya's malady was syphilis, for he was known to live a free-spirited life and had had a similar incident at age 32. Goya did not, however, experience the progressive deterioration associated with syphilis; he was robust, vital, and creative between his illnesses and remained active until his death following a stroke.

Because Goya's illnesses were well documented, medical historians have more recently attributed his sickness to lead poisoning, also known as plumbism. The symptoms may include paralysis of the limbs, tremor, coma, delirium, transient blindness, mental confusion, depression, paranoid thinking, and vertigo. Recurring depression, impaired hearing and vision, and personality changes are also observed, as are above-average rates of abortion, stillbirth, and fetal damage. While recuperating from illness, Goya was kept away from his paints, so he would gradually recover most of his strength. With the return to painting, however, the levels of lead would gradually accumulate, eventually provoking another period of prostration.

In the eighteenth century, use of highly toxic lead and mercury pigments—which today are known to be hazardous—was widespread. Twentieth-century analyses of Goya's paintings has confirmed his massive use of lead white—which gave his paintings their pearlescent luminosity—along with lead reds and mercury-containing cinnabar. Anecdotes from Goya's time suggest that he "scooped his color out of tubs, applied it with sponges, mops, rags, anything he could lay his hands on. He trowelled and slapped his colors on like a bricklayer, giving characteristic touches with a stroke of his thumb." Moreover, he ground the pigments himself and mixed them with solvents, "in short, spending his life in a storm of flying paint" (Karlen 1984).

When lead is ingested, less than 10 percent is absorbed by the adult intestines. When it is inhaled as dust, as from powdered pigments or dried housepaints used in old buildings, or inhaled as a volatile blend with linseed oil and turpentine, 30 to 60 percent of lead is retained by the body. Lead is stored in bones, which provide both temporary and permanent reservoirs. At some point, the bones can hold no more, and the lead enters soft tissues such as the aorta, brain, kidney, and liver. This body burden may lead to symptoms and signs of disease.

Goya is not the only historical example of health problems associated with art materials. Peter Paul Rubens (1577–1640), Pierre-Auguste Renoir (1841–1919), and Raoul Dufy (1877–1953) suffered from crippling arthritis. All were known to use pigments from toxic metals such as antimony, arsenic, cadmium, cobalt, chromium, lead, manganese, and mercury. Although only lead has been linked to vague musculoskeletal conditions, these metals, taken into the body, can inhibit enzymes, destroy proteins, and increase susceptibility to infections.

Paul Klee (1879–1940), who also used metal pigments regularly, suffered from scleroderma, a sometimes crippling disease of unknown cause that thickens the skin and can affect joints and internal organs. Some suggest that his disorder was aggravated by the pigments he is known to have used.

Vincent van Gogh (1853–1890) loved to drink camphor, turpentine, and absinthe. His cravings were so intense that he ate his paints and drank turpentine at the same time. These ingestions may have contributed to mental derangement and his early demise through suicide (Arnold, 1988).

## The Implications of
## Art Hazards

**L**earning that art can be hazardous hurts.

As children, many of us were introduced to art materials by parents and teachers who believed that creativity in the arts was personally, educationally, and culturally enriching. From this initiation, we concluded that making art is not only harmless but good for mind, body, and soul.

In high school, college, or art school, we were encouraged to be absorbed by art, taught to immerse ourselves in it mentally, spiritually, emotionally, and physically, to dig into it, become one with it, and become it. We engaged wholeheartedly, not knowing the potential risks.

It is difficult to acknowledge facts that contradict our ingrained sense of invincibility. Our rude awakening may understandably incite resentment, feelings of betrayal, and mistrust in basic art principles.

Just like other mid-life adults, artists develop cancer or other suspicious ailments such as joint pain, headaches, or slowed thinking. In a world committed to better living through chemistry, it is reasonable to ask whether excessive exposure to art materials may contribute to our health problems. If some art supplies contain potential health hazards, we may wonder why they have not been more appropriately labeled with warnings? Why is government regulation of substances developed only when there is an obvious menace, and sometimes not even then? Some of our anxiety may be reasonable.

Gradually, and with reluctance, we begin to question our assumptions and practices. Have we been too busy to read labels? If we read them, do we understand them? Are they relevant to our use of the material? What is meant by "adequate ventilation"?

Perhaps we opened a window that was better at blowing noxious vapors in our faces than removing them. If we used a fan, was it any more effective than opening a window?

Regarding protective masks, we may have used a readily available dust mask instead of the cartridge mask that would protect us from the harmful vapors of acids and solvents.

In the absence of easily accessible information, we wonder where to turn or what to do.

These are some of the issues addressed in *Making Art Safely*.

## Artist Vulnerability

**E**ven though we now know that harm to our health can arise from some of the art materials we use, *how* damage occurs is often unclear. As long as we remain ignorant of the ways toxins enter and affect our bodies, we will feel powerless to do anything about them. Facing the facts involves learning about:

◆ Forms toxic materials may assume

◆ How toxic materials can enter our bodies

◆ The reactions we may experience

◆ The risks we are taking

## THE PHYSICAL STATES OF ART MATERIALS

Like all physical matter, art materials exist in the three basic states: solid, liquid, and gas. Each state may convert from one to another.

While unlikely to be harmful in their natural state, **solids** can become hazardous if we change their form. When sanded or ground, solids become a dust, and if ground finely enough—particles smaller than ten microns in any dimension—the solid becomes a powder that is respirable: it can be inhaled into the respiratory tract and absorbed into the body. When heated, solids such as metals may become liquid, causing them to emit fumes (metal particles suspended in the air) or gaseous vapors. Fumes and vapors are both respirable and ingestible. A few solids, such as naphthalene (mothballs), naturally give off gaseous vapors with exposure to the air.

Some **liquids** may likewise emit gaseous vapors when exposed to air, or they may become gaseous through heating or chemical reaction. Through physical manipulation such as agitating, atomizing, foaming, splashing, or spraying, liquids may convert to a mist (liquid particles suspended in the air). Mists are both ingestible and inhalable. Liquids may be frozen or become dry by evaporation or oxidation, and thereby convert to a solid state, which can then be reduced to dust.

However they are created, **gases** are formless and will disperse at room temperature to occupy any space available. All gases readily enter the body through inhalation. While some are detectable by odor or color, the presence of odorless and colorless gases (such as oxygen and carbon dioxide) can be determined only by testing for their presence.

## HOW TOXINS
## ENTER THE BODY

Toxins can enter the body via three primary routes:

- **Ingested** into the digestive system
- **Inhaled** into the respiratory system
- **Absorbed** through the eyes and skin

Through all routes of entry, substances may enter our bloodstream to be circulated throughout the body. They may eventually be excreted with no effect and therefore be considered nontoxic, or they may deposit in one or more organs that respond adversely to specific chemicals. Either immediately or at some later time sufficient adverse effects may lead to recognizable illness or disease.

**Ingestion** Ingestion via the digestive system is the most controllable route of entry of harmful substances. Children are prone to ingestion, especially young children who instinctively put things in their mouths.

In the studio, adult artists working intensely can be poisoned by ingestion of contaminated food or drink. This often results from inadequate cleanup after working or before eating. In a recent survey of 300 artists in the New York City area, half of them said they ate while they worked. An artist in San Francisco, who ate while she worked, got paint on her sandwiches. She eventually developed mercury poisoning attributed to both inhalation and ingestion of mercury preservatives in the commercial paint she used (McCann 1989).

Policies that prohibit food and drink in the studio and art materials in the kitchen or dining facility can reduce the likelihood of inadvertent intake of harmful matter. For example, some foods may actually attract harmful vapors. Hand-to-mouth contact, pointing paint brushes with the lips, chewing pencils, or tasting to identify chemical solutions are undesirable habits in the studio. The threats of these practices to health are well documented (McCann 1979). For example, excess bladder cancer among Japanese silk kimono painters who worked with benzidine dyes and pointed their brushes with their lips sounded a tragic alarm about dangers in art materials. Ingestion can also inadvertently follow inhalation. For example, inhaled dust is trapped by the cilia and mucous membranes of the upper respiratory tract. As part of normal processes in the nose, throat, and mouth, mucus drains into the throat and is swallowed, carrying the particles into the digestive system. This is one of several important reasons to use a respirator when working with dusts and powders (see chapter 4). When dust is coughed up in phlegm, it should not be swallowed, but spit out. Because dust tends to cling to almost everything, including our clothing, hair, hands, face, and lips, it is a significant concern in making art safely. Licking lips and unwashed hands can carry dust readily into our digestive system.

**Do not eat, drink, or smoke in the studio.** That is the ideal. The reality is that we get hungry and thirsty while working. If you can relocate to eat, do so. When drinking water while working, for example, be conscientious about the cleanliness of both the container and your hands.

**Wash hands thoroughly after working and before eating.** Use soap and water and a scrubber if required. Avoid solvent cleaners such as turpentine or mineral spirits. Substitute a non-solvent-based waterless cleaner or baby oil, if necessary. Follow washing with a protective hand cream to prevent drying and cracking of the skin.

**Do not work with art materials in food preparation or dining facilities.** With a combination of sink and relatively impervious countertops, the kitchen can be convenient for art production activities. The dining room is also useful, especially for those continuing to create while raising children. If possible, do not mix food and art activities. Create a studio in the basement, attic, or a spare bedroom.

**Keep hands and tools away from your mouth while making art.** Pointing brushes with the lips and holding pencils, markers, and loaded paint brushes in the teeth have contributed to health problems in artists. In moments of contemplation, unconscious hand-to-mouth activities can also contribute to undesirable exposure.

Consider whether some of your work habits need to be altered. If so, tie a string on your finger, post signs, or do whatever is necessary to interrupt unsafe practice.

**Inhalation** Inhalation is the most common way that toxins are taken into the body in the workplace and studio. Dusts, fumes, gases, mists, and vapors are all easily inhaled. The respiratory system includes the nasopharynx (nose and throat), trachea (windpipe), bronchial tree, and the lungs, which are filled with cavities and ducts that absorb, exchange, and expel the gases and vapors we inspire. It is usually in the rapid, highly efficient exchange of oxygen and carbon dioxide—the fundamental breathing process upon which our lives depend—that toxic gases, vapors, microbes, and dust particles can enter the body. While many such substances may be expelled immediately, others may be trapped in the lungs or pass through the lungs to the bloodstream and to other organs. By this route hazardous chemicals in sufficient quantity over sufficient time may cause long-term health problems.

Strong odors may signal danger, but as exposure continues, olfactory fatigue can occur, reducing both sense of smell and awareness of potential harm. Inhalation, with or without odor, may lead to local irritation such as a burning sensation in the nose or throat or a suffocating feeling in the trachea or bronchial tubes (lower neck and upper chest area). This can be caused by irritating gases like ammonia and chlorine, which can also act as bronchial constrictors. The production and accumulation of fluid secreted from the irritated mucous membranes that line the respiratory system, temporary loss of the sense of smell, sore throat, hoarseness, and heavy or labored breathing are also symptoms of respiratory irritation. These reactions, known as primary and acute, are most often provoked by water-soluble compounds, such as hydrochloric acid and sulfur dioxide, that tend to react with tissues of the upper respiratory tract. Aggravations from primary irritants are generally reversible with treatment and prevention but occasionally result in chronic, long-term health effects.

Secondary irritants are less soluble chemicals that are more likely to penetrate deeply into the lungs, producing systemic effects after absorption from the lungs. Toxic gases, such as nitrogen dioxide and ozone, can cause cell damage that increases permeability of the lungs to other agents.

Respiratory allergy can be another manifestation of exposure to an irritating chemical. After one, and perhaps several, exposures, no effect is evident, but at some later time a subsequent exposure can trigger a reaction. Once sensitized to a substance, reexposure stimulates the allergic response. Sensitization often persists and, in some cases, smaller and smaller exposures may provoke reactions. Respiratory allergies may entail bronchoconstriction, labored breathing and coughing associated with asthma, hoarseness, sneezing, watery eyes, and weeping mucous membranes, not unlike hay fever. Those with hay fever or asthma are more likely to develop additional respiratory allergies.

Inhalable substances that represent the greatest threats to making art safely include:

♦ Aerosol mists from paints and fixative sprays

♦ Dusts and powders from dry pigments and chemicals

♦ Gases from many acid and alkali printmaking and photo chemicals

♦ Metal fumes from molten metals (usually not encountered in the two-dimensional art media)

♦ Solvent vapors from lacquer thinners and turpentine

Repeated exposure to these respirable substances has been associated with bronchitis, emphysema, and pulmonary edema. Tobacco smoke, a recognized carcinogen, also increases vulnerability to injury from airborne chemicals in two ways. Toxins in the work environment adhere to smoke particles and are inhaled into the lungs. Smoking also suppresses the activity of cilia (hairlike projections on the surface of cells lining the respiratory tree) that are responsible for moving mucus and foreign matter, such as dusts, tars, and other irritating substances, out of the lungs.

Artists may encounter serious respiratory disturbance from two other classes of chemicals:

♦ **Anesthetics,** such as acetylene, acetone, ether/ethanol, methyl chloroform and toluene, may cause dizziness, unconsciousness, and, with high exposures, death.

♦ **Asphyxiants** deprive the body of oxygen. Some asphyxiants, such as carbon dioxide, nitrogen, helium, methane, and ethane, may not be inherently toxic, but in sufficient concentrations can prevent oxygen from gaining access to pulmonary gas-exchange units. This can result in unconsciousness and death. Other asphyxiants, such as carbon monoxide and hydrogen cyanide, interfere with the utilization of oxygen already inside cells and/or decrease the oxygen-carrying capacity of the blood. In sufficient concentrations, these asphyxiants can also cause coma and death. While life-threatening, the adverse effects, if treated promptly, can be reversed.

Decreased alertness, diminished fine motor activity, and headaches are early symptoms of possible anesthesia or asphyxiation and call for immediate flight from the toxic environment to fresh air and medical assessment.

**Leave studio clothing in the studio.** Studio attire may be contaminated with dusts and other undesirable substances. Don't carry clothes into the living environment. Launder them promptly and dispose of contaminants properly. This issue is further addressed in chapter 4.

**Absorption** The skin is a highly effective barrier to environmental stresses and most foreign substances (xenobiotics). As the largest organ and most exposed part of the body, however, our skin often comes in contact with some chemicals that can penetrate this barrier. Furthermore, the skin is especially vulnerable when broken with cuts or sores. Environmental agents can also enter the bloodstream if injected through the skin by either accident or careless use of contaminated equipment.

The most frequent and immediate reaction to chemical exposure is skin inflammation, commonly called contact dermatitis. An irritant, known as a primary skin irritant, may alter skin chemistry, incite inflammation, or occasionally kill cells. This may be manifested by a mild rash with only visual evidence, a rash with itching and burning, or more painful and persistent reactions that include swelling, drying, and cracking of skin. Acids and alkalis, present in many printmaking and photography chemicals as well as in household and industrial cleaning agents, are especially irritating and corrosive to the skin. Solvents, such as alcohols, turpentine, toluene, and petroleum distillates, cause drying of the skin and may remove its natural protective waxy coating, thereby increasing the skin's vulnerability to absorption.

Many solvents and photographic developers can penetrate unbroken skin through hair follicles and sweat and sebaceous glands to get into the blood. As a result, solvents and photochemicals should be used only with protective gloves and local exhaust ventilation. In the absence of immediate irritation, we may not recognize that permeation of skin and absorption into blood is taking place, perhaps laying the groundwork for later reactions and manifestations of illness. These substances have been associated with a broad range of both acute and chronic

ill effects, including damage to the nervous system, reproduction problems, anemia, and kidney disease.

Skin that is damaged from dermatitis, cuts, abrasions, sores, or burns provides increased access to the bloodstream for substances that may otherwise be unable to pass the natural waxy barrier of the skin. Therefore, take extra precautions to protect damaged skin from exposure to art materials. Remember that the eyes and surrounding skin are also vulnerable to vapors, mists, and splashes.

A generalized rash or skin lesion located away from the site of contact with offending chemicals may signal an allergic reaction to substances that have been ingested or inhaled rather than absorbed through skin.

**Practice and teach safety in making art.** The earlier we learn respect for materials and procedures, the easier it will be to create art effectively and safely.

## REACTIONS TO TOXIC SUBSTANCES

Now that we know how substances enter the body, it is appropriate to look at what they can do to us.

Many art products contain acids, alkalis, dusts, dyes, pigments, solvents, and sprays that are, to a degree, toxic; that is, if not used properly, they may have adverse health effects, which can be acute or chronic.

An **acute** or short-term effect occurs as an immediate reaction, minutes to hours, after exposure to an irritating substance. The cause is often apparent and the effects generally short-lived, although permanent damage, such as scarring, may occur. Examples of acute effects are skin irritation from corrosive acid or alkali solutions and metal fume fever, a flulike illness from inhalation of metal fumes.

A **chronic** or long-term effect may result from low-level exposure to offending chemicals over an extended period of time (weeks, months, even years). Chronicity may involve subtle, cumulative, and often irreversible effects from which the body may not fully recover. Chronic effects may occur when offending agents accumulate due to incomplete transformation or excretion from the body. Examples of chronic health effects include emphysema (associated with exposure to nitric acid etching gases), pneumoconiosis (dusty lung disease), and cancer.

Unlike an acute effect, a chronic effect may not be obvious. The onset of symptoms may be gradual, and as a result the symptoms may go unnoticed or may be explained away as "getting old" or "being run down." It is difficult to trace the cause of a chronic effect; by the time the link between exposure and symptoms has been made, permanent damage may have occurred. In some cases, this **latency period** could be ten to forty years after exposure. Once symptoms do appear, it is often too late for adequate treatment or cure.

The difference between acute and chronic reactions may be more readily understood if we consider the effects of ingesting excess alcohol. The acute reaction would be drunkenness, with symptoms of slurred speech, slowed response time, unsteady gait, vomiting, and even unconsciousness. The chronic reactions to alcohol are addiction (alcoholism) and liver damage (cirrhosis) (New Jersey Department of Health, ongoing studies).

The effects of chemicals on the body can be local or general. An example of a **local** effect is irritation or injury following contact with acid or alkali. An example of a **general** effect is damage to heart, intestines, liver, kidney, and nervous system following chronic exposure to arsenic.

Because cumulative effects of hazardous chemicals are tough to detect, primary prevention becomes most important. We must learn the hazards in art materials, especially those we use, and adopt precautions and work habits that protect us as much as possible.

**Carcinogens** Cancer-causing materials are classified in two categories. **Suspected carcinogens** are agents that have been shown to increase the incidence of tumors in laboratory animals but not (yet) in humans. **Known carcinogens** are those agents shown to increase incidence of tumors in both humans and lab animals. Some authorities (and regulators) believe there are no known safe exposure limits for carcinogens of either kind. The effects of some carcinogens, moreover, may not be related to the dose of exposure, but may be affected by environmental and personal characteristics. Because of the long latency period, the time and expense required for conclusive evidence, and the painful, costly process of treating cancer, the best approach is avoidance of all carcinogens, both known and suspected. The following list spotlights carcinogens that have been used in art material manufacture. Fortunately, due to regulation and growing awareness of the dangers, some of them have been eliminated from art products, but not all. For example, lead is still permitted in the manufacture of artist paints. Remember that this list is not definitive. New carcinogens are discovered as cancer research progresses.

- Arsenic oxide, asbestos, zinc chromate, nickel compounds, and uranium oxide are known or believed to increase the risk of lung cancer.

- Benzidine-based dyes[1] increase the risk of bladder cancer.

- Chlorinated hydrocarbons and petroleum distillates, included in many solvents, are known to cause diseases of the nervous system and may be associated with increased incidence of some tumors.

- The photochemicals ammonium dichromate and potassium dichromate powders, pyrogallol, 6-nitrobenzimidazole, thiourea, hydroquinone, para-methylaminophenol, and para-phenylenediamine, as well as the preservative formaldehyde, are all suspected carcinogens.

- Ultraviolet light exposure is associated with skin cancer.

Some chemicals, including carcinogens, require only a very low dose to cause chronic effects and will produce an acute effect only at a much higher concentration. For example, vinyl chloride monomer (VCM), an ingredient used to make polyvinyl chloride (PVC), an ingredient in some plastic painting mediums, is known to have an acute effect of drowsiness, so it was limited in industrial environments to a threshold limit value (TLV) of 500 parts per million (ppm). Subsequent research on animals showed that VCM adversely affected the bones, kidneys, and liver. When, in 1974, several companies announced that a disproportionate number of workers exposed to VCM had died of liver cancer, the TLV was radically reduced to its current level of 1 ppm (New Jersey Department of Health, ongoing studies).

Because years may pass before evidence of chronic harmful effects is detected, it is essential for every artist to learn about potential hazards in art materials *before* using them. Rather than wait until results of studies of dubious chemicals are completed, eliminate suspected toxins from your work. Education and vigilance are your best means of protection.

When we know about the materials we use, safe art entails a triple-A list of basic principles:

- Avoid hazardous materials by substituting safer alternatives.

- Adopt the safest handling procedures for the toxic substances that we must use to realize our creative visions.

◆ Alert yourself to symptoms that may be associated with use of art materials. Early detection and treatment save lives.

## RISK FACTORS

Health risks linked to art materials depend on individual work practices (the environment), personal attributes (the host), and the degree of exposure (the agent). The following issues should be addressed to determine the potential harm of a toxic substance:

**Toxicity:** the intrinsic and relative harmfulness of the chemical. For example, although both toluene and ethyl alcohol are effective lacquer thinners, toluene is somewhat more toxic under the conditions in which most artists currently work.

**Total body burden** of any chemical: the accumulation of that substance in the body from all exposures, which may come from more than one source. If the total body burden exceeds the ability to eliminate or detoxify the substance, its accumulation may cause harm.

**Exposure:** the frequency, amount, duration, and conditions of use. "The dose makes the poison." (Even water can be toxic if administered in sufficiently high doses.) Based on the principle that the body is able to detoxify and eliminate chemicals if there is a sufficient interval between exposures, the American Conference of Governmental Industrial Hygienists publishes Threshold Limit Values (TLV), and the U.S. Occupational Safety and Health Administration (OSHA) establishes Permissible Exposure Limits (PEL), (discussed in chapter 2). When exposure overwhelms body defenses, however, injury, illness, and even death may result.

**Environment:** all surrounding physical factors such as temperature, humidity, ultraviolet and ionizing radiation, pressure differences due to altitude, magnetic fields, and the presence of toxic agents in the air, water, and food supplies. Several artists who submitted artwork for *Making Art Safely* wrote about environmental contamination where they live. These concerns reinforced the perceived need to work safely with art materials. Studio temperature and humidity, as well as toxins associated with smoking and industrial emissions, affect personal risk. Darkrooms, for example, should be kept cool and humid to reduce evaporation of photochemicals.

**Age:** children (not adolescents) are more vulnerable because of rapid breathing, lower body weight, smaller surface area, and (for neonates and some infants) a less mature capacity to detoxify chemicals. Pregnant and breast-feeding women should take special precautions to protect fetuses and infants as well as themselves (see chapter 4, "Art Hazards and Human Reproduction"). Immaturity and an inappropriate sense of invincibility among some adolescents are good reasons to postpone introduction of some art procedures, such as lithography and color photography. The elderly are vulnerable because their ability to transform and detoxify chemicals is likely to have declined.

**Susceptibility:** individual variation in genetic traits and overall health. Susceptibility is influenced by heredity, temporary and chronic health problems, and allergies. Persons under stress, smokers, and heavy drinkers may also have greater susceptibility to harmful effects of chemicals.

**Synergistic effects:** the combined result of simultaneous exposure to two or more chemicals. Such effects can be multiple rather than merely additive; that is, the combined effect is greater than the sum of the potential harmful effects of each agent by itself. Potentially harmful synergistic effects may also occur following simultaneous exposure to art materials and some medications (see chapter 4, "When to Seek Medical Attention").

**Precautions:** the safeguards and cautious habits one brings to the studio. We can greatly reduce the risk of possible injury by being **alert, avoiding** hazardous substances, and **adopting** safe procedures.

## Art Hazard Classification and Identification

The toxicity of chemicals has been determined by toxicologists who have measured the oral dose for ingestion, the concentration of an airborne chemical for inhalation, and the amount applied to the skin for absorption that are required to kill 50 percent of a group of laboratory animals.

The terms:

◆ Highly toxic

◆ Moderately toxic

◆ Slightly toxic

◆ No known toxicity

will be used to classify the toxins in art materials identified in subsequent chapters and Appendix I, "Art Material Chemicals." These classifications, drawn from several sources,[2] represent a consensus at the time of this writing. Due to continuing research, it is possible that classifications may change. In recent years, changes have been toward addition of newly recognized hazardous chemicals and, in some cases, classification changes toward greater degrees of potential harmfulness. For example, when we began writing this book, acetone was considered one of the safest solvents. Recently, however, new evidence of adverse effects on some people prompted a reduction in the threshold limit value.

In this book, a chemical is declared **highly toxic** if it

- May cause major damage or fatality from a single exposure,
- Has been clearly associated with substantial chronic health problems following long-term or repeated exposures, or
- Is reported to cause significant numbers of life-threatening allergic reactions.

Materials declared **moderately toxic** are those that

- May cause minor damage from a single exposure,
- May cause an allergic reaction in a large percentage of people,
- May have chronic health effects, or
- May be lethal if exposure involves a massive dose.

A substance is considered **slightly toxic** if a single exposure

- May cause a minor injury that is readily reversible,
- Has no known long-term health repercussions,
- May cause an allergic reaction in a small percentage of people,
- May cause a more serious injury if exposure involves massive overdoses.

Many art materials are **not known to be toxic (NKT):**

- relatively nontoxic
- not significantly toxic
- are not associated with any short- or long-term health hazard
- have a toxicity believed to be negligible

Further classifications include:

- **Carcinogen:** an agent that is known to increase tumors or induce cancer in both humans and laboratory animals.
- **Suspected carcinogen:** a substance that has been demonstrated to cause cancer in laboratory animals, but is as yet unproven for humans.
- **Reproductive toxin** (includes teratogens, agents that cause birth defects): a material that is associated with adverse effects on reproduction, fertility, and pregnancy; may cause spontaneous abortion, fetal death, retarded growth, premature birth, low birth weight, birth defects, and other complications.
- **Anesthetic:** a substance that causes dizziness, unconsciousness and, with high exposures, death.
- **Asphyxiant:** a substance that deprives the body of oxygen and causes unconsciousness, coma and death.

When relevant, materials are also identified as:

- **Oxidizer** (oxidant): any substance that contains oxygen and gives it up readily to support combustion.
- **Combustible:** a material that burns easily, with a flashpoint higher than 100° Fahrenheit (38° Celsius).
- **Flammable:** a material that burns easily, with a flashpoint lower than 100° Fahrenheit (38° Celsius).
- **Explosive:** a material that, under specified conditions, will explode.

**Systemic toxins** are those chemicals that adversely affect internal organs such as the kidneys, liver, or lungs and/or systems such as the nervous system, circulatory system, or blood-forming system. Examples include effects of aniline and benzene on blood and its formation, effects of ethyl ether on the nervous system, effects of mercury on the kidneys and nervous system, and effects of phenol on the kidneys. Systemic toxins may gain access to the body through multiple routes of entry, cause a variety of responses—including cancer—and affect more than one organ. Simultaneous exposure to several chemicals may elicit synergistic effects.

We recommend that you do not use materials that are carcinogenic. For materials that are highly toxic we recommend that you avoid them if possible, and use less toxic alternatives. For products that are highly and moderately toxic that you must continue to use, we recommend that you adopt stringent precautions and reduce exposure as much as possible. For all art materials we recommend alertness to any and all adverse health effects. If necessary, seek medical attention from providers qualified to recognize and treat occupational disorders (chapter 4).

## How, Why, and Which Materials Are Hazardous

The materials we use in making art can be classified by basic physical and/or chemical properties, which we list here as an overview of the most common hazards in art materials.

### SNEAKY AGGRESSORS

We have dubbed the acids and alkalis, used in many photochemicals and printmaking processes, "sneaky aggressors." While some are noxiously pungent, many are odorless and give scant warning of insidious effects on the respiratory system.

**Acids** are gases or liquids that can corrode metals. Used in printmaking and photographic processes, acids can be highly toxic, especially in concentrated solutions, and can have serious effects through all paths of entry. Inhaling acid vapors and mists, which may be odorless and seemingly innocuous, can irritate the mucous membranes of the eyes, nose, mouth, trachea, bronchial tubes, and lungs. Acute inhalation can cause chemical pneumonia and pulmonary edema, while prolonged inhalation can contribute to chronic bronchitis. Persistent coughing and shortness of breath call for medical attention.

Ingesting even small amounts of acid can severely damage the stomach, alter digestive function, and even cause death. Seek immediate medical attention for ingestion of known acids.

Acids are also primary skin and eye irritants that cause damage on contact. In concentrated solutions, they can cause serious chemical burns that warrant medical attention. Mild solutions may cause contact dermatitis.

Most acids are purchased as concentrates that are diluted by the artist. When diluting acids, use local exhaust ventilation, gloves, and goggles. Few acid solutions have an odor to warn against accidental drinking, so keep them in clearly labeled containers, and never use food containers to store acids.

**Alkalis** are potentially caustic chemicals also known as bases. Most are formed from solids that have been dissolved in water and are used to neutralize acids. Alkalis are found in household cleaning solutions, dye baths, paint removers, and photographic processes. Most alkali solutions should be considered extremely toxic.

Inhalation of alkali dusts or gases can cause pulmonary edema. Ammonia is perhaps the most common culprit.

Ingestion of alkali solutions, even in small quantities, can cause painful damage to the mouth and digestive system and can be fatal. Many alkali solutions do not have odor to warn against accidental drinking, so keep them in labeled containers. As with acids, never store alkali in food containers.

Burns are frequently observed after skin and eye contact with alkalis; ammonia in the eyes is particularly worrisome because it can increase intraocular pressure (glaucoma) and cause corneal ulceration and blindness.

### DECEPTIVE FRIENDS

The materials we use frequently in making art become familiar friends. Although familiar, however, some are deceptive, for they are inherently toxic and may pose significant hazards to health.

**Adhesives and binders** serve many purposes in the visual arts. They are used to suspend pigments in solution until they are bound to the surface to which they are applied. Binding agents, such as acrylic polymer, gum arabic, and various oils, are generally considered nontoxic, but the solvents required to make them workable can be highly toxic. Acrylic polymers, used as both glue and pigment binder, contain small amounts of ammonia and formaldehyde. Either can be an irritant to some individuals. Avoid rubber cement, which may contain n-hexane and other toxic solvents. Epoxy resins, model cement, "instant" cyanoacrylate glues, and plastic adhesives also contain toxic solvents. Some glues contain glycidyl ethers, which in high doses have been found in animal studies to alter blood-forming and testicular function.

**Dusts** are particles generated by handling, crushing, grinding, or sanding stone, wood, metals, and other solid materials. They are chemically identical to the native compound; physically, they are tiny particles that may disperse in air. Chalk, charcoal, and pastels are common sources of dust in the two-dimensional arts and require appropriate safeguards addressed in chapter 5. Many liquid materials, such as paints, if not cleaned up when spilled but are instead allowed to dry, can also become a source of dust.

Dust also results from growth of bacteria, fungi, and molds in contaminated art materials, especially those that are water-soluble and made from plant and animal substances. This dust is known as *biological dust* or organic dust, and may contain intact infectious microbes. In some people, proteins and other small-molecular-weight substances can stimulate allergic reactions such as tearing eyes, runny nose, and labored breathing.

Inhalation of dust presents a well-known health threat. Particles smaller than ten microns, also known as powders, are more easily trapped or retained by the lungs, may aggravate respiratory infections, and, in special cases, may lead to pneumoconioses (dusty lung diseases) or even tumors. Very small powder particles may actually cross alveolar (lung) membranes to the bloodstream.

Ingestion of dusts, while not as problematic as inhalation, can occur when phlegm carrying inhaled dust is swallowed. Ingestion of dusts should be avoided, and dust-containing phlegm expectorated.

Skin and eye irritation may erupt from exposure to specific dusts, some of which may also pit and scratch contact lenses. Dust can be caught under the lenses and corneal abrasions may follow.

**Dyes** are coloring agents that attach or bind to the material being colored (as contrasted with pigments that will not attach without a binding agent). Most commercial dyes are produced in dry forms (see the comments about dusts and powders), and many are synthetically created from petrochemicals. Some artists create their own dyes in liquid states from animal, mineral, and vegetable products (Dadd, 1984). The few dyes that have been studied for safety are those used in foods and cosmetics. Little is known about exposure patterns or the long-term health effects of the many dyes available to artists. We do know that benzidine-containing dyes are carcinogenic and should be safely discarded and avoided. Cold-water (fiber-reactive) dyes have been associated with severe respiratory allergies. All dyes should be handled with extreme care and safeguarding attire (chapter 4). Avoid all inhalation, ingestion, and skin contact of dyes in both liquid and dry states.

**Fumes** are very tiny solid particles formed above molten metal or by chemical reactions such as oxidation. They should be considered as toxic as the substance from which they are made because their toxic components are more likely to be inhaled. In recent years, some artists (e.g., Anselm Kiefer) have poured molten lead onto their paintings, with dramatic visual effect. We hope that all who work with molten metals do so with appropriate ventilation and protective attire because inhalation and/or ingestion of lead fumes is highly undesirable.

**Gases** are compressible formless fluids that occupy a space or container, and upon release dissipate into the air. In two-dimensional art, gases arise primarily from aerosols and from chemical reactions and evaporation in etching, lithography, and photoprocesses. Ammonia, chlorine (from bleach), and nitrogen dioxide (from nitrous oxide etching) are common examples of irritant gases. In their presence, use local exhaust ventilation and goggles. If ventilation is not adequate, use approved respiratory protective equipment such as a face mask with an appropriate cartridge. Some gases are called *nonirritating* because their predominant mechanism of toxicity does not involve irritation of the respiratory tract. This does *not* mean they are harmless! For example, nonirritating gases include the systemic poisons carbon monoxide and hydrogen cyanide (formed in some photoprocesses) and the propellant vinyl chloride (banned in the 1970s), a known carcinogen that causes liver injury. Contact lenses may be damaged by exposure to gases that are corrosive or capable of dissolving the material from which the lenses are made. Some gases, known as asphyxiants, cause problems by limiting the amount of oxygen available for breathing. Finally, remember that under the right conditions, gases can burn or explode, while other gases react with one another to form toxic products (e.g., ammonia reacts with chlorine to form toxic chloramine gas).

**Metals** such as cadmium, lead, manganese, mercury, selenium, silver, palladium, and platinum are used in a variety of art materials, especially in pigments, photochemicals, and emulsions. Exposure to some forms of metals can pose a major health hazard to the artist. Lead, mercury, and manganese, among others, can cause illnesses ranging from psychological problems at low doses to extensive retardation and paralysis at higher doses. Some forms of antimony, arsenic, cadmium, lead, manganese, mercury, and selenium are associated with problems in human reproduction. Occasionally, metal exposure causes allergies, respiratory problems, or multisystem adverse effects.

**Mists** and **sprays** are droplets that become airborne through aerosol spraying, air brushing, aquatinting with spray and similar activities. They are as toxic as the substances from which they are made, though the hazard may vary with effectiveness of ventilation and the dose and duration of exposure. Mists and sprays may contain pigments, lacquers, and paints mixed with undesirable solvents. Artists may also use paints and lacquer fixatives containing easily inhaled and occasionally ingested solvents. Therefore, we prefer spraying only in an appropriately ventilated area such as a spray booth (see chapter 3). An early common propellant, vinyl chloride, is a recognized human and animal carcinogen, and it has not been used in art material manufacture since the mid-1970s. If your studio has spray cans more than ten years old, discard them. Before doing so, however, check with local recycling centers or county hazardous waste unit offices for advice on proper disposal. Propane, butane, and carbon dioxide are common propellants used today. Fluorocarbon propellants such as freon have also been widely used. With excessive misuse or abuse, fluorocarbons may initiate irregular heart rhythms in a few susceptible individuals. In addition, some authorities believe the release of fluorocarbons contributes to the destruction of the upper atmospheric ozone layer that protects us from harmful ultraviolet radiation. In response to this concern, legal restrictions have been placed on the use of fluorocarbons in new products.

**Pigments** are coloring agents created from various substances that have been ground to a powder. In addition to ground metals, minerals, and synthetic materials, plant and animal substances are also used to create pigments. Few pigments have been fully tested, so knowledge of human health hazards associated with pigments in art materials is limited. We do know, however, that some of the metals in pigments pose serious health threats. Cadmium, lead (flake white, chrome, and lead white), manganese, and mercury have been used in pigments and may cause skin allergies, irritation, and ulceration. Inhalation and ingestion of these metal-based pigments can also cause, in extreme cases, death. Their use in manufacture, therefore, has been greatly curtailed, and one should carefully avoid them in the studio and classroom. Other heavy metals such as nickel, iron, and chromium (chromium oxide green, chrome yellow) are suspected human carcinogens.

For a number of years, the phthalocyanine colors were manufactured using polychlorinated biphenyls (PCBs), probable animal carcinogens that may cause chloracne (a skin condition) as well as liver and kidney damage. Pigments produced in the 1990s, however, should no longer contain PCBs. Although we are not aware of any artists who have developed chloracne from PCB colors, we suggest you discard any phthalocyanine pigments that may have been produced before 1982.

When using pigments, avoid skin contact by wearing gloves or using barrier creams. Wear protective clothing and wash it frequently, separating it from other clothes. Because dry or powdered pigments can be inhaled or ingested, we discourage their use unless you work in a glove box or use a particle filter mask together with local exhaust. Most pigments, however, are produced in bound formulations as part of art products such as inks, crayons, and paints. Many binders in solution, such as acrylic polymer, gum arabic, oils, and water, have not been associated with significant human health effects.

**Powders** are finely ground and refined substances. In specific art techniques, powders are found in dry pigments, dry photochemicals, rosin, talc, and whiting. Inhalation of some powders can cause or aggravate respiratory infections and lead to pneumoconioses (dusty lung diseases). Chronic coughing and/or shortness of breath, especially when accompanied by weight loss, call for immediate medical attention. Gum arabic (acacia), used in food products and as a dispersing agent and binder in watercolors, gouache, and pastels has, in its powdered form, been linked to respiratory allergies in some people. Respiratory protection is recommended when using pastels, which are bound with gum arabic, and when using powdered gum arabic in printmaking.

Ingestion of powders is another potential route of exposure that may contribute to long- or short-term illness. Skin irritation, which is often increased if moisture is present, may develop from contact with any of these powders.

**Preservatives** are chemicals added to art materials to prevent deterioration from increased acidity or contamination by fungi or molds. They include common household bleach, boric acid, formaldehyde, naphthol, mercuric chloride, phenol (carbolic acid), phenyl mercuric chloride, and sodium fluoride. Formaldehyde is a likely animal carcinogen and, for some people, an agent that can trigger respiratory distress, asthma, and/or skin irritations. It is commonly used in water-based paints, resin wood glues, hide and animal glues, and hardeners and stabilizers used for photography. Formaldehyde can also be found in the home in plywood, tobacco smoke, cosmetics, and hair products (must be on label); as a textile finish in clothing; and as an emission of gas and wood- and coal-burning stoves.

The low concentrations of preservatives in most art materials have not been associated with significant health hazards. However, since information on preservatives is likely to be withheld because manufacturers have proprietary rights over trade secrets, we recommend that only preservative-containing products that bear the AP and CP labels be used (see chapter 2, "Labeling"). This is especially relevant for all art materials in water solutions.

## EVIL SPIRITS

Evil may lurk in the art studio in the form of toxic solvents or their vapors. To appreciate (and effectively communicate) the potential relationship of solvents to adverse human health effects, artists must understand the basic terminology of solutions. In general, a solution is a liquid consisting of a mixture of two or more substances dispersed through one another in a homogeneous manner. A solution consists of a solute and a solvent, in which the solute is the substance dissolved, and the solvent is a liquid that dissolves or is capable of dissolving. Water is the best-known solvent, but many substances do not dissolve in water. So we use a variety of nonaqueous fluids called organic solvents, which are usually mixtures, have different trade names, and often contain similar chemicals. Everyone is exposed to these solvents. Billions of pounds (excluding water) are manufactured and used every year. Secretaries use correction fluid, attendants or drivers pump gas, hobbyists glue model airplanes, some of us change oil in the family car, and others use paints. Solvents are used to dissolve inks, lacquers, oils, paints, varnishes, thinners, plastics, resins, and waxes. Solvents are found in aerosol sprays, felt-tip pens, photographic chemicals, textile dyes, glues, and adhesives. Solvents are everywhere, so why all the fuss? Why should artists learn about solvents?

Clearly, mere exposure to solvents should not be equated with toxicity, and both exposure and toxicity should be distinguished from harm or adverse effect. The fundamental principle of toxicology—the dose-response relationship—requires both exposure and a toxic potential to create harm or adverse effect. It is important to understand this crucial concept: both significant exposure and significant toxic potential are needed for a solvent to cause a health problem. Exposure to solvents varies; likewise, the inherent toxicity of solvents varies. Therefore, to lessen the possibility of adverse health effects, the artist has two goals: the first is to minimize exposure; the second is to choose a safer solvent.

We suggest that artists learn, teach, and constantly update a hierarchy of solvents. To do so, make a list of all solvents useful in your work. Use this book (and other sources) to rank solvents in terms of inherent toxicity, intensity of exposure, and fire hazard. List the safest solvents first; water will be at the top of the list. Denatured alcohol should be a close second, followed by acetone. The task becomes more difficult at this point. Based on current knowledge, we will offer some guidelines later in this book for completion of your list. Unfortunately, exposure to mixtures of solvents raises the possibility of unpredictable additivity, synergism, or potentiation of effects. We cannot address this issue in a rank ordering because the human health effects of solvent mixtures are not known. We therefore rely on current knowledge of the toxicology of individual solvents and the relationship between the structure of solvents and their toxicity within chemical classes.

Solvents can be ingested, inhaled, or absorbed through skin. Immediate symptoms from excessive solvent ingestion or vapor inhalation include headaches, dizziness, feelings of intoxication or "being high," fatigue, mental confusion, increased heart rate, nausea, and loss of coordination. At higher doses, symptoms can progress from "drunkenness" to unconsciousness and death. Some authorities believe that years of chronic exposure may contribute to memory loss and psychological problems.

Inhalation of vapors from most organic solvents also causes eye, nose, and throat irritation and occasionally breathing problems, convulsions, kidney damage, and even death. Although the mechanisms of solvent toxicity are poorly understood, there is a growing consensus that adverse health effects can occur with prolonged, low-dose inhalation exposure. (Incidentally, constant exposure to solvent vapors can also shorten the life of contact lenses by slowly dissolving the material from which they are made.) Liquid solvents making contact with the eyes can cause serious damage.

The presence or absence of odor has no relation to the toxicity of a solvent. For example, benzene and toluene have fairly pleasant odors, while acetone, which is less toxic, has a strong, disturbing odor.

Solvents such as the aromatic hydrocarbons (benzene, toluene, xylene), chlorinated hydrocarbons (carbon tetrachloride, trichloroethylene, trichloroethane), aliphatic hydrocarbons (petroleum distillates such as kerosene and mineral spirits), and turpentine are also skin irritants that can also be absorbed through cuts and abrasions. Most solvents are combustible or flammable and must be kept away from sources of heat, sparks, flame, and static electricity. (Table 6-3 alphabetically lists the uses, precautions, and alternatives for solvents employed by artists.)

**Alcohols** are found in paint, shellac, rosin, and varnish thinners and removers. Whether ingested or inhaled, alcohol depresses the nervous system. Methyl alcohol (methanol, wood alcohol) is highly toxic by ingestion. Less than two ounces has led to blindness, and larger doses have resulted in death. Inhalation and skin absorption can contribute to injury of internal organs and the nervous system. Ethyl alcohol (grain alcohol or ethanol) is less hazardous through inhalation and skin absorption, but will cause drunkenness if ingested. That sold as denatured alcohol often contains methyl alcohol, which makes it more toxic. Isopropyl alcohol (rubbing alcohol) is usually toxic only by ingestion. Small amounts can cause coma, shock, gastritis, liver damage, and kidney damage. Coingestions of rubbing alcohol and a hydrocarbon (petroleum distillates) warrant immediate medical attention. Alcohols have anesthetic qualities, are irritating to the eyes and mucous membranes, and also dry the skin, making it vulnerable to cracking and permeation. Significant long-term adverse effects of alcohol inhalation have not been well documented.

**Aliphatic hydrocarbons,** also known as petroleum distillates, include gasoline, kerosene, mineral spirits, naphtha, and petroleum ether. Benzine, also known as VM&P Naphtha, is less toxic than benzene but may also be a skin irritant. n-Hexane, a solvent of rubber cement, can be highly neurotoxic. It can cause axonopathy, which may be manifest as short-term and reversible damage to the peripheral nerves (arms and legs), and long-term, irreversible damage to the central nervous system. Early symptoms include cramping, numbness, and/or weakness in hands, legs, and feet; loss of appetite and weight; and fatigue. The symptoms have been mistakenly diagnosed as multiple sclerosis (McCann, *AHN* 2:6). n-Hexane–containing products should be used only with appropriate local exhaust ventilation (chapter 3). Aliphatic hydrocarbons are capable of dissolving a number of substances, including the natural waxy coating of skin. These substances are irritating to mucous membranes and the skin, will defat the skin and be absorbed, and are central nervous system depressants with narcotic effects both by ingestion and inhalation. Petroleum distillate solvents typically contain up to 25 percent aromatic hydrocarbons, which give them their distinctive odor. We believe that deodorized or odorless mineral spirits, which have had the aromatic hydrocarbons removed or reduced, are safer to use. However, all petroleum distillates are flammable, and some explosive. Turpentine, sometimes designated a petroleum distillate, is actually an oleoresin, and is described later in this chapter under its own name.

**Aromatic hydrocarbons** are potentially dangerous solvents because they are easily absorbed and especially irritating to skin. Benzene (benzol) is believed to cause some forms of leukemia and is known to damage bone marrow, where red and white blood cells are formed. Toluene (methyl benzene, toluol) is used to thin and dissolve lacquers, varnishes, and paints. If large amounts are inhaled or ingested, toluene may cause dizziness, unconsciousness, and with extreme exposure even death. Large amounts irritate the skin, and significant chronic exposure may cause liver and kidney damage, abnormal heart rhythm, and nervous system damage. Xylene (dimethyl benzene, xylol) is also a lacquer, paint, and varnish solvent that has adverse effects similar to toluene. In massive, acute exposures, styrene is a skin and respiratory irritant. Long-term, low-dose exposure to styrene has been associated with persistent and premature dementia in some workers.

**Chlorinated hydrocarbons,** including carbon tetrachloride, trichloroethylene, ethylene dichloride, methylene chloride, and methyl chloroform are anesthetics that depress the central nervous system. Effects range from mild dizziness to unconsciousness and, with extreme exposure, death. PCBs (polychlorinated biphenyls) are suspected carcinogens and are not allowed to be used in the manufacture of art materials. All chlorinated hydrocarbons can dissolve the fatty layers of the skin to increase skin absorption and cause local irritation. Most are not flammable, but they decompose in heat or ultraviolet light to give off phosgene, a poisonous gas.

**Glycol ethers,** including butyl Cellosolve, Cellosolve and methyl Cellosolve, represent several classes of solvents commonly used to disperse pigments in solution. They are mild skin irritants and are easily absorbed through the skin. Their vapors can be moderate irritants to mucous membranes and the upper respiratory tract. If inhaled, ingested, or absorbed through skin, methyl Cellosolve (also known as ethylene glycol monomethyl ether or 2-methoxy ethanol), used in photo etching, can cause central nervous system depression; cardiac, lung, and kidney problems; anemia; birth defects; and testicular damage in animals. The combination of glycol ethers and organic solvents in offset lithographic printing has been associated with bone marrow damage. Glycol ethers are used in photoresists and color photography; as solvents of resins, paints, and cellulose inks; as an ingredient in a variety of coatings such as wood stains, epoxies, and varnish; and in cleaning compounds. Many are relatively safe (the propylene glycol ethers are preferred) except in extreme quantities, and some are approved as an ingredient in cosmetics. The possibility of an association between glycol ether exposure and reproductive toxicity remains a serious and open question as of this writing.

**Ketones** are organic solvents that include acetone, methyl ethyl ketone, methyl butyl ketone, cyclohexanone, and isophorone. They are used to dissolve lacquers, oils, plastics, vinyl inks, and waxes. Methyl butyl ketone has been associated with neuropathy and mucosal irritation.

**Turpentine,** the solvent artists most commonly use, is an oleoresin distilled from coniferous (pine) trees. Gum turpentine is distilled from the sap, and wood turpentine is distilled from the wood. Labels on turpentine containers identify it as a "petroleum distillate" because that designation is an alert to the medical profession that vomiting should not be induced if the solvent is accidentally ingested. It may be applied to products that are not, in fact, derived from petroleum (Rossol, *AHN* 8:4). Wood turpentine is more hazardous than gum, but both are potentially toxic, especially upon ingestion. Vapors from turpentine are irritating to the eyes and mucous membranes and represent a significant threat to making art safely. Repeated exposure has been associated with pulmonary edema, chronic respiratory diseases such as bronchitis and emphysema, and kidney disease.

In response to growing concern about the volatility and toxicity of turpentine, manufacturers have produced odorless paint thinners such as Turpenoid. The aromatic hydrocarbon content of these products is less than turpentine (McCann, *AHN* 11:9). The lack of pungent sensation to the eyes and nose does not, however, mean total absence of hydrocarbon. (Remember, there is no correlation between odor and toxicity of any solvent.) If any of the common symptoms of solvent inhalation are present, such as headache, dizziness, feelings of intoxication or "being high," mental confusion, increased heart rate, nausea, or loss of coordination, consider the possibility that inhalation of an odorless solvent may be contributing to the problem.

Some **waterless cleaners** for removing paint, inks, and other materials from skin are safer than others. Kerosene, alkalis, and abrasives, however, can be almost as irritating as the agents they replace (McCann 1979). D-Limonene, a paint thinner ingredient in Citrus Clean, Citrus Turp, Grumtine, Really Works, and other products, is natural terpene extracted from orange peel. D-Limonene is "an EPA registered pesticide ingredient" (Rossol, *ACTS* 4:6). Testing has demonstrated significant acute toxicity only with massive ingestions. Therefore, the presence of D-limonene does not legally require a warning label. However, some D-limonene–containing products include a warning on the label. We advocate labeling for D-limonene because a recent government study provided preliminary evidence of kidney tumorigenicity in one strain of rats. However, some authorities discount any risk of tumorigenicity in humans exposed to D-limonene. They argue that the kidney tumors were noted only

in male rats that produce a protein called alpha-2-microglobulin. Because humans, mice, and female rats do not produce the protein, they may not be expected to develop tumors due to D-limonene exposure.

**Vapors** represent the gaseous phase of liquids. The vapors of most significance to artists have been identified previously in the gases and solvents sections. Vapors can be more hazardous than the substance from which they evaporate because vapors enhance the likelihood of inhalation. Artists encounter potentially harmful vapors most often from printmaking solutions and photochemicals that, when used in open trays, have large surfaces from which evaporation easily occurs. Therefore, keep trays covered as much as possible. Other sources of toxic vapors are solvents such as lacquer thinners, turpentine, and mineral spirits. Vapors vary in toxicity, flammability, and capability to react with other gases.

## Facing the Future

**M**ost artists want their work to thrive and survive. The materials from which art works are made affect their durability in important ways. The chemicals, inks, paints and papers used may require preservation and restoration someday. For example, works on paper are vulnerable to moisture and hence contamination by molds or fungi. Biodegradability is a desirable and noble quality for throwaway art, but not for that we wish to preserve.

The conservation of art objects in all media is an issue to be addressed at the time of their creation. By using dangerous or deteriorating materials, artists subject future conservators to hazards and frustrations that could be avoided if artists were sufficiently conscious of potential consequences. If not stored properly in an acid-free, dry environment, art works may have to be treated with antifungal and antivermin fumigants, such as ethylene oxide, methyl bromide, and carbon disulfide. If the wood pulp papers used are acid rather than pH neutral, they will discolor, become brittle with age, and eventually crumble to dust. They may have to be deacidified, a less toxic but potentially expensive process.

Environmental consciousness today focuses on hazardous industrial emissions and commercial products, but as those issues are addressed, the ecological implication of increasing print media and art cannot be far behind. Let us not put our heirs, customers, and conservators at risk from potentially toxic materials, such as lead paint. Further, we live in a time when it is necessary to deal with the toxic wastes of our shrinking planet. As leaders and anticipators of change, artists, as they often have, can lead the way.

Chapter 1 is a wake-up call and an overview of potential hazards in the products used in making art. We have suggested a continuing need for a more enlightened approach among artists to health and safety issues. To assist the concerned artist, we provide additional information about specific materials in the chapters devoted to drawing, painting, printmaking, graphic design, illustration and photography.

Continued growth of knowledge and the development of new art materials preclude preparation of a definitive reference. Fortunately, there are a number of individuals, institutions, organizations, and government agencies who are responsible for keeping abreast of developing information and making it available to the public. So that our readers may avail themselves of these resources, the next chapter provides information and channels by which to remain current on the topic of safety in art.

Material Safety Data Sheet
May be used to comply with
OSHA's Hazard Communication Standard.
29 CFR 1910.1200. Standard must be
consulted for specific requirements.

**U.S. Department of Labor**
Occupational Safety and Health Administration
(Non-Mandatory Form)
Form Approved
OMB No. 1218-0072

IDENTITY (As Used on Label and List)

Note: Blank spaces are not permitted. If any item is not applicable, or no
information is available, the space must be marked to indicate that.

**Section I**

| Manufacturer's Name | Emergency Telephone Number |
|---|---|
| Address (Number, Street, City, State, and ZIP Code) | Telephone Number for Information |
| | Date Prepared |
| | Signature of Preparer (optional) |

**Section II — Hazardous Ingredients/Identity Information**

| Hazardous Components (Specific Chemical Identity; Common Name(s)) | OSHA PEL | ACGIH TLV | Other Limits Recommended | % (optional) |
|---|---|---|---|---|
| | | | | |
| | | | | |
| | | | | |
| | | | | |
| | | | | |
| | | | | |
| | | | | |
| | | | | |
| | | | | |
| | | | | |
| | | | | |

**Section III — Physical/Chemical Characteristics**

| Boiling Point | | Specific Gravity ($H_2O$ = 1) | |
|---|---|---|---|
| Vapor Pressure (mm Hg.) | | Melting Point | |
| Vapor Density (AIR = 1) | | Evaporation Rate (Butyl Acetate = 1) | |
| Solubility in Water | | | |
| Appearance and Odor | | | |

**Section IV — Fire and Explosion Hazard Data**

| Flash Point (Method Used) | | Flammable Limits | LEL | UEL |
|---|---|---|---|---|
| Extinguishing Media | | | | |
| Special Fire Fighting Procedures | | | | |
| Unusual Fire and Explosion Hazards | | | | |

(Reproduce locally)

OSHA 174, Sept. 1985

A rtists must know about potential hazards in art materials and practices because our health and well-being are ultimately our personal responsibility.

Art educators must know about latent dangers in art supplies and techniques because we are, to a degree, re-

# Staying in the Know

sponsible for the protection of our students and for teaching them safe art-making practices.

Leaders of educational institutions and professional studios must be informed because we have legal responsibility for the health and safety of our students and employees in the workplace.

Whatever our perspective, we can no longer leave our

◆ ◆ ◆

heads buried in sand. At each stage of making art—from purchasing products and using them, to cleaning them up and preserving them—artists of all genres have many opportunities to be cognizant and enlightened. While we hope that *Making Art Safely* will provide accessible and useful information, the issues are so vast that no single reference can provide all the facts and knowledge pertinent to the topic. Therefore, we include the following information sources for artists and recommend that pertinent telephone numbers be posted with other emergency numbers near studio telephones.

## Resources

**P**oison information centers operate 24 hours per day, year round. The telephone number is listed in telephone directories, often with emergency listings. Poison information centers, also known as poison control centers, have a toxicologist on call 24 hours a day, every day of the year, a role that one of us (JWS) fulfills periodically for our center in the Delaware Valley. In our experience, calls have been answered immediately and returned promptly by the toxicologist, within a few minutes in an emergency. In addition to handling emergencies, poison centers can help identify hazards of specific chemicals and decipher chemical abbreviations and identification numbers that may be used on product labels.

In some areas, the poison control center has not had the personnel or the latest information on file regarding art materials and they turn to our primary resource and advocate:

**Center for Safety in the Arts (CSA)**
**5 Beekman Street, Suite 1030**
**New York, NY 10038**
**212-227-6220**

Founded and directed by Michael McCann, Ph.D., a pioneer in art hazards, the center (previously known as the Center for Occupational Hazards) maintains an extensive resource and information file, distributes pertinent publications, provides workshops and seminars, and maintains a computer bulletin board and database. *Art Hazards News*, CSA's newsletter published ten times per year, reports new information, legislation, case studies, and a broad range of issues relevant to hazards and safety in the arts, including an annual resources directory. The Center for Safety in the Arts is open Monday through Friday, 10:00 A.M. to 5:00 P.M.

**Arts, Crafts and Theater Safety (ACTS)**
**181 Thompson Street # 23**
**New York, NY 10012**
**212-777-0062**

Founded and led by Monona Rossol, ACTS publishes a monthly newsletter, *ACTS Facts*. Along with articles relevant to art and theater safety, it contains a review of pertinent items from the Federal Register, a compilation of regulations and public notices issued by federal agencies.

We recommend that artists, especially those who find it necessary to work with any hazardous art materials, subscribe to both *Art Hazards News* and *ACTS Facts*, which are available for a nominal yearly rate.

**Art and Craft Materials Institute (ACMI)**
**100 Boylston Street, Suite 1050**
**Boston, MA 02116**
**617-426-6400**

ACMI tests the safety and quality of products submitted by art materials manufacturers. Certification seals then appear on the labels of tested products. ACMI answers questions about specific products and periodically publishes a list of evaluated materials.

**Center for Hazardous Materials Research (CHMR)**
**University of Pittsburgh**
**Applied Research Center**
**320 William Pitt Way**
**Pittsburgh, PA 15238**
**412-826-5320**
**Hotline: 800-334-CHMR (2467)**
**FAX 412-826-5552**

CHMR provides a full range of environmental services and offers a confidential contact for answers to environmental and workplace questions. Their free publication, *The Minimizer*, updates environmental issues which may affect one's business.

**National Institute for Occupational**
**Safety and Health (NIOSH)**
**4676 Columbia Parkway**
**Cincinnati, OH 45226**
**800-356-4674**

NIOSH funds educational resource centers. To get the name and address of the resource center nearest you, call NIOSH, open Monday through Friday, 9:00 A.M. to 4:30 P.M. If you are concerned about the health effects of a substance or occupational hazards in the workplace, NIOSH will provide information needed.

**Occupational Safety and Health**
**National Office**
**Information and Consumer Affairs**
**200 Constitution Avenue, NW**
**Washington, DC 20210**
**202-523-8151**

The federal OSHA administers the Occupational Safety and Health Act, which applies to all nongovernmental employees throughout the country and, by executive order, to all federal government employees. In 25 states[1] with OSHA-approved plans, health and safety responsibilities have been delegated to the state. Since one of the conditions for federal OSHA approval is that state plans cover state and municipal employees, many schools and teachers are regulated by OSHA. Local OSHA offices are listed in telephone directories under U.S. Government Department of Labor. OSHA has issued permissible exposure limits (PELs) for hundreds of toxic and hazardous substances; the 1989 issue, *Air Contaminants—Permissible Exposure Limits* (OSHA 3112) is 2100 pages in length. Each local office will have this document (as should poison control centers), from which you may obtain information about how long and under what conditions you can safely work with a given chemical. Individual states also maintain an OSHA consultation service that will provide free inspection and advice on safety to workplaces and studios that have employees (they may not advise individual artists working alone, however). If not listed in your local telephone directory, you may locate your state office by calling the above number or the Center for Safety in the Arts.

**National Fire Protection Association (NFPA)**
**1 Batterymarch Park**
**P.O. Box 9101**
**Quincy, MA 02269-9101**
**800-344-3555**

As part of its mission to protect lives and property from fire, NFPA produces the *National Fire Codes* as well as hundreds of other fire safety materials, including training programs, educational brochures, textbooks, references, films, videos, and more. Through products and publications, NFPA makes the latest fire safety information available to the public.

**National Safety Council**
**444 North Michigan Avenue**
**Chicago, IL 60611**
**312-527-4800**

The National Safety Council produces a general educational materials catalog that includes films, videotapes, and instructional slide sets on safety in the home and workplace.

**Hazardous Substance Fact Sheets**
**New Jersey Department of Health**
**CN 368 Trenton, NJ 08625-0368**
**609-984-2202**

As part of the state's Right to Know Program, the New Jersey Department of Health prepares hazardous substance fact sheets that contain information about a specific chemical. (Right to know programs are also available in other states.) Approximately 2,000 chemicals are covered. Similar to Material Safety Data Sheets (described later in this chapter), they are organized clearly and written to be understood by the lay reader. Fact sheets also include information from the New Jersey Occupational Health Service and answer questions commonly asked about exposure to a hazardous substance. The fact sheets are available in both printed form and on electronic media. In printed form, the first ten data sheets are provided at no charge, and thereafter a small charge is made. Call for information.

**Chemical Referral Center**
**2501 M. St., N.W.**
**Washington, DC 20037**
**800-262-8200**

An information line for the chemical industry; provides data on chemicals and manufacturers.

**Canadian Centre for Occupational Health**
**and Safety**
**250 Main Street East**
**Hamilton, Ontario L8N 1H6 Canada**
**416-572-2981**

An information service on workplace health and safety; produces a CD-ROM known as CCINFO disc.

**Ontario Crafts Council**
**35 McCaul Street**
**Toronto, Ontario M5T 1V7 Canada**
**416-977-3551**

Affiliated with the Canadian Occupational Health and Safety Authority, the Ontario Crafts Council has produced a workplace safety training kit that consists of a book and videotape, "The Split Second."

Ontario Crafts Council
35 McCaul Street
Toronto, Ontario M5T 1V7 Canada
416-977-3551

Affiliated with the Canadian Occupational Health and Safety Authority, the Ontario Crafts Council has produced a workplace safety training kit that consists of a book and videotape, "The Split Second."

Environmental Research Foundation
P.O. Box 73700
Washington, DC 20056-3700
202-328-1119

The foundation offers technical assistance for fighting toxic problems; it publishes a weekly newsletter and maintains a database called RACHEL (Remote Access Chemical Hazards Electronic Library).

**Committees on Occupational Safety and Health (COSH)** are coalitions of labor unions and health professionals that lobby for occupational health and provide health and technical services. For the one nearest you, contact the Center for Safety in the Arts.

## Nontoxic Art Materials:

## A First Choice

**N**ontoxic means that a substance in a reasonable quantity can be introduced into the body:

- Through the mouth and digestive tract
- By the way of the respiratory system
- Through the skin

and pass through or be absorbed with no known short- or long-term ill effects.

In recent years, through the efforts of artist and art material producer advocacy groups,[2] progress has been made toward better testing and labeling of art materials. These efforts contribute to greater safety in the arts and, as our knowledge and interest grow, to an expansion of the quantity and quality of safe art materials. It should become increasingly easier for artists to distinguish nontoxic art materials from those that contain dangerous ingredients.

A list of 2500 identified nontoxic art materials, "Acceptable Children's Art and Craft Materials," is periodically prepared by the California State Department of Health Services.[3] Because it is for children, it includes the safest art materials known. Although this is relevant for all artists, it is especially important for young people and other high-risk groups, such as the retarded, handicapped, and elderly. This list can be obtained from the Center for Safety in the Arts or from Arts, Crafts and Theater Safety.

The Art and Craft Materials Institute (ACMI), supported by art material manufacturers, periodically publishes a list of the products that have been reviewed by their certification program.

While helpful, and an important first choice in art materials, a nontoxic label can be misleading for several reasons. Some manufacturers of art products have taken it upon themselves to designate their products as nontoxic. How reliable are these designations? Perhaps some are very reliable. The skeptical consumer, however, knowing how long it takes for misleading or erroneous labeling to be discovered, challenged, and verified, will note that huge quantities may be sold before health problems are reasonably linked to a particular product. Then, more often than not, all it takes is a pledge to stop the practice, and the manufacturer may continue to sell whatever it wishes.

If a product does not bear an Art and Craft Materials Institute label, responsible artists should either obtain the information needed to protect themselves or decline to purchase the product.

A nontoxic label may also be misleading because it is difficult and expensive to assess the risk of cancer and other ailments posed by a single chemical. There are thousands of chemicals and products marketed as art materials, and despite the 1988 legislation to improve testing and labeling, it may be years before a clear understanding of actual risks is attainable. Even with the materials designated as safe by the Art and Craft Materials Institute—identified with an AP or CP seal or the health label—the distinction between toxic and nontoxic can be, in some cases, narrow and based upon limited research. Most tests are effective only for identifying single-dose, acute reactions, and not chronic, long-term health hazards that may take years to be manifested. Research on chronic hazards has been required only since 1988, and because of the time lapse between exposure and illness, it is safe to assume that information on long-term health effects will only gradually become available over the next few decades.

As a general rule, young children should be kept out of studios. Art materials, except those known to be safe for children, should be kept out of their reach. Parents and educators have a responsibility to prepare children to respect art materials, to help them learn the potential hazards, and to develop safe art-making work habits.

The nontoxic label (ACMI approved or otherwise justified) means *only* that the art materials:

- Are not known to cause adverse reactions for *most* people when used for the intended purpose *only*

- Are *probably* (more likely than not) nontoxic

It is still *possible*, however, that individual sensitivity and susceptibility may lead some persons to experience discomfort and ill effects.

One should read designations such as nontoxic as *relatively safe, probably safe,* and not as absolutely safe.

Significant questions and variables remain. We must be careful and alert. Caution and moderation are always justified. Independent artists and art educators must handle art materials more attentively, with constant vigilance for suspicious symptoms. With initiative, responsibility, and careful attention to labels and information, we can work safely.

## Labeling

### BACKGROUND

The labeling of art materials was once a simple matter that included little more than the name of the product and the company that manufactured it. Art supplies were treated as consumer products, where users were assumed to have only occasional and therefore insignificant exposure. In the late 1940s, when some children became ill from eating crayons, the Crayon, Watercolor and Craft Institute, an organization of manufacturers of children's art materials, created a safety certification program. It eventually became the Art and Craft Materials Institute (ACMI), which today determines the safety of art and craft materials and products used by both children and adults.

The economic and product advances of the 1950s and 1960s were accompanied by growing interest in consumer safety, a concern that culminated in the Federal Hazardous Substances Act (FHSA) in the late 1960s. The FHSA required that ingredients in consumer products known to be hazardous or to cause acute (single exposure) reactions, must be indicated on the label of that product.

These label warnings were general. For example:

**Caution: Flammable Mixture**
**Do Not Use Near Fire or Flame**
**Contains Petroleum Distillates**
**Keep Away from Children**

Before long, the inadequacy of this kind of labeling became apparent. This warning does not indicate, for example, that petroleum distillates can be harmful if ingested, inhaled, or absorbed through the skin or that the harmful effects of one petroleum distillate are different from another. Further limitations include recognition of only acute reactions (omitting long-term effects) and the type of testing used to determine the toxic properties that cause acute effects.

One testing method typically involved feeding ten rats five grams of a suspected ingredient per kilogram of body weight. If more than five rats died within 14 days, the substance would be declared toxic. By the same standard, if five (or fewer) rats died, the product could be labeled nontoxic. This procedure permitted substances now known to be toxic—for example, asbestos, which is incapable of killing animals in two weeks, but is a carcinogen (in some forms)—to be declared nontoxic (Rossol, *AHN* 8:4). The significant shortcoming of this testing method is that it fails to investigate long-term chronic effects, those that result from gradual absorption of a substance over a period of time, often years, before any ailment is noticed.

Although labeling of consumer products is generally regulated by the Consumer Product Safety Commission, the Occupational Safety and Health Administration (OSHA) got involved in the 1970s and early 1980s. Following a series of hearings, OSHA generated the 1983 Hazard Communication Standard, known as HAZCOM,[4] to reduce risk to workers by requiring evaluation of hazardous substances and conditions in the workplace and education of employees about hazards and safety. HAZCOM requires that the presence of a hazardous ingredient, with its name and its precautions, be stated on the product label. In conformance with HAZCOM, a label must include:

- Identity of hazardous chemicals

- Appropriate hazard warnings

- Name and address of the manufacturer, importer, or other responsible party

For example, a brush cleaner label states:

**CONTAINS XYLOL**
If swallowed do not induce vomiting. CALL PHYSICIAN IMMEDIATELY. Keep away from heat, sparks and open flame. Avoid prolonged contact with skin and breathing of vapor or spray mist. Close container after each use. Do not transfer contents to bottles or other unlabeled containers. Use with adequate ventilation.
**KEEP OUT OF THE REACH OF CHILDREN.**
Name and address of distributor.

This warning, while it contains more information than the first example, is still plagued by vagueness, by reference only to immediate dangers and acute reactions, and by lack of information about long-term, chronic effects, which may include possible liver and kidney damage.

## LABELING TODAY

In 1979, congressional hearings on art were held to identify and explore the insidious threat of chronic hazards in art materials. This led to the development of a chronic hazard labeling standard for art materials, ASTM (American Society for Testing and Materials) Standard D-4236. As a result of encouragement from arts advocacy groups, many art product manufacturers voluntarily increased labeling information.

In 1988, Congress passed the Labeling of Hazardous Art Materials Act[5] that amended the Federal Hazardous Substances Act (FHSA) with the ASTM Standard D-4236. It required that by November 1990 manufacturers indicate on all labels both the chronic and acute effects of chemicals in their product, to the extent known. As of November 1990, it is illegal to sell products not labeled in conformance with the ASTM standard D-4236, although some manufacturers have received extensions to delay conformance. The Hazardous Art Materials Act (HAMA) also prohibits use of hazardous art supplies in schools by children in grades six and lower and preempts state and local labeling laws so that labels will be consistent throughout the United States.

A significant additional development in the 1988 Labeling Act is that, unlike previous legislation that depended on manufacturers to identify hazardous substances and their precautions, it directs the Consumer Product Safety Commission (CPSC, a federal agency) to develop scientific criteria by which manufacturers can evaluate their products for chronic health hazards. The American Society for Testing Materials; Artists' Equity Association; Art and Craft Materials Institute; Arts, Crafts, and Theater

Safety; the Center for Safety in the Arts; and the National Art Materials Trade Association are working with the CPSC and art materials manufacturers to establish testing criteria and standards that are informative and protective for the artist. They include:

- Identification of the chronic hazards in art materials
- Differing reactions of children, adults, and the elderly
- Synergistic effects of exposure to multiple chemicals
- Bioavailability of specific ingredients

CPSC will also develop and distribute educational materials about art products and is empowered by HAMA to sue art materials manufacturers that do not conform to the act. Retailers as well as manufacturers who sell products that are not labeled properly may be penalized by CPSC.

The Art and Craft Materials Institute (ACMI) operates a certification and labeling program for a broad spectrum of art and craft materials (more than 17,000). Products presented to ACMI for testing are either designated as nontoxic or are otherwise labeled with health warnings. According to ACMI's fact sheet, a consulting toxicologist reviews the formulas of all products included in the certification process, and an advisory board of toxicologists reviews the consultant's criteria and decisions. ACMI also conducts random testing of products bearing its certification labels to ensure continuity of the product's content.

Only one of three seals may be used on a given product. Whenever the square health label is used, always be sure to read the print below it. ACMI devised the phrase "No Health Labeling Required" to avoid having to label differently domestic and exported Institute-certified nontoxic products because some European countries prohibit the nontoxic designation in labeling.

As artists increasingly choose to buy only products with adequate labeling, a greater number of art materials manufacturers respond by subscribing to ACMI to have their products evaluated. Even so, we must be responsible and willing to learn by reading labels before purchase. We must be sure that the product contains only substances we and our studios are equipped to handle safely and responsibly. Guidelines for safe handling can be found in chapters 3 and 4.

Under the law, only recognized hazardous sub-

Figure 2.1.–The Art and Craft Materials Institute certification labels for art materials. "AP NONTOXIC" circular seal designates an approved product, one that conforms to ASTM D-4236 as having no acute or chronic health hazard.

"CP NONTOXIC" circular seal indicates a certified product, one that, in addition to conforming to ASTM D-4236, also has been found to meet quality performance standards as set by the American Society for Testing Materials (ASTM D-4302) or the American National Standards Institute (ANSI Z356.1-5). Wording below the seal should indicate which of these performance standards the product meets.

The "HEALTH LABEL" is a square seal that indicates a product is certified to be properly labeled. By itself, it is not informative, for it is the words below that must be read carefully. This information ranges from "nontoxic" and "no health labeling required" (each of which may also, if qualified, include the ASTM or ANSI performance standard) to specific labeling for toxic ingredients as required by ACMI's toxicologist.

over, recommend legislation to require the same degree of ingredient information and listing as is found on food and cosmetic labels. When we consider the intimacy with which we use art materials, inclusive labeling is a development we hope will decrease the number of artists suffering from materials-related illness. It may be impossible, however, to list all required information on labels, especially on small items, so we can expect to see package inserts with the required facts and references from which to obtain additional information. Any details that are separate from a product container, as found on a package or package insert, should be retained in the studio in a product information and material safety data sheet file, as described later in this chapter.

Such information is useful only if it is read carefully before using the product, preferably even before purchasing it. Just because a product is on the market does not mean that it is safe or that the label tells the whole story. Government safety standards are often a compromise between the health of human beings, environmental issues, and the economic pressures of manufacturers. If a product fails to list its ingredients or does not conform to the present laws, do not buy it. Moreover, discard products in your studio that do not have sufficient ingredient and safety information on the label (see in chapter 3, "Safe Disposal of Toxic Materials").

## LABEL TERMINOLOGY

Effective reading of labels requires knowledge of how terms are used and what they mean. According to the Federal Hazardous Substances Act:

- **Caution** indicates the product is toxic and a chronic hazard.
- **Danger** means the product is highly toxic.
- **Poison** indicates the product is highly toxic by ingestion.
- **Warning** means the product is toxic with short-term effects.

stances and their appropriate precautions and antidotes must be listed on the label. Artists' rights groups, however, have convinced some manufacturers to voluntarily include more complex information on the label. These additional data conform to the American Society for Testing and Materials standard ASTM D-4236 and include identification of the pigment composition, binders, and other less toxic chemicals. Some artist advocacy groups, more-

These words indicate the existence of a toxic substance or mixture which is usually identified by "contains _____," followed by appropriate precautions. Do not be misled by statements such as "Contains no _____" or "_____-free" (lead, formaldehyde, hexane, etc.). Other potentially harmful ingredients may be present.

Different brands of the same product may have different formulas. For example, lacquer thinner can be made from a number of different chemical mixtures, some of which are safer than others (see chapter 7).

Brand, trade, and product names can look the same on a label over a period of years, but the ingredients may have been changed since your last purchase. Modifications, like dangers, may be in the smallest print, so it is essential to read the entire label each time you buy a product (McCann, *AHN* 8:4).

"Use only with adequate ventilation" is a nebulous directive found on many labels. Opening a window, circulating air with a room fan, or taking work outdoors to be sprayed may, in fact, be counterproductive: an unexpected breeze can send the spray in your face instead of onto the work. What **adequate ventilation** means is to provide sufficient ventilation to keep the airborne concentrations of mist, dust, fumes, or vapors away from your personal intake area and below the levels considered hazardous to health. Information on ventilation can be found in chapter 3.

The term **biodegradable** does not mean that a product is safe or nontoxic, but only that it decays into other substances that do not pollute the earth. In sufficient quantities, these substances may be harmful to humans despite their biodegradability.

Precautions and antidotes must be included on product labels of harmful materials. The words **petroleum distillate** are an alert to the medical profession that vomiting should not be induced if the solvent is accidentally ingested. It is a designation that may be applied to a range of products, only some of which are, in fact, derived from petroleum. Read the precautions and antidotes *before* purchasing the product to be sure you and your studio are equipped to handle it safely. Read all precautions and antidotes before using the product to be informed in case of emergency.

Flammable, combustible, and nonflammable will be indicated on labels of liquids. The Federal Hazardous Substances Act (FHSA) classifies:

– **Extremely flammable** liquids are those that can be ignited by a spark at temperatures below 20°F (− 7°C).

– **Flammable** liquids (classified as **IA** or **IB**) have a flashpoint (at which they will be ignited by a spark) between 20° and 80°F (− 7° and 27°C).

– **Combustible** liquids (classified as **II, IIIA,** or **IIIB**) have a flashpoint over 80°F (27°C).

The National Fire Protective Association has slightly different classifications in which:

– **Flammable** has a flashpoint under 100°F (38°C).

– **Combustible** has a flashpoint which starts at 100°F (38°C) or higher.

In *Making Art Safely* we have adopted this last designation. See chapter 3 for additional information on fire hazards and prevention in art studios.

As of 1990, labels on newly manufactured materials contain more detailed information and include warnings about the product's possible chronic health effects. These developments, efforts by artist advocacy groups, the OSHA HAZCOM standards, and the Labeling of Hazardous Art Materials Act of 1988 are important allies of the artist in making art safely. The responsibility is ours, however, to **read labels** and know what they mean. They are the first source of information about potential hazards and handling requirements. Typically, labels today also carry the notice, "It is a violation of Federal law to use this product in a manner inconsistent with its labeling." Adherence to label directions is essential for the safety of ourselves and those close to us.

The label or product insert may also contain **warranty** information that may be a veiled disclaimer of responsibility, so be sure to read it.

It is our responsibility to know what we are using and how to use it safely.

## LABELS OF THE FUTURE

While progress has been made in recent years, there is still much to be done to improve labeling. Standardization in the use of symbols would help: we all know that a skull and crossbones designates poison. Why not develop symbols that would designate hazard via inhalation, ingestion, or skin absorption or standardize different shapes or colors of labels to communicate relative safety or danger? These are possibilities that we hope to see addressed soon.

## Material Safety Data Sheets

When product labels do not contain sufficient information, artists may be uncertain if their studios are properly equipped to handle a substance. To obtain more detailed information on product ingredients, request a Material Safety Data Sheet (MSDS) from the manufacturer. The purpose of an MSDS is to provide, in a concise format, information about hazards of materials so that one can take appropriate action when necessary. The MSDS can help artists and studios plan safe handling of materials, the kind and level of protective attire needed, first aid precautions and treatment, and the best management of accidental mishaps such as spills and fires. If you are not employed by an OSHA-covered entity, you have no *legal* right to obtain the Material Safety Data Sheet.

We recommend that artists obtain and keep a notebook or file of MSDSs for all hazardous materials in the studio. Employers and institutions are required by law to maintain such a file.

MSDSs must be written in English and provided to product distributors and commercial customers. Manufacturers must enclose MSDSs with shipments to all employers, including schools. An MSDS for a product may be procured from the manufacturer, its importer, its local distributor or retailer who, if cooperative, can provide artists with a copy upon request. If they don't have an MSDS file, most art retail outlets will have the National Art Materials Trade Association (NAMTA) Directory of Art Materials Producers and Manufacturers and may allow you to consult it for a product manufacturer's name and address.

As an overview, an MSDS, when complete, will include:

- Identity of the material with its brand, chemical (and chemical index number), and common name

- Ingredients that are hazardous that amount to 1 percent or more of the product's composition

- Cancer-causing ingredients in amounts of 0.1 percent and more

- Physical and chemical hazard characteristics such as corrosive, explosive, or flammable

- Health hazards, whether they are acute (occur immediately) or chronic (accumulate and are manifest after a period of time)

- Primary entry routes (ingestion, inhalation, skin absorption)

- Organs likely to be damaged following exposure

- Medical conditions that can be aggravated by exposure

- Worker exposure limits

- Precautions and safety equipment

- Emergency and first aid

- The organization that created the MSDS and its date

While the information in MSDS is standardized, the formats, order of sections, and writing styles may differ. The following outline is representative.

**Section I** includes:

Name, address, and telephone number (important in case of an emergency with the product) of the company that produced the material. Pay special attention to hotline numbers that may be toll-free and/or answered around the clock.

■■

MSDS date of issue (or most recent revision). The date on the MSDS should reflect a time relevant to the production of the product. This may require obtaining an updated MSDS whenever new supplies are purchased.

■■

Name of the product. The product name must be spelled *exactly* on the MSDS as it is on the container you have. If it is not, you may not have the correct MSDS (Accrocco 1988).

**Section II,** Hazardous Ingredients/Identity Information:

Identifies the product and its ingredients. Substances may be identified by their chemical name, chemical family and formula, trade name, and in some cases by a special code name or number which the company uses internally. The chemical family enables one to learn about general properties of the product. (Specific information may be obtained from the Chemical Referral Center hotline listed in Appendix II of *Making Art Safely.*)

■■

Lists the product ingredients for which the U.S. Occupational Safety and Health Administration (OSHA) has set health and safety standards. This may include the product's relative percentage of concentration, although it is not required by OSHA.

■■

Includes threshold limit values (TLV) and permissible exposure limits (PEL) based on time-weighted averages (TWA), and where applicable, ceiling limits (TLV-C) and short-term exposure limits (STEL) for each ingredient.

■■

Material Safety Data Sheet
May be used to comply with
OSHA's Hazard Communication Standard,
29 CFR 1910.1200. Standard must be
consulted for specific requirements.

**U.S. Department of Labor**
Occupational Safety and Health Administration
(Non-Mandatory Form)
Form Approved
OMB No. 1218-0072

| IDENTITY *(As Used on Label and List)* | *Note: Blank spaces are not permitted. If any item is not applicable, or no information is available, the space must be marked to indicate that.* |
|---|---|

**Section I**

| | |
|---|---|
| Manufacturer's Name | Emergency Telephone Number |
| Address *(Number, Street, City, State, and ZIP Code)* | Telephone Number for Information |
| | Date Prepared |
| | Signature of Preparer *(optional)* |

**Section II — Hazardous Ingredients/Identity Information**

| Hazardous Components (Specific Chemical Identity; Common Name(s)) | OSHA PEL | ACGIH TLV | Other Limits Recommended | % *(optional)* |
|---|---|---|---|---|
| | | | | |

**Section III — Physical/Chemical Characteristics**

| | | | |
|---|---|---|---|
| Boiling Point | | Specific Gravity ($H_2O$ = 1) | |
| Vapor Pressure (mm Hg.) | | Melting Point | |
| Vapor Density (AIR = 1) | | Evaporation Rate (Butyl Acetate = 1) | |

Solubility in Water

Appearance and Odor

**Section IV — Fire and Explosion Hazard Data**

| | | | |
|---|---|---|---|
| Flash Point (Method Used) | Flammable Limits | LEL | UEL |

Extinguishing Media

Special Fire Fighting Procedures

Unusual Fire and Explosion Hazards

(Reproduce locally) OSHA 174, Sept. 1985

**Figure 2.2.—U.S. Department of Labor, Occupational Safety and Health Administration (OSHA) *Material Safety Data Sheet*, front.**

**TLVs** represent the airborne concentration of substances to which nearly all workers can be repeatedly exposed: eight hours per day, five days per week, without adverse effects. TLVs have been established by the American Conference of Governmental Industrial Hygienists (ACGIH) for fewer than 600 of the known most hazardous chemicals and are based upon human and animal studies and "the best available information from industrial experience" (ACGIH 1988). Their list alerted OSHA to study hundreds of additional substances for their 1989 report, *Air Contaminants: Permissible Exposure Limits.* It must be kept in mind that TLVs are established for a hypothetical worker who is male, between the ages of 18 and 65 years, healthy, and with no prior medical conditions or risks. Persons at higher risk, such as children, pregnant women, smokers, the disabled, the elderly, those with a medical condition such as asthma, those already overexposed to hazardous materials, and those with smaller stature than the average male, should exercise greater precautions appropriate to their individual risk factors. TLVs, therefore, are intended only as guidelines or recommendations in the practice of industrial hygiene and offer no guarantee that exposure below the stated limits will not result in injury to especially susceptible individuals. As the most stringent standards, however, TLVs are the ones (when both TLVs and PELs are provided) artists should follow.

■■

Permissible exposure limits (**PEL**) for air contaminants have been established by OSHA for thousands of chemicals in the workplace—the 1989 list is more than 2100 pages—and are similar in principle to TLVs. TLVs and PELs are based upon parts per million (ppm) for gases and vapors and milligrams per cubic meter ($mg/m^3$) for fumes and dusts. Since no artist we know has the equipment to measure and monitor airborne contaminant concentrations, the measurement can be useful only relatively. Generally speaking, over 1000 ppm is not a hazard, down to 500 ppm is slightly toxic, down to 100 ppm moderately toxic, and below 100 ppm highly toxic and all precautions should be observed. Dusts of ten $mg/m^3$ and higher are considered nuisance dusts, and below ten $mg/m^3$ become increasingly toxic as the numbers decrease. Greater precautions and safeguards are required the lower the number of $mg/m^3$.

■■

These criteria also take into account the duration of exposure using the concept of time-weighted averages (**TWA**). The recommended exposure limit is averaged over an eight-hour shift, and the limit may be exceeded during the work period as long as it is compensated with a period of lower exposure. PELs also include a "skin" notation for harmful substances, such as chlorinated hydrocarbons, hydrogen cyanide, petroleum distillates, phenol, and picric acid, that may be absorbed through the skin.

■■

Some chemicals (e.g., chloroform, dichlorobenzene, and methyl bromide) act so fast that time-weighted averages are irrelevant. Instead, ceiling limits (**TLV-C**) are established. These are limits that should not be exceeded during any part of the work period.

■■

Short-term exposure limits (**STEL**) are based upon exposure periods limited to 15 minutes, with a maximum of four STELs permitted during a work shift and at least one hour separating each exposure.

Use the lowest limit as a guideline for the safest exposure limit, and if possible never exceed the TLV or PEL, TLV-C, or STEL for any chemical. Additionally, use all appropriate studio and personal precautions to keep exposure as low as possible. When limits are exceeded, provide compensatory time free from exposure to allow your body to recover.

Some ingredients will not have limits. This may be because they are rarely present as particulate, vapor, or other airborne contaminants or because their toxicity is believed to be negligible.

**Section III,** Physical/Chemical Data list certain chemical data for the material such as:

- Boiling point: A lower number means less heat is needed to evaporate the material.
- Vapor pressure: The higher the number, the greater the tendency for vapors to escape at a given temperature.
- Solubility in water.
- Appearance and odor: Help to identify the material. If the actual physical appearance of the chemical you are working with does not match the description on the MSDS, there may be some discrepancies to be investigated further. Be aware that the absence of odor does not indicate safety, and some substances can depress the sense of smell.
- Specific gravity: Above 1.0 means heavier than water; below 1.0 means lighter than water.
- Melting point.
- Evaporation rate: The higher the number, the greater the volatility and availability of vapors for inhalation.

**Section IV,** Fire and Explosion Hazard Data:

The flash point of the material is the temperature at which the material gives off sufficient vapor to ignite in the presence of a spark or flame. The lower the flashpoint, the more hazardous the material. A flashpoint at or below 100°F (38°C) is designated as flammable, and is particularly dangerous: a cigarette, hot weather, or static electricity could set off a fire or explosion. A flashpoint above 100°F (38°C) is designated combustible rather than flammable. For example, among paint thinners, use mineral spirits, which have a relatively high flashpoint of 107°F (McCann 1989).

■■

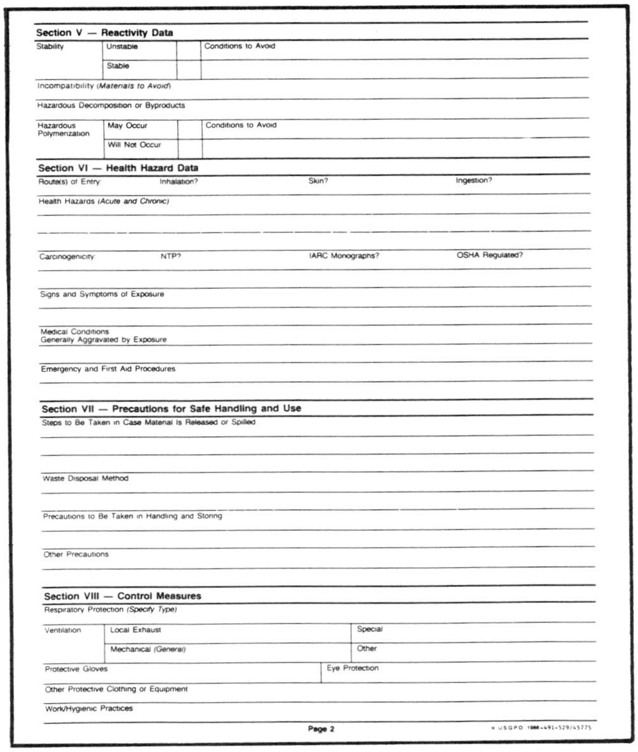

**Section V — Reactivity Data**

Stability: Unstable / Stable — Conditions to Avoid

Incompatibility (*Materials to Avoid*)

Hazardous Decomposition or Byproducts

Hazardous Polymerization: May Occur / Will Not Occur — Conditions to Avoid

**Section VI — Health Hazard Data**

Route(s) of Entry: Inhalation? Skin? Ingestion?

Health Hazards (*Acute and Chronic*)

Carcinogenicity: NTP? IARC Monographs? OSHA Regulated?

Signs and Symptoms of Exposure

Medical Conditions Generally Aggravated by Exposure

Emergency and First Aid Procedures

**Section VII — Precautions for Safe Handling and Use**

Steps to Be Taken in Case Material Is Released or Spilled

Waste Disposal Method

Precautions to Be Taken in Handling and Storing

Other Precautions

**Section VIII — Control Measures**

Respiratory Protection (*Specify Type*)

Ventilation: Local Exhaust / Mechanical (*General*) — Special / Other

Protective Gloves — Eye Protection

Other Protective Clothing or Equipment

Work/Hygienic Practices

Page 2

* USGPO 1986-491-529/45775

**Figure 2.3.—*Material Safety Data Sheet*, back.**

Extinguishing equipment and materials, such as water, foam, dry chemicals, or dry powders, needed to put out a fire.

■■

Special fire fighting instructions on appropriate protective clothing and respiratory equipment. Sections III and IV together can give a picture of how carefully a material should be handled under certain circumstances and of whether your studio is appropriately equipped for optimum safety.

**Section V,** Reactivity Data, provides information on:

Chemical stability and safety during normal handling and storage. Unstable chemicals require special care and precautions.

■■

Materials with which the product is incompatible and from which it should be isolated during handling, storage, and disposal.

■■

Hazards that might result from change in temperature, decomposition, or polymerization (changing to a more complex compound) of the material. For example, photographic fixer, which typically has no hazardous ingredients listed (Section II), may decompose (Section V) with release of sulfur dioxide, which can cause chronic bronchitis. This is the major hazard of fixers.

**Section VI,** Health Hazard Data, provides information on:

- Routes of entry: inhalation, ingestion, skin absorption.

- Short-term and long-term hazards.

- Exposure limits: threshold limit values (TLVs) set by the American Conference of Governmental Industrial Hygienists (ACGIH), permissible exposure limits (PELs) established by the Occupational Safety and Health Administration (OSHA), and limits recommended by the manufacturer.

- Effects of overexposure: the most common symptoms or sensations associated with overexposure to the product and the likelihood of synergistic interaction with other substances (the mixture may be more hazardous than its separate ingredients).

- Emergency and first aid procedures for overexposure (what should be done until medical assistance is obtained).

**Section VII,** Precautions for Safe Handling and Use, provides information on:

- Procedures for handling and cleaning up chemical spills and leaks, including ventilation, protective attire, equipment, and other safety practices.

- Appropriate and safe handling, storing, and waste disposal.

- Other precautions such as removal of sources of accidental ignition of flammable and combustible products.

You can learn from this section the procedures, equipment, and safeguarding attire needed for handling and cleaning up materials you use.

**Section VIII,** Control Measures, provides special protection information:

- Respiratory protection required.

- Ventilation to be used to reduce exposure during normal handling of the material.

- Impermeable gloves, eye protection, and other protective measures.

- Work/hygienic practices such as for the care and disposal of contaminated clothing and equipment and for personal cleansing.

Information regarding the safe use of a chemical product is essential for both the artwork and your health. For a complete guide to reading MSDSs, *The MSDS Pocket Dictionary*, edited by J. O. Accrocco, is available from:

**Genium Publishing Corporation**
**1145 Catalyn Street**
**Schenectady, NY 12303-1836**
**518-377-8854**

## Artist Power

Information on MSDSs has two limitations: one allowed by HAZCOM and the other associated with facts of life.

The right-to-know state and federal regulations requiring manufacturers to provide information to the consumer clearly aid the safety-conscious artist. As a result of lobbying by materials manufacturers, however, HAZCOM (Hazard Communication Standard) dilutes the real power of right to know by allowing companies to impose confidentiality restrictions on the release of product composition data. In the area most germane to artists, claims of trade secrecy for all chemicals, no matter how hazardous, are permitted by the HAZCOM standard.

*According to OSHA, a trade secret may consist of any Formula, pattern, device or compilation of information which is used in one's business and which gives the employer an opportunity to obtain an advantage over competitors who do not know or use it. Chemical manufacturers, importers or employers must be able to support or validate the trade secret claim.*

**(PROCTOR 1987)**

As of this writing, this confidentiality aspect of HAZCOM impairs the use of product composition information for artist and worker health education. (Trade secret claims are also permitted by many *state* right-to-know laws.) Thus far, no challenge to the federal law has been successful.

Acquisition of information may be easier for health professionals than for employers, workers, and artists. Only licensed medical persons, in the context of medical care, can obtain proprietary information about a product's ingredients. Information may be granted only under a contractual arrangement whereby the physician agrees not to divulge that information to anyone, including the artist-patient.

Complete ingredient identification on labels is especially critical for persons who have known sensitivities or allergies to substances and for those who develop reactions so they may be treated properly. Some artists, particularly in photography and several printmaking processes, use a number of different chemicals in the studio. They should know the chemical composition of products to avoid inadvertent chemical reactions.

By purchasing only products in which all ingredients are identified, artists inform manufacturers that we take our right to know seriously.

The second (fact-of-life) reason that information on MSDSs is limited is that testing of materials lags far behind their production. Many substances are never tested until, and unless, a significant suspicion of hazard has emerged. Moreover, the vast majority of MSDSs do not contain information on potential long-term adverse health effects of particular ingredients because of the limits of medical and scientific knowledge.

Until all labels on products conform to the Labeling of Hazardous Art Materials Act of 1988, MSDSs, with their limitations, still provide a useful source of information on potentially hazardous ingredients in any product. Even though a manufacturer has no legal obligation to provide artists or workers with trade secrets, they do have a moral obligation to do so. If you suspect that an ingredient in an art product you are using is affecting your health, you or your physician may press for complete ingredient listing. To do so, you may obtain product manufacturer names and addresses from the National Art Materials Trade Association (NAMTA) Directory of Art Materials Producers and Manufacturers, available at most art retail outlets.

If the MSDS states that users should wear gloves or maintain a certain level of ventilation, it is the employer's responsibility to provide the particular equipment or safe working conditions required by the amount of material and exposure to it. It is the employee's responsibility to utilize them (Accrocco 1988, p. 15).

This means that the MSDS may provide a standard by which employer action may be judged. Furthermore, if individual artists use a product in ways or for durations that are contraindicated by appropriate labeling, the manufacturer may not be held responsible for ill effects thought to be attributable to the product.

As long as artists work with chemicals, there is potential for hazardous conditions. If we request MSDSs and insist upon maximum information, or switch to products that do not keep trade secrets, manufacturers may get the message. We are convinced that improved labeling requirements and the use of MSDSs will enhance our protection and aid in making art safely.

## Health and Safety Programs

We cannot *stay* in the know unless, at some point, we *get* in the know. Education of young people in art, both at home and at school, should include observations and advice about health and safety. Artists who teach or work in an art studio have a multifold and ongoing task: to inform themselves, their students, their colleagues, and the administrators of their school or business about coping effectively with art hazards.

Educating ourselves and our students may be easier than educating our colleagues and administrators. Recognition of the need to change to safer studio practices may be accompanied by resentment toward the necessary inconvenience and habit breaking. Thus we should help our colleagues to recognize the importance of these issues and support attempts to acquire relevant knowledge of preventive measures. Administrators must recognize that it is in their interest to implement health and safety programs. Institutions and businesses may be liable for injury attributed to hazardous conditions, whether caused by fires, accidents, or overexposure to chemicals. Prevention of injury through competent instruction and responsible supervision carries legal as well as safety incentives, for the HAZCOM law is intended to reduce chemically related illness and injury in nonmanufacturing as well as manufacturing workplaces.

Educational health and safety programs in art studios and schools, as in all workplaces, have reduced the likelihood of work-related illnesses, accidents, negligence lawsuits, and workers' compensation claims (McCann, *AHN* 9:2). Moreover, such programs provide a means to comply with OSHA regulations regarding establishment of health and safety educational guidelines. Training courses should be presented at the beginning of the term of employment or study because NIOSH statistics show that the highest percentage of accidental injuries and exposures occur within the first six months after a person enters a new work situation (GAIU 1989, p. 9). The program should discuss labeling and Material Safety Data Sheets and provide training for all employees and students in the uses and hazards of the chemicals and equipment they are likely to encounter.

Most large universities have an industrial hygienist on staff to oversee worker safety in maintenance and in science, engineering, and medical laboratories. This individual is a resource for art departments and should participate in decisions that affect safety in the studio. Autonomous art schools, colleges, and professional studios may find it useful to have an industrial hygienist on staff or under contract as a consultant to advise administration, staff, faculty, and students about optimal safety in classrooms, studios, and the workplace.

Guidelines for setting up health and safety training programs, as required by the Hazard Communication Standard, are available from the National Institute for Occupational Safety and Health (NIOSH). Procedures for establishing a safety educational program in both schools and multiperson art workplaces are much the same:

◆ Organize or appoint a taskforce or committee at both teacher/supervisor and student/worker levels

◆ Assess the art program to identify hazardous materials and activities

◆ Develop strategies to cope with existing hazards

◆ Replace dangerous materials and activities with safer alternatives

◆ Seek expert advice, such as from the school or workplace nurse and industrial hygienist, the local OSHA representative, or the fire department, on technical matters. If possible, do so with the administration's knowledge

◆ Keep the administration informed of your findings and changes

◆ Engage your colleagues in the health and safety program

The ultimate goal of a health and safety program is to prevent, reduce, or eliminate occupational illnesses and injuries. To that end, specific objectives are useful:

◆ Identification and recognition of the actual hazards in the studio

◆ Evaluation of those hazards and understanding of their implications

◆ Specific steps to eliminate and/or reduce those hazards [6]

Plan changes in small bites so they are likely to succeed. For example, safely dispose of hazardous materials and substitute safer ones before designing and installing a new ventilation system. This action will clarify what materials must be used only with appropriate ventilation, thereby helping to convince administration of the necessity of its installation. Learn the way your administration implements change and improvement; as far as possible, work within those procedures (and politics). Find out how projects are funded, and think creatively about how you may contribute to obtaining assistance and support.

Once hazards have been identified and addressed, a basic student or employee training class on health and safety in the studio should be implemented and offered regularly to incoming students and workers. To be effective, funding and time are needed for such an educational program, requiring the support and recognition of the principals in the institution or workplace. Certain individuals should be designated, recognized, and held accountable for the program, in which there is a clearly defined curriculum of informative activities toward actualization of the goals. Course content should include:

◆ Chemical hazards likely to be encountered and their potential health effects

◆ Chemical identification

◆ Proper labeling, handling, storage, and disposal of art material chemicals

◆ Obtaining and using Material Safety Data Sheets and ongoing access to MSDSs in a readily available, clearly organized file

◆ Safe use of equipment and hazards likely to be encountered

◆ Using equipment instruction materials and ongoing access to these instructions in a readily available, clearly organized file

◆ Personal protection with protective attire, equipment, and safe work procedures

◆ Fire and accident prevention

◆ Emergency and hazard communications and procedures for fire and accident

◆ Employee/student rights and responsibilities

In colleges and art schools it may be appropriate to hold faculty development seminars to enable each faculty member to address potential hazards and safety procedures in his or her discipline. Once they are in the know, faculty should integrate that information into the classroom or studio instruction of their students. Methods to evaluate and implement safety compliance in studio environments are provided in the last section of chapter 3, "Evaluating Classroom and Professional Studios for Safety and Ventilation." Let us join together with all who can help to lock the gate on potential hazards before tragedy strikes.

Staying in the know about proper use of hazardous materials is fundamental to protecting our health and prolonging our creative lives. We may nevertheless come to realize that using only safe art materials and procedures may restrict our creative imperatives. This prompts us to acknowledge that there are techniques and visual qualities that cannot be achieved without employing chemicals and practices that contain certain hazards. Thus, in chapters 3 and 4, we present the general precautions and methods that will optimize making art safely despite the presence of potentially hazardous materials.

*3*

**F**or independent artists, the ideal studio is a personal haven designed to support creativity in space clearly separated from normal living space. By contrast, studios in academia, business, or artist co-ops are shared and must be designed to foster collective work without collision or

## The Ideal Studio

friction. Shared studios should support artists in spontaneously choreographing activities toward satisfying creative results. For all users and uses, the ideal studio is function-specific, yet flexible, well lighted, appropriately ventilated, and maintained for continuing health and safety.

Safe materials, good personal hygiene, an appropriate environment, good housekeeping, and proper storage and

◆ ◆ ◆

**Figure 3.1.—S**tudent wearing appropriate eye protection while sharpening an etching tool on a grindstone. (Ontario College of Art, Toronto. Photo: Bruce Green)

handling of all materials and equipment (including personal protective gear, addressed in chapter 4) are key aspects of the successful studio.

Choice of space for the studio requires both intuition and logic. In the selection of desirable space and light, evaluate electricity, structure and building materials, and water and plumbing. Can the floor withstand spills, or will it need to be covered by a protective surface? (Nonskid, easily cleaned surfaces are preferred.) Will the floor and other work surfaces support the weight of any equipment? Is the layout designed for task-oriented work areas as well as storage of materials?

Is the electricity supply sufficient? Are safe outlets (including ground fault circuit interrupters) appropriately located for generous lighting? Suboptimal

vision results from lighting that is either too intense or too dim. Light that is adequate for close work may be insufficient for the studio as a whole. A relatively dark studio predisposes to accidents such as bumps and falls.

The physical layout of the studio has an ergonomic component: the interrelationship between individuals and equipment in the work environment. All tools should be evaluated with regard to comfort and efficacy to prevent undue fatigue, inflammation, cuts, pinches, or other accidents. Locate all equipment with an eye toward safe and efficient use.

Is there sufficient water and plumbing in good repair? Water is required in the studio for making art, cleaning up, and emergency first aid. Its presence, location, and dispersal for each technique and medium are unique and therefore addressed, as relevant, in chapters 5 through 9. In general, artists need access to a toilet, hot and cold running water, and a utilitarian sink, especially to flush eyes or skin, should accidental contamination or injury occur. Ideally, a faucet fitted with a flexible sprayer (or an eyewash fountain) is preferred for all studios. OSHA requires eyewash fountains and deluge showers that quickly deliver large quantities of water in workplaces where chemical exposure may occur. These should be placed near acids, alkalis, and corrosives for immediate washing when necessary. Water in the studio, of course, should also be fit to drink in case of accidental ingestion of harmful substances.

What is the air tempering system? Will it be warm enough in winter and cool enough in summer? Is current ventilation effective? If needed, can a ventilation system be installed? Is the studio big enough to store years of work, without fostering acquisition of a junk heap? Spaciousness is attractive, and our hunger for it can seem limitless. Space must also be maintained, however, so choose a studio that will maximize creativity and minimize the chores of housekeeping.

## Diligent Upkeep of the Studio

**T**o make art safely and to reduce the risk of accident and fire, keep the studio clean and tidy. The most common circumstance for accidents anywhere is carelessness. When we are intensely involved in the process of creating, we often let things fall where they may. At some point, however, the responsible artist recognizes when it is time to stop and clean up. The best approach is continuous monitoring and clearance of workplace hazards. This is that "ounce of prevention" that avoids the "pound of cure."

Today the cure may not only weigh more than a pound, but also may be elusive, if it even exists.

The following are basic precautions for optimum safety in the studio:

**1.** Maintain all equipment and tools in good working order.

**2.** Keep the studio clean and free of hazards such as objects, debris, or wet, slippery floors. Clean studio floors and work surfaces regularly with a damp mop, followed by vacuuming. (This is especially important immediately following work with toxic chemicals and dusts.) Home vacuum cleaners and small hand vacuums are not adequate for most studio materials because the filters do not prevent fine powder particles from reentering and recirculating in the air. If you work with dry pigments (including pastels and chalks), resins, gum arabic, or gum dichromate, use a high-efficiency particulate air (HEPA) filter vacuum cleaner to draw dust-contaminated air through a filter that traps microscopic particles. (To purchase this equipment, call your local industrial supply shop.) Do not allow such particles to accumulate in the studio; vacuum after each use or at the end of the day. Do not dry sweep; it stirs up dust that can be inhaled or ingested.

**3.** Clean up small liquid spills (one quart or less) immediately to prevent further contamination by evaporation. For spills of any size damp mop or vacuum with a designated wet pickup vacuum that has an explosion-proof motor. Wear protective clothing, especially gloves and a mask, when cleaning up spills of toxic materials. Preferred materials for cleanup of spills include activated charcoal, amorphous silicate, diatomaceous earth, cat litter (free of deodorizing chemicals), or vermiculite. Once liquid has been absorbed by these nonflammable materials, dispose of them in a secured heavy-duty trash bag. If, in an emergency, rags, sawdust, or paper is used to mop up spilled flammable substances, place soaked materials in a sealed fireproof container that will not pollute the environment, in accordance with local regulations. Such regulations may be obtained from your township or municipal waste-disposal administration.

Wet pickup vacuums can be used to clean dust accumulations that have first been sprayed with water. Spray gently above the dust; don't disturb dust by spraying directly into it. In both dry and wet vacuuming, the filtered contaminants must be discarded regularly and carefully, in accordance with manufacturer's recommendations as well as local, state, and federal regulations.

Spill-control pillows with hazardous waste disposal bags are available from safety supply outlets and should be in any studio where hazardous liquids are used. The pillows contain amorphous silicates that absorb hazardous liquids quickly. Neutralizing acid spills by adding sodium bicarbonate (baking soda) or alkali spills with citric acid or dilute acetic acid (vinegar) is an option but often creates a more difficult mess to clean up (McCann, *AHN* 4:11).

**4.** Designate and minimize the area where potentially hazardous materials are used; conversely, maximize hazard-free space in the studio.

**5.** If you must use extension cords or hoses in your work, do not cause accidents by leaving them trailing across the floor. Keep them in retractable reels that hang above the work space, or coil them onto a hook when not in use.

## Appropriate and Adequate Ventilation

**V**entilation[1] is the primary environmental issue that affects artists who work with hazardous materials. Poor ventilation means that airborne toxic materials in the form of dusts, fumes, gases, mists, or vapors can be inhaled. Many agents make it to the nose and mouth, some get into the lungs, and a few are absorbed into the circulatory system. Health effects can range from negligible to irritation, immediate reaction, serious illness, chronic health problems, and, in extreme cases, death. The accumulation of vapors from flammable liquids, moreover, can create a potentially explosive environment. Appropriate ventilation provides protection so that many art materials can be used without threat to our well-being. Effective ventilation is essential to making art safely.

The purpose of ventilation is simple: to supply clean air while removing contaminated air from a space. Depending upon the system, it may

- Control heat and humidity for comfort
- Remove toxic dusts, fumes, gases, and vapors
- Prevent fire and explosions

A common recommendation on product labels is USE ONLY WITH ADEQUATE VENTILATION. Such a nonspecific phrase is of little help. Not knowing what it really means, we might take work outdoors to be sprayed, which may feel safer but in

Figure 3.2.—Thomas J. Cutter: *OK Clean Air Past Face*, 8″ × 9″ pencil drawing showing the advantage of lateral slot ventilation.

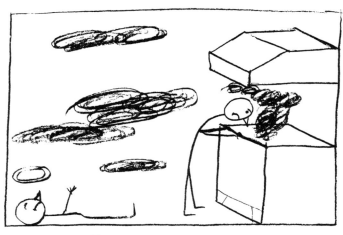

Figure 3.3.—Thomas J. Cutter: *No-No*, 7″ × 9″ pencil drawing depicting the dangers of a canopy hood over chemicals.

fact be counterproductive; an unexpected breeze can send the spray to our faces instead of onto the drawing. "Even in the winter I blow vapors out an open window with a strategically placed fan," proudly states an artist. However, opening windows or blowing with a fan may not be sufficient. Airflow varies with the wind, temperature, and air pressure in quantity, quality, and direction. A blowing rather than exhausting fan in the studio may reintroduce toxic chemicals into one's personal intake area instead of carrying them away. Ceiling-mounted down-draft or reversible fans likewise do more to stir dusts, fumes, gases, mists, and vapors into our personal intake area than to provide effective ventilation. In many instances, they may cause more harm than good.

Some artists believe that airborne substances they encounter are heavier than air, and are therefore found primarily close to the floor. A floor-level fan is not appropriate because gases, vapors, and fumes enter the air in all directions. Some portion of airborne mists and particles does fall as a result of gravity, but spraying, pouring, and blowing by a floor-level fan can bring substantial quantities of undesirable substances to one's personal intake environment. To improve ventilation significantly, floor fans must effectively dilute or remove contaminated air from the workplace.

"Use with adequate ventilation" requires provision of sufficient ventilation to keep airborne concentrations of mist, dust, fumes, gases, or vapors away from one's personal inhalation intake area and below the levels considered hazardous to health

(McCann, *AHN* 8:4). Once air is in motion, the pressures created can maintain clean, fresh, and continuous supplies of air. Appropriately designed ventilation is that which will move air in, around, and out of the workplace.

The two basic ventilation systems are general or dilution ventilation and local exhaust ventilation. **Dilution ventilation** is an air circulation system, commonly used in air conditioning. By contrast, **local exhaust ventilation** clears a specific area by drawing contaminants and venting them to the outside or to a filtered system. In the next sections we describe these two systems, discuss the procedures for selecting appropriate ventilation, and provide resources for obtaining ventilation systems and advice. When adequate ventilation is not available, personal protection may be provided with respiratory masks, which are described in chapter 4.

Basic rules for ventilation include:

♦ Use local exhaust ventilation whenever possible.

♦ Enclose (contain) contaminating processes in the smallest area possible, e.g., in a glove box, spray booth, or small room.

♦ Be sure that the air entering your breathing zone is clean, not already contaminated.

♦ Know the source of replacement air; be sure it is clean.

♦ Make certain that exhausted, contaminated air does not enter or reenter work or living spaces.

## GENERAL OR DILUTION EXHAUST

General (dilution) ventilation operates on the principle of diluting the concentration of undesirable substances in the air by mixing them with uncontaminated air. Dilution is provided by a central air system with or without heating or cooling (air conditioning) that constantly circulates air throughout a space via fans, ducts, and filters. These systems are expensive to install, and in an older building they may require visually and physically intrusive ductwork. While there is an intake from outside air, the air that is moved about is primarily recirculated and constantly passed through filters to dilute the concentration of contaminants. The filters must be frequently cleaned and replaced for the system to work effectively.

The intake vent must be located to attract the cleanest air possible. The exhaust outlet should be placed where it will not expose others through open windows or intake vents. A system in balance between intake and exhaust will allow doors to open and close easily. Undue pulling or slamming of doors, both within the ventilated space and between it and outside locations, suggests airflow imbalance. Adjustments may be required.

General or dilution systems are not without limitations. Few circulate the air evenly throughout a space; many create drafts and dead spots. Dilution ventilation does not always provide adequate protection from harmful gases and vapors that disperse in all directions; moreover, these systems occasionally stir up dusts and powders into the air that one breathes. Contaminants may be carried out of a darkroom or studio into other rooms. This possibility, however, does not eliminate the overall value of dilution to decrease airborne concentrations of potentially hazardous substances. To decrease the likelihood of dust agitation and contamination of other areas, adjustments can be made in air pressures and exchange rates without total elimination of dilution ventilation.

## LOCAL EXHAUST

Local exhaust works on the principle of capturing toxic materials at their source before they contaminate the general air in the room. The simplest, most direct, and most effective method of ventilation is an exhaust device located immediately behind the work area to draw contaminated air away from the artist. Local exhaust ventilation is preferred when

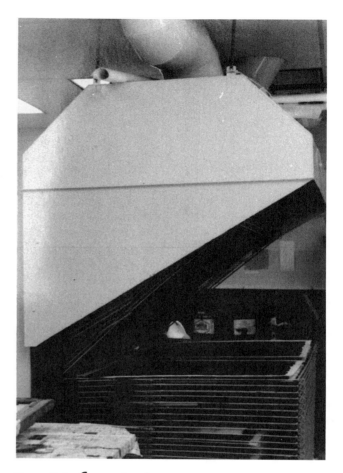

Figure 3.4.—Canopy hood used to exhaust vapors directly from prints in drying rack, University of Texas, Austin.

contaminants are highly toxic or when large amounts of toxic materials are produced. Use of the following art materials requires local exhaust ventilation:

- Acid and ammonia solutions such as those used in photography, printmaking, and etching

- Aerosol sprays of fixatives, paints, and airbrush

- Gases and vapors from solvents and photo and printmaking solutions and processes

- Powders from dyes, pigments (including pastels and charcoal), gum arabic, gum dichromate, and all photography and printmaking chemicals in powdered form

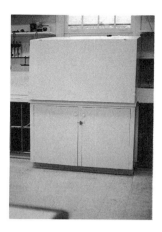

Figure 3.5.–An enclosed compartment that allows etching acids to be vented directly to the outside. A, closed; B, open. University of North Carolina at Wilmington.

The Occupational Safety and Health Act of 1970 requires that local exhaust ventilation be provided in workplaces that use paints; spray finishes; open-surface containers of solvents, acids, and alkalis; and flammable and combustible liquids in storage rooms and enclosures. These guidelines are appropriate for individual artist studios as well.

The local exhaust-ventilated work area should be only as large as necessary to handle any hazardous materials used in your work. Generally speaking, dusts and powders require a stronger exhaust system to clear the air than do vapors and gases. Exhausting contaminated air from the work area with a window panel or wall-mounted fan is both safer and more efficient than blowing it out with a fan, which will disperse some of the contaminants back into the work space. When arranging a studio, recall that an exhaust fan mounted in a window at work level (with the work table against the window) is the easiest and least expensive local exhaust installation. Second in expense is a wall fan that vents directly to the outside because it avoids costly duct length.

Place the exhaust opening (fan or outtake duct) of the ventilation system as close as possible to the source of contaminants. To reduce the expense of ductwork, limit the distance air must travel from contaminants to the fan or outlet. When working, face the outtake device so that clean air from behind you comes across and away from the work, to be expelled by the exhaust. If you work with combustible materials such as solvents, lacquer, and fixative sprays, make sure the fan and wiring are explosion-proof. Electrical sparks from the fan could ignite flammable substances.

An intake of clean air from an open window, door, or vent should be located on the side of the work space opposite the exhaust device. The intake must permit airflow related to the draw of the exhaust. Excessive intake sources create conflicting and self-defeating cross drafts, while insufficient intake creates negative pressure and occasionally backflow of contaminants from the fan. If there is a chimney or exhaust flue in the work space for oil, gas, kerosene, or wood-burning space heaters or kilns, safety devices—such as a valved flue—must be used so the exhaust fan does not draw toxic fumes and gases, such as carbon monoxide, from the flue into the work environment.

There are three basic types of local exhaust systems: enclosures, receiving hoods, and exterior hoods. Fume hoods, ventilated compartments, spray booths, and glove boxes (see chapter 4) are environment-containing or self-contained enclosures available for limited use of specific noxious chemicals and sprays. They can augment local exhaust ventilation

Figure 3.6.–Using exhaust ventilation, Diane Goodman works on a monotype at the Cheltenham Center for the Arts.

Figure 3.7.—**U**se of a respirator accompanies lateral slot exhaust ventilation during printing with oil-based screenprinting inks at Ontario College of Art, Toronto.

and facilitate making art safely.

A **spray booth** is an excellent method for containing contaminants produced by spraying lacquer, paints, fixatives, adhesives, and other flammable and toxic materials. Not to be confused with an inexpensive spray shield (sometimes also called a spray booth) that merely prevents sprays from falling on surrounding areas, a spray booth contains and vents particulates with a filtering device that removes particulates before discharging air into the environment. Whether custom-made or a commercial product, select a spray booth large enough to accommodate your work and a fan powerful enough to draw at least 150 cubic feet of air per minute (cfm) per square foot of the spray booth opening. Locate your spray booth along an external wall vented directly to the outside with minimum duct length. Vented booths are safer than recirculating booths with a particulate filter and an activated charcoal filter to remove odors. One pound of activated charcoal becomes saturated with 11 ounces (0.315 liter) of liquid, requiring frequent replacement. High humidity, moreover, reduces the amount of solvent vapor that can be trapped. Neither particulate nor charcoal filters trap some gases and vapors that may be encountered in art and photography studios (e.g., ammonia, formaldehyde, nitrogen dioxide, ozone, or sulfur dioxide). Recirculated air from a spray booth exhaust is not desirable in either a multiemployee workplace or an otherwise unventilated private studio.

Flammable propellants or solvents require a spark-proof spray booth located away from possible sources of ignition such as electric switches, motors, lighted cigarettes, or flame. When working with a spray booth, avoid buildup of flammable vapors to explosive levels. Use spray booths only as recommended by the manufacturer. Clean and replace filters regularly, as advised by the instructions, to improve safety, efficiency, and energy conservation.

Receiving overhead canopy hoods, similar in principle to those used over kitchen stoves, are not recommended for most two-dimensional art making. As we lean over the work area, these hoods may draw contaminated air upward through our personal inhalation space. Canopy-hooded local exhaust systems can be made safer, however, by installing a compartment or Plexiglas enclosure over the work space to prevent vapors from rising into the personal intake area. A see-through shield is preferred, with a slot at the bottom for handling artwork in trays that contain photochemicals and printmaking acids. Installed correctly, with an intake to provide sufficient clean air, such compartments provide efficient, direct, local exhaust.

Receiving lateral slot ventilation hoods are safer than overhead canopy hoods. Lateral slot exhaust systems have horizontal slots behind the work area that continuously draw contaminated air away at multiple levels, from the work surface up and above the personal intake level. All local exhaust systems require regular inspection, maintenance, and cleaning in accordance with the manufacturer's recommendations.

Exterior hoods are capture devices with a flexible duct (elephant trunk) that may be placed close to the contaminant. They can be useful in lithography and for solvent cleaning of printmaking plates.

All exhaust systems require one or more fans, which may be axial flow or centrifugal. Axial fans, which have standard propeller blades that rotate around an axis, are appropriate for most two-dimensional art work. Centrifugal fans operate like a waterwheel or hamster treadwheel, with blades that are perpendicular to the outtake surface. They are most effective for dusts formed by grinding, mixing, and sanding procedures used in three-dimensional arts.

Learn where your ventilation exhaust leaves the building. Be certain that the escape port is not located near open windows or intake vents that may circulate contaminants back into buildings. In one art school, the ventilation system in the printmaking studio exhausted out the window on the sixth floor. The contaminated air blew right back into the windows of the seventh-floor painting studios. Not unexpectedly, the painting students expressed concern.

Protect the environment. Many communities regulate what can and cannot be vented to outside air. Be sure vented substances are permitted under all codes and regulations. If they are not, find safer substitutes and/or install filtered ventilation that will provide you, others, and the environment with nonpolluting air.

In large studios where many artists work with hazardous materials, industrial local exhaust ventilation systems with outtake hoods (preferably lateral slot), filters, and ducts must be installed. Be sure that the devices you plan to install are in compliance with local, as well as federal and state, regulations.

## SELECTING A VENTILATION SYSTEM

Formulas are available to calculate the appropriate exhaust volume for an effective ventilation system (see "Darkroom Safety and Ventilation" in Chapter 9). Fans can be classified on the basis of size and capacity to move cubic feet of air per minute (cfm). The choice of fan depends upon the size of the room and characteristics of the contaminants to be vented, including their threshold limit values, evaporation rates, molecular weights, and specific gravities. The details of calculation are beyond the scope of this book, but the following suggestions should help readers to intelligently acquire a ventilation system.

Setting up a ventilation system requires time, patience, common sense, and a willingness to ask questions. If necessary, complicated calculations and procedures can be undertaken by a ventilation specialist. Several factors impact on the choice of appropriate ventilation. To establish the kind and size of exhaust device you need, note the following:

◆ List the identity and quantity of each potentially contaminating material used on a typical day.

◆ Identify the physical forms that each material may take (e.g., dust, fumes, gas, liquids, mists, vapors; compounds).

◆ Identify the route, frequency, and duration of exposures.

Then review your working procedures:

◆ Note where, when, how, how often, and how long you use each material.

◆ Determine which procedures need ventilation, and decide how the relevant contaminants can best be contained in a local exhaust area.

◆ Establish the best ventilation combination for your needs: local exhaust, general dilution, and/or containment (e.g., a spray booth).

To acquire and install an effective ventilation system:

◆ Measure the room and the immediate local exhaust area. Diagram the space, including location of fresh air intake, windows, doors, mixing tables, sinks, storage, and other designated areas.

◆ Consult your local yellow pages under air conditioning equipment and systems. Ask the vendor if they install ventilation systems; if so, visit them with the materials list, working procedures, and plans.

Your discussion with the ventilation system supplier should focus on the most logical and affordable system for your needs. If you have any doubts about the suggested system, consider the services of a professional engineer who specializes in industrial ventilation. Up-front costs vary and can be high, but this expertise can save money in the long run. For example, if you live in a cold-winter location, fresh air intake will involve cold outside air. A ventilation engineer will calculate heating, circulating, and cooling costs for your particular environment and can advise you on effective, economical, long-term procedures. The more complex your needs, the more likely a ventilation engineer will be able to help. Alternatively, consultants from state OSHA offices, the Society of Toxicology (SOT), and the American Conference of Governmental and Industrial Hygienists (ACGIH) may also be available.

Some air purification systems claim to purify air without the expense of a ventilation system. Do not be fooled. Ion generators may remove some particles from the air, but greasy residues may be left on walls and other vertical surfaces. The particles eventually lose their negative charge and can reenter the air. Ionizers also "deodorize and freshen" by producing ozone, a respiratory irritant that can cause breathing problems and lung damage. Electrostatic precipitators do not remove vapors, gases, or large particles such as sawdust, pollen, lint, or bacteria. They may remove some very small particulates (e.g., from smoke), depositing them on special collector plates that must be cleaned frequently. Activated charcoal filters, cyclones, fabric collectors, and HEPA (high-efficiency particulate air) filters have their place in some workplaces but are inappropriate for most two-dimension artists' studios and workplaces. Do not waste your money on ineffective or unnecessary equipment.

## VENTILATION RESOURCES

For guidelines on ventilation system installation, "American National Standard Fundamentals Governing the Design and Operation of Local Exhaust Systems" is available from the American National Standards Institute (ANSI) (see Appendix II). Universities also offer courses in the practical application of ventilation systems.

If you suspect your studio is not ventilating properly, consider measuring the airborne concentration of chemicals. Colorimetric tubes with hand-held pumps are relatively inexpensive, but different chemicals require different tubes, and cross-reactions with closely related chemicals complicate the interpretation of the readings. If you are seriously concerned about potential health hazards, employ an accredited laboratory that measures airborne chemical concentrations. A list of these laboratories can be obtained from the American Industrial Hygiene Association at 216-873-2442. Appropriate ventilation is essential for making art safely. Its importance cannot be overemphasized.

## Safe and Appropriate Storage

In the development of a studio, do not forget issues of storage. Proper storage techniques and habits are crucial for making art safely. Take time to organize, assess, label, and discard chemicals and materials that have accumulated (see "Safe Disposal of Dangerous Materials" later in this chapter).

To store potentially hazardous art materials safely:

Know your local regulations. Some areas limit the amount of hazardous materials, especially flammable solvents, that may be stored.

■■

Designate storage space that is well ventilated, cool, and protected from direct sunlight. Store liquids and solids in separate areas.

■■

Prohibit mixing or dispensing chemicals in the storage area.

■■

Store materials and chemicals in reasonable quantities for a reasonable time, preferably in amounts that will be used within two months.

■■

Keep materials in tightly closed, original, or unbreakable containers. Do not use food product containers. This prevents accidental ingestion of chemicals that do not have warning odors. Ingestion of chemicals stored in soda bottles and other food containers has led to severe illness and death.

■■

Clearly, prominently, and appropriately label each container with permanent ink, noting the date received and date opened.

■■

Limit storage areas to above-the-floor and below-shoulder height for more convenient access.

■■

Separate materials that may generate undesirable reactions if inadvertently mixed. For example, bleach mixed with acid or ammonia may create highly toxic, potentially lethal gases. Acid concentrates used in photography and printmaking, such as hydrochloric, phosphoric, and sulfuric, should be stored in a special, distinct, and distant cabinet. In addition, highly reactive nitric acid should be isolated from other acids.

■■

Store flammable and combustible materials such as acetone, ethyl alcohol, petroleum distillates, toluene, turpentine, and other solvents in specially designed, OSHA-approved fireproof cabinets. Keep only a minimum practical amount on hand. Keep the cabinets closed; vent if required by local regulations. Explosion-proof refrigerators also decrease the risk of fire or explosion. Solvents retained for reuse should be stored in metal safety cans that have spring-loaded caps to prevent evaporation.

■ ■

Store oil- and solvent-soaked rags loosely in specially capped waste cans that permit air to circulate around them to dissipate heat buildup and prevent spontaneous combustion.

■ ■

Store chemicals that are highly toxic if ingested (e.g., mercuric chloride and sodium cyanide) in a locked cabinet.

As you reassess and clarify your storage needs, think about fire prevention as well as general safety. Make safe storage and disposal an ongoing part of your work, with seasonal cleanings and reevaluations.

Figure 3.9.—This waste can is foot-operated for easy access when hands are occupied; it can be used for disposal of solvent-soaked rags and other flammable waste.

Figure 3.8 .—This self-closing, leak-tight, and spill-proof polyethylene bottle features a spring-loaded plunger and brass nozzle that closes automatically, allowing one-handed dispensing. Bottles of this type can be purchased from safety supply stores.

Figure 3.10.—Lateral slot exhaust system, metal safety cans with spring-loaded cap for solvent storage, and warning signs at Ontario College of Art, Toronto.

## Safe Disposal of Dangerous Materials

Toxic waste disposal is a global problem. Responsible artists will help to solve rather than exacerbate the problem by avoiding hazardous substances, and substituting safer materials whenever possible. If hazardous materials are necessary for your work, learn to handle and dispose of them safely. "Hazardous Waste Minimization Manual for Small Quantity Generators" and *The Minimizer,* a free quarterly newsletter, are published by the Center for Hazardous Materials Research (CHMR) at the University of Pittsburgh (see listing in chapter 2 and Appendix II, Resource Directory). If you must regularly use, and hence discard, hazardous materials, the newsletter can keep you informed. CHMR maintains a directory of state agencies for waste management and also advises small and medium-sized firms on hazardous waste management. Their hotline is **800-334-CHMR (2467)** for instructions about safe disposal.

If you have any of the following substances in your studio, call the EPA about disposing of them safely: asbestos, pentachlorophenol, lead, uranium, PCBs, and benzene. Call your local waste management office or health department as well, for they may have a hazardous waste system or procedure that follows local as well as state and federal law. Consult your fire department for advice on disposal of flammable and combustible substances. Universities, hospitals, service stations, reclamation centers, and independent waste disposal contractors may also help solve your hazardous waste problems.

The following guidelines are offered for disposal of materials artists are likely to use. If a product label or Material Safety Data Sheet makes specific recommendations, follow those directives.

Aqueous liquids, such as acid and alkali solutions used in photography and printmaking, should be neutralized (baking soda for acids; citric acid for alkalis) before they are slowly poured down the sink with lots of running water. If you have a septic tank instead of a municipal sewage system, do not pour highly toxic chemicals, such as bleaches or cyanide compounds, down the drain because they may destroy the bacterial culture in the tank. Photographers with a home septic tank should severely restrict disposal into the system. If you must use color processing chemicals, we suggest disposal through an approved chemical disposal service.

Organic solvents, such as turpentine and mineral spirits, should not be poured down the drain because they kill bacteria that break down other waste products. Some solvents can be recycled, which greatly reduces the amount to be disposed:

1. Slowly pour used or dirty solvents through fine steel mesh, or a coffee filter, inserted in a metal funnel and into a receiving can.

2. Cap and clearly label the container as to what it contains and when it was retrieved.

3. Package the dirty (greatly reduced!) solvents and chemicals in the filter in a sealed container, and dispose of it through regular trash pickup procedures.

For small quantities (less than a liter), let solvents evaporate. Prevent inhalation by placing evaporating solvents under a fume hood or outdoors where the material is secure from children. For large quantities of solvents and solvent-containing materials, contact your local fire department for regulations and recommendations.

Solvent-soaked rags and papers should be set outdoors to allow evaporation prior to discarding them in sealed, flame-proof containers. Do not use plastic containers because many solvents will dissolve them.

All refuse containing chemicals, whether in packages, cans, or bottles, along with sharp-edged objects and broken glass, should be discarded in a separate container that is visibly labeled to warn and protect refuse collectors. This is especially important where compactor trucks are used.

If you live in an area served by a local dump, do not discard hazardous materials there without consulting the local health department or waste management agency for regulations and recommendations. Private contractors for waste disposal may be listed in your yellow pages under garbage, rubbish, or trash removal.

## Fire Prevention

Some art materials, especially organic solvents and materials containing them, such as paints, lacquer, and fixative sprays, can be serious fire hazards. As a result, studios that use or store these materials should conform to local and federal laws concerning storage and handling of flammable liquids. Flammable liquids are identified by the Federal Hazardous Substances Act (FHSA) as:

- Extremely flammable: a flashpoint below 20°F, meaning that a spark will ignite the substance or its vapors even in very cold weather

- Flammable: a flashpoint between 20° and 80°F

- Combustible: a flashpoint above 80°F

The National Fire Protection Association (800-344-3555) classifies liquids with a flashpoint below 100°F (38°C) as flammable, and combustible if 100°F (38°C) and above. These are the standards we have adopted in *Making Art Safely*.

Every studio should be equipped with smoke alarms and fire extinguishers appropriate to the size of the space and use of flammable materials. Much of the equipment mentioned here is available at hardware stores, with specialty items available from the safety equipment suppliers listed in Appendix III, Safety Equipment Suppliers. **Smoke alarms** and **fire extinguishers** are especially important for artists who smoke or use flammable solvents, materials, or equipment. Remember that flammable vapors, though they dissipate in air, may still be hazardous several feet from their source.

Smoke alarms are readily available and easily installed. Because they are more sensitive and less likely to be triggered by nonsmoke dusts and vapors, photoelectric alarms are preferred over ionization detectors. Be sure the battery works; occasional beeps signal that it is time for a replacement.

There are four kinds of fire extinguishers, classified by a letter in a symbol, that are effective for four kinds of fire. A combination A-B-C type covers most hazards artists are likely to encounter. Be sure the kind(s) you have in your studio are appropriate to the types of fires you could have, and place them in areas appropriate to their possible need:

- Class A Natural Materials: paper, wood, and textiles that require cooling, quenching, or smothering, can be put out with water.

- Class B Flammable or Combustible Liquids: alcohol, printing inks, and solvents must be smothered or blanketed. Water is not effective.

- Class C Electrical: no matter what material is burning, if it has been caused by electricity and the current is on, using water can cause electrocution. Electrical fires require a nonconducting extinguishing agent.

- Class D Metals: magnesium and titanium, used in the printing industry, are combustible, and require professional firefighting.

Small office fire extinguishers are available at art and office supply outlets.

**Foam generators** will cover small toxic and/or combustible chemical spills, such as acids, alkalis, and hydrocarbon liquids, to reduce the chances of vapors being released into the air. Special, noncombustible **fire blankets** are also available for smothering fires; they should be located near the work area where flammable solvents and inks are used.

Be sure that you, and others working in your studio, know how to use fire extinguishers, foam generators, and fire blankets properly **before** an emergency arises. If at all uncertain, call your local firefighter company for instruction.

Should a fire break out:

First, sound the alarm and have all persons not trained to use a fire extinguisher leave the building.

■■

Second, call the fire department.

■■

Third, if you have the right kind of extinguisher and know how to use it, put out the fire. If the fire is too big for the extinguisher, leave the building **immediately** and leave the fire to professional firefighters. Smoke is the volatilized product of combustion consisting of toxic gases and particulate matter. Smoke inhalation is one of the principal causes of death in fires, sometimes because the victim fails to recognize its lethality.

To be sure that your studio is as fire-safe as possible:

Store only the minimum amount of flammable and combustible liquids needed for your work.

■■

Substitute noncombustible materials whenever possible.

Store flammable and combustible liquids in nonbreakable containers, preferably safety cans with spring-loaded caps. Never store them in plastic or glass containers.

Store large quantities of flammable and combustible liquids in OSHA-approved safety storage cabinets. Do not exceed cabinet capacity. Keep the cabinets capped or closed; vent if required by local regulations.

■■

Remove and discard safely old containers of flammable and combustible liquids.

When handling flammable or combustible liquids, observe the following safeguards:

Pour or distribute the contents of one container at a time. Do this at a safe distance from any source of flame or spark, including cigarettes, electric equipment, and static electricity.

■■

Handle liquids in an area set apart from other materials and activities.

■■

If transfer of materials involves use of air pressure, be sure the nozzle and receptacle are properly connected and grounded.

■■

Keep containers capped when not in use, and promptly clean up spills with spill-control pillows or amorphous silicates manufactured for the purpose. Wear protective clothing for cleanup, such as boots, impermeable suit, gloves, goggles, or respirator. If rags or paper is used, dispose of them promptly in sealed containers in accordance with local regulations. Do not store them except in special metal waste disposal cans with self-closing lids or in sealed containers that exclude air, to minimize evaporation.

■■

In the event of large spills, immediately extinguish any flame or pilot light in the area. Call the fire department and report the spill. Cut off electric power from a remote circuit breaker or switch. Open windows and evacuate the building. Allow professionals to clean up the spill.

■■

For your safe escape during a fire, do not place trash containers that may contain flammable or combustible materials near exits from the studio.

■■

Periodically assess your studio for fire hazards and make corrections as needed.

■■

Hold fire drills as often as necessary to keep all who use the studio informed about the best emergency procedures.

## Evaluating and Establishing Safety in Small and Individual Studios

**O**SHA does not regulate self-employed persons or workplaces already covered by statutes of other federal programs. Furthermore, because work settings with few employees may carry low priority with government agencies, we think it is important to outline procedures for individual artists to evaluate the safety of a personal studio as well as those in the workplace. If you have not already done so,

Inventory all products used in the studio. Include consumer products, especially those used in greater quantities or in atypical ways. Include soaps, cleansers and maintenance products because some of these contain toxic materials.

■■

Discard materials no longer used and old materials for which ingredient information is not available.

■■

Read the labels to locate hazardous materials, and note them on the inventory list.

■■

When a label states, "For professional or industrial use only," it is a significant caution that should be heeded. It means that the product requires special handling and should never be used by children or untrained adults. A Material Safety Data Sheet (MSDS) should be obtained from the manufacturer so that guidelines for use of the product may be followed.

■■

Set up and maintain a file or looseleaf notebook of Material Safety Data Sheets, preferably in alphabetical order by product name so you can refer to it quickly if necessary (see chapter 2 for how to use MSDSs). By law, MSDS files must be maintained by employers and made available to all employees. Know how to access the file; update it annually. Art teachers (as employees) and their students must also have access to the MSDS file.

■■

Use the inventory process to organize, store, and discard hazardous materials.

Figure 3.11.—**D**eluge shower and eye wash fountain at Auburn University printmaking studio.

Once aware of the identity and location of hazardous substances in the studio, determine, acquire, and install the appropriate

♦ Ventilation system

♦ Smoke detection and fire fighting devices

♦ Eye wash, shower, and water stations

♦ Appropriate identification and warning signs

By law, signs are required in public spaces for exits, fire fighting equipment, and other safety and danger topics. A variety of signs are available from safety equipment suppliers (listed in chapter 4 and Appendix III, Safety Equipment Suppliers). They include such admonitions as KEEP CONTAINERS COVERED, FIRST AID STATION, USE ONLY AFTER INSTRUCTION, and WEAR YOUR SAFETY GLASSES. Astute placement, selective use, and periodic changing of signs can contribute to greater safety in individual as well as shared studios.

While cleaning and sprucing up your studio, you might wish to add something both esthetic and healthful. If you have space and light to accommodate plants and the inclination to care for them, a few well-chosen ones may contribute not only to your decor but to your health as well. In particular, corn and dracaena (dragon) plants, golden and

Figure 3.12.—**V**arious signs are available from safety supply companies.

green pothos (the common variegated, heart-shaped leaf "devil's ivy"), philodendrons, and spider plants may capture small amounts of such contaminants as formaldehyde, carbon monoxide, and solvent vapors from the air. A study conducted by the National Aeronautics and Space Administration (NASA) determined that these plants, placed in sealed containers in which formaldehyde was introduced, removed 80 percent of the formaldehyde molecules within 24 hours. Although plants are no substitute for appropriate ventilation, one medium-sized plant for every 800 cubic feet of space adds color and may improve air quality (Danneberg 1989; Caplan 1990).

## Evaluating Classroom and Professional Studios for Safety and Ventilation

**W**e recently spotted an art school advertisement showing a printmaking student wearing a respiratory mask around her neck. While the school most likely wanted to show serious concern for health and safety in the studio, we speculated that use of the mask could mean the ventilation system at that school is inadequate or nonexistent. The advertisement highlights a fundamental, often ignored principle of oc-

cupational safety and health. Engineering controls such as effective ventilation are almost always preferable to use of respiratory protective gear. This is reinforced by federal law, which states that respirators are not to be used as primary protection except under rare and brief circumstances. When a mask is necessary, both mask and instruction for its proper use must be supplied by the employer (Rossol 1982).

The Occupational Safety and Health Act of 1970 created government authority to investigate and prevent workplace accidents and illnesses and set up three federal organizations to implement it:

- Occupational Safety and Health Administration (OSHA) to enforce health and safety standards in the workplace
- National Institute for Occupational Safety and Health (NIOSH) as a research center
- Occupational Safety and Health Review Commission (OSHRC) to settle disputes that arise as a result of efforts to enforce the Act

The Act also grants employees the right to inspect and evaluate workplaces for conformance to safety and health regulations.

HAZCOM (hazard communication standard)[2] requires information dissemination and gives artists significant opportunities to learn about materials

Figure 3.13.—Signs can remind artists of dangers and precautions in the studio. (Ontario College of Art, Toronto. Photo: Bruce Green)

they encounter on the job. Also, employers (including educational institutions and art studios) must develop and present a hazard communication and education program (see in chapter 2, "Health and Safety Programs").

Under HAZCOM law, the employer or institution must keep an up-to-date MSDS file on all hazardous chemicals and materials in the workplace or studio, and any employee has a right to refer to it at any time. Each hazardous chemical in the school or workplace must be labeled as to:

+ Its identity

+ The degree of health, flammability, and reactivity hazard

+ Special handling information

Every time a new chemical is introduced into the studio or workplace, workers are to be informed and trained in its handling.

The Occupational Safety and Health Administration (OSHA) is a federal law enforcement agency with a number of regional offices throughout the United States. It is listed in telephone directories under the U.S. Government Department of Labor. The purpose of OSHA is to ensure that employers:

+ Reduce workplace hazards

+ Implement safety education programs

+ Maintain job safety and health standards

Typically, basic fire and safety procedures are enforced through periodic inspections by the local fire marshal and building, licensing and inspection officials. It is relatively easy to check exit and warning signs, fire extinguishers, smoke alarms, water sources and eye wash stations. In addition, electrical hazards such as frayed wires, scorching around switches and receptacles, and improper use of extension cords should be checked.

Air quality is not addressed as easily as fire safety. Air is always changing, and what is problematic for one person may be imperceptible to others. When an artist experiences symptoms such as burning in eyes, nose, and throat passages, difficult breathing or chronic cough, headaches or dizziness, or numbness or tingling in extremities, there may be reason

to check for possible contamination of workplace air. If more than one person experiences symptoms, expert evaluation may be required.

A student or entry-level artist may find it difficult to suggest the possibility of an environmental or health hazard in the studio. If you have questions or suspicions about the safety of your work environment, share them with peers to see if anyone else has similar concerns. Discuss safety issues with those who should know and/or be responsible. Find out what they know and what they are doing for your safety. If you are an employee and are unconvinced by their knowledge, sincerity, or eagerness to establish a safe workspace, contact your local OSHA State Consultation Service for further information and advice (U.S. Government, Department of Labor listing in the telephone directory). It is possible that an inspection and evaluation of the studio-workplace by an OSHA representative would be considered appropriate.

Any employee or employee representative (such as an attorney) who perceives a threat of physical harm or imminent danger from a possible health or safety violation may request an inspection. A written request for an inspection must include the reasons for the request and the signature of the employees or employee representative (Tell 1988). To protect employees from inappropriate (and illegal) reprisal, the written request to OSHA for an inspection can specify that the employee's name be withheld. Alternatively, the request to OSHA for an inspection can be made by a lawyer who represents the employee(s). Names of "whistle-blowers" do not have to be included or divulged to the employer. While the Occupational Safety and Health Act of 1970 provides protection against discharge and discrimination for employees who initiate or testify in complaint proceedings, efforts to prevent possible harassment can be prudent and time-saving. If, like many entry-level artists, your financial resources are limited, a local COSH (Committee on Occupational Safety and Health) group, a legal aid society, or a Volunteer Lawyers for the Arts group may be able to help.

If a hazard is thought to exist, an OSHA representative will make a preliminary visit to determine whether there is reason to believe a threat to health does exist; if so, an inspection will be undertaken. A copy of the request must be given to the employer by the time of the inspection, but the name(s) of the initiators may be held in the confidence of an attorney. The inspection may include measurement of airborne contaminants to determine whether concentrations exceed standards set by the American Conference of Governmental Industrial Hygienists'

Figure 3.14.—Elaine Cohen Nettis works with water-soluble crayons on paper. Her studio has an exhaust fan and windows that can be opened according to the weather and ventilation needs. (Photo: Joseph Nettis.)

threshold limit values (TLVs) or OSHA's permissible exposure limits (PELs), based upon time-weighted averages (TWAs).

Despite understaffing and time-consuming bureaucratic procedures, federal and state OSHA officials have clout. They are authorized to question privately any employer, owner, operator, or employee. Testimony may be required under oath, and failure to comply may result in a contempt of court citation. An employer cited for a violation must agree to take steps to comply with the law. When notified of a health or safety violation, the employer has 15 days in which to contest the finding and may have to pay a fine.

OSHA standards are predicated on healthy males between the ages of 18 to 65, with exposure limited to eight hours per day, five days per week. OSHA standards and evaluation, however, may not be entirely appropriate for schools and studios where the majority of artists are other than healthy adult males. Greater restrictions on exposure may be warranted for youngsters, for the elderly, and for women, especially those who are pregnant or nursing. Restricted exposure may also be appropriate for individuals with allergies or other chronic health problems.

These factors should be remembered and placed in perspective with the OSHA report. Ultimately, each of us individually is responsible for our personal health, safety, and well-being.

# 4

In relentless pursuit of the best possible odds for minimizing the perils of major diseases, many elements of society have recently addressed the role of environment in health. Most artists have heard the current prescription for a healthy life: Do not smoke or chew tobacco; use

# *The Healthy Artist*

alcohol only moderately if at all; exercise regularly and control your weight; and eat a diet low in fat, high in fiber, low in salt and pickled foods, and high in fruits and vegetables. However, most artists have not seriously considered that the art-making environment might also influence quality of life and contribute to risk of harm.

Developments in politics, law, economics, and medicine suggest an increased societal demand for control of

◆ ◆ ◆

many sources of risk. This book focuses on control of *health risks* associated with one's occupation (art-making). By sharing the statements of some practicing artists, we hope to persuade others that health risk control in artmaking is possible, desirable, reasonable, and worthwhile.

In this chapter we focus on personal hygiene, protective attire, safe handling of toxic substances, and medical intervention.

## Attentive Personal Hygiene

**A**ttentive personal hygiene is a key aspect of making art safely. To decrease oral, dermal, or inhalational exposure, we suggest the following:

Wash skin and hands with water and mild, pH-balanced soap. If soap and water is inadequate, use baby oil or a waterless cleaner.[1] Follow with soap and water and a soothing cream to prevent dry skin. Never use toluene, turpentine, kerosene, or other solvents to remove paints, inks, or stains on your hands.

■■

Before leaving the studio, wash under your fingernails. Keep a nailbrush and file near your studio sink.

■■

Keep fingernails trim and clean; do not bite or pick nails.

■■

Use skin lotion to prevent dryness and chapping; wear gloves when it is cold.

■■

Keep eating and living areas separate from working areas. If you must use a home sink (kitchen, bathroom, or laundry) for washing brushes and art materials, scour and rinse the sink thoroughly to remove all traces of art materials. Do not mix food and art materials; eat (and smoke) outside the studio.

■■

Never hold brushes or tools in your teeth or mouth.

■■

Launder studio attire frequently and separately from other clothes to eliminate an additional source of contamination.

To prevent mechanical accidents, recognize your physical, emotional, and mental limits:

Periodically evaluate your workspace and equipment; look for ways to prevent bruises, bumps, burns, cuts, ear and eye strain, falls, muscle fatigue, strains, sprains, trips, and other mishaps.

■■

Eat and sleep appropriately because decreased alertness, often due to hunger or fatigue, is a major cause of accident and injury.

■■

Get an annual medical checkup with a physician trained to recognize work-related health problems.

## Personal Protective Attire

**T**he cornerstone of safe artmaking is a thoughtfully equipped, effectively ventilated studio where artists use the safest materials available. For some activities, however, engineering controls and substitution measures are not enough. Potential health hazards may warrant use of personal protective equipment.

We recognize that artmaking is a personal and intimate process that relies upon touch, smell, and vision. Not surprisingly, some artists feel that per-

**Figure 4.1.—René Michele Joseph** of Minneapolis working in her studio. I wear a jumpsuit that covers most of my body, and found this one at a flea market for 50¢. It is used only in the studio, so pigment dusts and wet paint are not transferred to other living areas. I also use specific shoes for the studio.

Having developed severe asthma, which can be aggravated by dusts and other pollutants, I have replaced my youthful attitude that I am bulletproof and am taking more health precautions. For example, I use an air-purifying respirator, heavy gloves, and exhaust ventilation when cleaning with solvents.

sonal protective gear interferes with immediate, direct, and sensitive use of art materials. Nevertheless, responsible artists must know that goggles, masks, shields, gloves and aprons can be used to decrease exposure to most hazardous art materials.

It should be noted that long, unsecured hair, baggy clothing, flowing scarves, and neckties can be hazardous in the studio.

## SUPPLY SOURCES

If your art-making environment requires special personal protective equipment not available at most art and photo supply stores, consult the yellow pages under "Industrial Equipment and Supplies" or "Safety Equipment." In large cities, check the business yellow pages for the listing. If there is no local source of safety equipment, you can obtain safety products, catalogs, and price lists from the firms listed in Appendix III.

## EYE, FACE, AND RESPIRATORY PROTECTION

Vision is central to the creation of art; loss of vision or eye injury is particularly tragic for artists. Eyes and face must be protected from abrasives, acids, airborne particles, alkalis, splashes, volatile solvents, and harmful radiation (carbon arc lamps, glare, infrared, and ultraviolet). Various types of eye and face protection are available that meet standards set by the American National Standards Institute (ANSI) or the National Institute for Occupational Safety and Health (NIOSH) for specific hazards. Equipment labeled with ANSI and/or NIOSH seals may cost more, but the buyer gets a product that meets acceptable standards. We recommend only products bearing one or more of these seals of approval.

Eyeglasses are better than no coverage at all, but glasses provide no barrier to gases and vapors, provide inadequate security against particles or fumes, and may have inadequate impact resistance. However, many eye-care professionals today routinely

**T A B L E  4.1**     RECOMMENDED EYE, FACE, AND RESPIRATORY PROTECTION

| Chemical | Goggle Type | Respirator (Mask) | Comments |
|---|---|---|---|
| **ACIDS** | Nonventilated, flexible-fitting, hooded goggles, with a face shield | Full-face respirator with acid-gas cartridge | Always wear goggles when mixing acid; add acid to water, never water to acid, to prevent an exothermic reaction |
| **ALKALIS** | Nonventilated flexible-fitting, hooded goggles, with a face shield | Full-face respirator with an ammonia cartridge | Always wear goggles when mixing alkalis |
| **DUSTS AND NONCORROSIVE CHEMICALS** | Flexible-fitting, hooded goggles, with vents | Respirator with toxic dust filter; half- or quarter-face dust mask | Wear a mask as long as contaminants are present |
| **FORMALDEHYDE** | Nonventilated flexible-fitting hooded goggles | Most organic vapor cartridges provide protection, as do cartridges specifically for formaldehyde (and pesticides) | Wear goggles when handling and use local exhaust hood |
| **SOLVENTS** | Nonventilated, flexible-fitting, hooded goggles are necessary to protect both eyes and contact lenses from damage by solvent vapors | Full-face respirator with an organic vapor cartridge | Use local exhaust ventilation |
| **VAPORS** (irritating or harmful) | Flexible-hooded goggles | A face shield or full-face respirator with acid-gas cartridge | Use local exhaust ventilation |

Figure 4.2.—**S**afety spectacles protect eyes from flying particles produced by chipping or grinding. Polycarbonate lenses provide greater impact protection than regular eyeglasses or contact lenses. Replace lenses when scratched or pitted to maintain proper vision. (Courtesy Clement Safety Equipment, Inc., of Mississippi. Photo: AO Safety, Cabot Safety Corp.)

recommend ultraviolet (UV) coatings on eyeglasses for daily protection from the sun's ultraviolet rays.

**Safety goggles** that fit over eye glasses are available, as are prescription lens goggles. Goggles that are hooded fit snugly over the eyes like a mask. Use of vented or nonvented goggles depends upon chemicals and conditions of exposure. Nonvented goggles with a seal around the mask provide the best protection against acid, alkali, and solvent vapors that may irritate eyes and damage contact lenses.

Vented goggles are sufficient protection against dusts, airborne particles, and noncorrosive chemicals. Choose the right goggles for the task, and keep goggles clean by washing periodically with a mild detergent, rinsing thoroughly, and hanging to dry before storing.

**Face shields** of high-impact Plexiglas attached to a headband or cap offer limited protection from splashes of acids and alkalis and from flying particles. They do not prevent inhalation of fumes, gases, mists, solvents, or vapors, however, for which a respirator is preferred.

**Hearing protection** is advised for anyone who uses noisy equipment such as electric saws or printing presses. Headphones, muffs, and corded ear inserts are capable of reducing noise levels by about 25 decibels. In addition, hearing protection can also

Figure 4.3.—**S**afety goggles can prevent eye irritation from dusts, corrosive and noncorrosive chemicals, and impact hazards. Two mask types are available: (1) impact (direct vented) for general impact applications and (2) chemical splash (indirect vented) for impact and liquid splash hazards. (Courtesy Clement Safety Equipment, Inc., of Mississippi. Photo: AO Safety, Cabot Safety Corp.)

Figure 4.4.—**A** face shield protects against chemical splash from corrosive acids or alkalis. (Courtesy Clement Safety Equipment, Inc., of Mississippi. Photo: AO Safety, Cabot Safety Corp.)

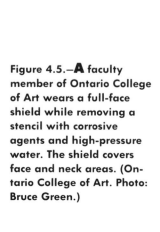

Figure 4.5.—**A** faculty member of Ontario College of Art wears a full-face shield while removing a stencil with corrosive agents and high-pressure water. The shield covers face and neck areas. (Ontario College of Art. Photo: Bruce Green.)

reduce stress associated with excessive exposure to chronic noise.

**Respirators** There are two basic types of respirators: air supplying and air purifying. Both are used to prevent inhalation of hazardous airborne substances.

**Air-supplying respirators** provide a source of clean air from compressed air tanks or compressors. The most common designs are the familiar self-contained underwater breathing apparatus (SCUBA)

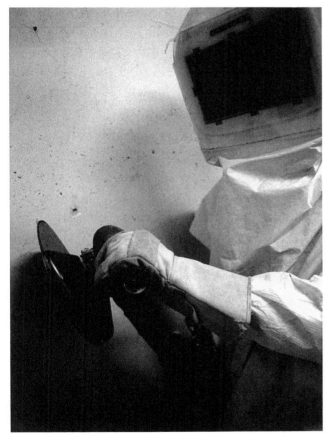

Figure 4.6.—Edgar Buonagurio wearing protective gear while working in his New York City Studio. To contain the acrylic dust generated as a by-product of my work, I grind only in a specially outfitted room isolated from the larger studio space and equipped with a filtered, negative pressure blower system. I also use an industrial-quality collection unit with a specially designed pick-up to extract as much dust as possible close to the source. To protect myself directly, I wear a Bullard air-supplied, positive-pressure respirator hood—designed for sandblasting—over a full-body disposable Tyvek jumpsuit. I also have regular medical checkups; so far, no problems. (Photo: Toby Buonagurio.)

and those worn by people with lung diseases. These devices are expensive and cumbersome but offer protection for those who work in very hazardous environments. An example of the use of an air-supplied respirator is shown in Figure 4.6, where E. Buonagurio employs a unit as protection against eight-hour-per-day sandblasting of acrylic paint. His ventilation system is also described; the ventilator and air-supplied respirator augment one another. In general, however, air-supplied respirators are not needed by artists working in the "typical" studio environment.

By contrast with air-supplying systems, **air-purifying respirators** are comparatively inexpensive and much less cumbersome. These units do not use an external supply of air but instead attempt to reduce the concentration of potentially harmful substances in the air we breathe. The breathing process pulls "contaminated" air inside the respirator mask through a chemical cartridge, a filter, or both. The cartridge reduces the level of gas and vapor in the air delivered inside the mask; filters block small particles arising from paint spray, mists, fumes, and dust.

Respirators, however, are not without limitations and problems. Before using a respirator, try substituting a less hazardous agent or provide effective exhaust and dilution ventilation. If a respirator must be used, make sure the unit is approved by the National Institute for Occupational Safety and Health (NIOSH). The Occupational Safety and Health Act (OSHAct) requires employers to furnish employees with NIOSH-approved respirators to protect against specific hazards in the workplace. Furthermore, OSHA permits respirator use as primary protection only during brief periods when existing or planned ventilation is impaired, during emergencies, or when evidence shows that a well-engineered ventilation system will not significantly reduce exposure.

There are three classes of air-purifying units for respirators. In ascending order of cost they are:

The quarter-face **dust mask,** which covers the nose and mouth with a filter, is available at hardware stores and must be checked for NIOSH approval. We do not recommend use of generic "nontoxic" or pollen masks.

■■

The half-face (covers nose, mouth, and chin) or full-face (eyes as well) **dust/mist** respirator, which essentially filters out dust and spray mists. Some models have replaceable filters, while others must be discarded entirely when clogging occurs.

■■

**Figure 4.7.–S**tudent using a toxic solvent to wash oil-based ink from a silk screen. Note the use of an air-purifying respirator and lateral slot (exhaust) ventilation. (Ontario College of Art. Photo: Bruce Green.)

and asbestos (Hepa). Examples of substances and processes for which protection is provided by specific cartridges of interest to artists include:

### Acid Gases

| | |
|---|---|
| chlorine | hydrogen sulfide |
| hydrofluoric acid | polyvinyl chloride |
| hydrogen chloride | sulfur dioxide |

### Organic Vapors

| | |
|---|---|
| aerosol spray mists | plastic cements |
| formaldehyde | plastic resins |
| lacquers and fixatives | plastic sanding, |
| lacquer thinners | grinding, and cutting |
| leather dyes | silkscreen washups |
| paint strippers | solvents |
| photo and | spray adhesives |
| printmaking | spray mist with organic |
| solvents | solvents |

The **gas/vapor** (half-face or full-face) respirator, which absorbs gases and vapors, can be equipped with a filter to block particulate matter and comes in removable cartridge or completely disposable models.

These classes of air-purifying respirators utilize human breathing to pull air through the cartridge and filter. This process creates negative pressure inside the facepiece or mask. Negative pressure, however, may permit external "contaminated" air to enter the facepiece at perimeter points where it does not closely fit facial contours. (Hair, beards, bony facial structure, and skin disorders are most likely to interfere with proper fit.) Therefore, the artist must select and regularly check the air-purifying respirator for secure fit to be maintained during likely motions of the head, arms, and torso.

The fourth class of air-purifying respirator is the **powered air filter,** which delivers air through a belt assembly containing a small pump and a filter/cartridge. Unlike negative-pressure devices, this positive-pressure system does not readily permit entry of external air through the small spaces in the mask fitting.

Those who wear respirators must use the filter or cartridge appropriate to the exposure. As mentioned, filters trap particulate matter including dusts, paint spray, mists, and fumes. By contrast, various cartridges protect against ammonia and methylamine, acid gases, and organic vapors as well as protecting against dust and vapors, dust and mist,

When lacquers, fixatives, adhesives, or mists containing organic solvents are sprayed with aerosol cans or airbrush, additional protection in the form of special filters and appropriate clothing is recommended.

Chemicals within cartridges remove unwanted gas or vapors. In some cases, as the protective chemical is depleted with use, one may detect the odor of the hazardous substance, indicating that the cartridge can no longer remove incoming gas or vapor. By contrast, odorless substances cannot provide a warning. A surrogate odor-bearing agent (e.g., nail polish remover) may help you decide if an organic vapor cartridge is functioning properly. The challenge of deciding when a cartridge has lost its function rests upon the user. Thus it compares unfavorably with the use of exhaust ventilation or substitution of less toxic agents.

A respiratory mask is a personal item that should be used by only one person (who is responsible for its maintenance). Check your mask frequently for signs of wear, cracks, tears, stiffening, or accumulations of dirt. Replace cartridges and filters when they become clogged, when breathing becomes difficult, or when odor can be detected through the cartridge. Some chemical cartridges do not show signs of wear; these should be tested per manufacturer's instructions. Chemical cartridges may also deteriorate with time, even if they are not used. Do not use them beyond the expiration date. Cartridges are not designed to handle massive concentrations; under such

Figure 4.8.—**A** half-face air-purifying respirator with replaceable filters and cartridges; mouth, nose, and chin are covered. (Courtesy Clement Safety Equipment, Inc., of Mississippi. Photo: AO Safety, Cabot Safety Corp.)

Figure 4.9.—**A** full-face air-purifying respirator with replaceable filters and cartridges. Used to prevent ocular (eye), dermal (skin), and inhalational (airway) exposure to airborne substances. (Courtesy Clement Safety Equipment, Inc., of Mississippi. Photo: AO Safety, Cabot Safety Corp.)

circumstances, they may lose effectiveness within minutes. Once respirator cartridges are opened, they should be replaced after ten calendar days, after eight hours of cumulative use, or if odor can be detected (odor breakthrough), whichever is sooner.

Periodically, remove the cartridges and clean the cartridge parts only as recommended by the manufacturer. Disassemble the rest of the mask, and wash it with warm water and a mild detergent. Use a hand brush to remove dirt. Rinse thoroughly. To disinfect, soak the mask for ten minutes in a solution of two tablespoons of bleach in a gallon of water, rinse thoroughly, and allow to dry without distortion of the contours of the mask. Store the entire apparatus in a closed container or sealable plastic bag away from potential contaminants, sunlight, temperature extremes, and moisture.

**Some Additional Issues** Respirators are not a panacea in our effort to minimize exposure to hazardous substances. Breathing through cartridges and filters requires effort, which may cause difficulty for those with cardiovascular or respiratory disorders. These individuals should not use respirators without medical advice. Moreover, odor detection is difficult for those with allergies or loss of olfaction (sense of smell).

Selection of an appropriate respirator may require the services of an industrial hygienist, who can identify and measure many hazardous substances in the air. Each type of respirator cartridge/filter combination is assigned a numerical "protection factor" rating. The magnitude of the protection factor determines whether the respirator combination is sufficient to protect the user against an assumed external concentration. If the measured concentration is too high for a given respirator, another device with a larger protection factor should be used. To purchase an appropriate respirator, engage a reliable dealer, read package inserts carefully, and ask questions to clarify information on specifications.

Remember, a respirator protects only the wearer. By contrast, exhaust ventilation protects all who enter the work space. Effective ventilation is a primary preventive measure; use of respiratory protective gear is a last resort.

# HAND AND BODY PROTECTION

A **glove box** is an apparatus that provides face, respiratory, hand, and body protection. Most sealed glove boxes have see-through tops and holes on the left and right sides where gloved hands are inserted.

Figure 4.10.—**V**arious types of respirator cartridges and filters are available. Some are designed to trap organic vapors; others trap fumes. Some prevent inhalation of dusts, while others keep out mists (including mist from paints, lacquers, and enamels) (Courtesy Clement Safety Equipment, Inc., of Mississippi. Photo: AO Safety, Cabot Safety Corp.)

Glove boxes are particularly useful for mixing photo and printmaking chemicals. Glove boxes are easy to make and are commercially available.

1. Choose a box large enough to accommodate both your hands and the vessels and chemicals you will be mixing.

2. Cut holes on each side just large enough to let your hands through; the opening should create a snug fit near the end of your elbow-length gloves.

3. Remove the top of the box and coat the interior with an easy-to-clean sealer such as acrylic medium or gesso.

4. Replace the top with glass or clear plastic, which you may hinge to one side of the box. If the edges of the glass or Plexiglas are rough, tape them to prevent cuts and scratches.

To use the glove box, place the mixing vessel and closed containers of agents to be mixed into the box. Close the transparent top, taping it shut if necessary. Insert your gloved hands into the holes, and mix the chemicals according to manufacturer's directions. Recap all containers and wait for powders to settle and vapors to dissipate. Then open the lid slowly and remove the mixture. Wipe the interior of the glove box with a damp cloth before storing it.

**Protective Gloves** When handling materials that may irritate skin or can be absorbed through skin, wear protective gloves. Some glove types are available at hardware stores, while others must be ordered from safety equipment companies or scientific supply houses. Gloves may feel clumsy and annoying at first, but with proper fit and the lightest weight consistent with adequate protection, you can incorporate use of gloves into most art-making procedures. We believe that the incidence of stains, absorption, chafing, cracking, and other skin irritations can be significantly decreased by regular use of gloves.

Glove composition must be matched to the chemical nature of the substance; that is, different glove materials protect against different hazards. For example, leather, fiberglass, or quilted cotton protect against heat and abrasion but are inappropriate when using chemicals. Chemical-resistant plastic and rubber gloves, which may be thick or thin, long or short, lined or unlined, are preferred, but appearances can be deceptive and offer no clues as to permeability. Gloves that can withstand strong acids may be dissolved by turpentine. Those that are resistant to liquid chemicals may be penetrated by vapors. For example, some painters use inexpensive disposable (single use) vinyl gloves from the local pharmacy, which have limited barrier effectiveness. Certain solvents degrade and dissolve this glove material, and glycol ethers, used as wetting agents in many water-soluble inks and paints, pass directly through gloves, giving a wet feeling inside the glove. At the time of this writing, there is no one glove that will protect against all chemicals. It is very important to make sure that your glove is impervious to the solution being used, especially one containing organic solvents.

Thus selection of appropriate gloves requires knowledge of the options. First, identify the agents you commonly use. Second, learn which glove materials will decrease your exposure to those agents. Finally, recall that gloves made from the same material by different manufacturers may have different qualities.

The glove materials, attributes and applications listed below are general. More specific evaluations can be obtained from glove manufacturers, who sometimes include chemical resistance and performance charts in the packaging. (We have also found them responsive to requests for copies of these charts.) Tables 4.2 and 4.3 summarize glove needs for materials most often encountered by artists working with two-dimensional media.

Once you know the composition of the gloves you need, consider the following when selecting the best glove for the task:

How much dexterity is required? Glove thickness varies; use the heaviest (most protective) glove that will allow you to work effectively.

Figure 4.11.—Ansell Edmont Industrial Inc. provides Sol-Vex® NBR Nitrile gloves for most solvents, including aromatic hydrocarbons and petroleum distillates, claiming that they are superior to neoprene and rubber. Gloves are available in a number of lengths, thicknesses, and sizes.

■■

**TABLE 4.2**    GLOVE MATERIALS AND THEIR ATTRIBUTES

These attributes and uses are provided as general information. Each glove manufacturer has its own formulations of density and thickness for each glove material, resulting in variation of degradation and permeability times. Obtain performance information from the manufacturer if it is not provided with the packaging. Use all gloves with observant caution. If irritation or wetness inside the glove occurs, or the glove surface becomes gummy or sticky, discontinue use of both the gloves and the material until safer handling procedure is obtained.

| Glove Material | Attributes | Uses |
| --- | --- | --- |
| BUNA-N | Resistant to cuts, abrasions and punctures; good gripping qualities | For alkalis, petroleum distillates, turpentine, formaldehyde, and mineral and organic acids |
| BUTYL RUBBER | Good flexibility and resistance to abrasion, cuts, and punctures; fair gripping qualities; poor heat resistance | For organic acids, alkalis, alcohols, ketones (acetone), and formaldelhyde |
| LATEX | Puncture and heat resistant; good gripping | For acids, alkalis, alcohols, petroleum distillates, formaldehyde, acetone, Cellosolve, and glycol ethers Please note: there is increasing evidence that latex gloves are not as impermeable as once thought. Skin reactions have also occurred. |
| NATURAL RUBBER | Puncture and heat resistant; good gripping | For acids, alkalis, alcohols, formaldehyde, acetone, Cellosolve, and glycol ethers |
| NBR RUBBER | Cut resistant; good abrasion and puncture resistance and gripping qualities | For alkalis, petroleum distillates, turpentine, formaldehyde, alcohols, lacquer thinner, toluene, and xylene |
| NEOPRENE | Resistant to cuts, punctures, and heat; good flexibility and dry grip | For mineral acids, alkalis, alcohols, petroleum distillates, formaldehyde, glycol ethers, and hexane |
| NITRILE | Resistant to abrasions, cuts, and punctures; good flexibility and dry grip | For acids, alkalis, alcohols, D-limonene, petroleum distillates, toluene, turpentine, chlorinated hydrocarbons, toluene, and xylene |
| POLYETHYLENE | Abrasion and puncture resistant; good flexibility and dry grip | For alkalis, alcohols, toluene, petroleum distillates, formaldehyde, xylene, and phenol (carbolic acid) |
| POLYVINYL ALCOHOL (PVA) | Resistant to abrasions, cuts, punctures, and heat; excellent gripping qualities | For acetic acid, chlorinated hydrocarbons, lacquer thinner, benzene, toluene, xylene, turpentine, and VM&P naphtha |
| POLYVINYL CHLORIDE (PVC) | Excellent gripping; good abrasion resistance | For alcohols, alkalis, formaldehyde, and potassium hydroxide |

These recommendations are provided as general information. Each glove manufacturer has its own formulations of density and thickness for each glove material. Because degradation and permeability times for each glove material may vary from one manufacturer to another, we advise obtaining performance information from the manufacturer if it is not provided with the packaging. With this variability in mind, the recommendations are given generally and in order of preference for each substance.

| Chemical | Recommended Glove | Comments |
|---|---|---|
| **ACIDS** (*acetic, carbolic [phenol], hydrochloric, etc.*) | Neoprene and Buna-N | Natural rubber, butyl rubber, polyvinyl chloride (PVC); polyvinyl alcohol (PVA) gloves will also protect from organic (e.g. acetic) acids; and nitrile from mineral (e.g., hydrochloric) acids |
| **ADHESIVES, GLUES, AND BINDERS** | Natural rubber, neoprene, Buna-N, butyl rubber, and PVC | |
| **ACRYLICS** | Natural rubber, neoprene, Buna-N, butyl rubber, and PVC | |
| **ALKALIS** (*ammonia, sodium hydroxide*) | Natural rubber, neoprene, Buna-N, butyl rubber, and polyethylene | |
| **OIL PAINTS** (*turpentine, mineral spirits*) | Neoprene and polyethylene, Buna-N, polyvinyl alcohol, nitrile | |
| **SOLVENTS** | | Butyl rubber not recommended for aromatic and chlorinated hydrocarbons |
| **ALCOHOLS** (*ethyl, isopropyl, methyl*) | Natural rubber, neoprene, butyl rubber, PVC, polyethylene, and nitrile | |
| **ALIPHATIC HYDROCARBONS** (*petroleum distillates: gasoline, kerosene, mineral spirits, naphtha, petroleum, ether, benzine/VM&P naphtha*) | Nitrile, natural rubber, neoprene, Buna-N, and polyethylene | |
| **AROMATIC HYDROCARBONS** (*benzene, toluene, xylene*) | Nitrile; polyethylene for toluene; PVA for benzene and xylene | Most likely to degrade materials; permeates all glove materials in time |
| **CHLORINATED HYDROCARBONS** (*carbon tetrachloride, ethylene dichloride, methyl chloroform, methylene chloride, perchlorethylene, PCBs, 1,1,1-tricholoroethane*) | PVA, nitrile, and polyethylene | Will degrade PVA in 17 minutes, all others within moments; avoid these chemicals, and use only with respiratory protection as well |
| **GLYCOL ETHERS** (*cellosolves*) | PVA | |
| **HEXANE** | Nitrile and neoprene | Substitute heptane |
| **KETONES** (*acetone, methyl ethyl ketone*) | Butyl rubber, natural rubber, neoprene, and polyethylene | |
| **LACQUER THINNER** (*may include alcohols, aliphatic, and aromatic hydrocarbons*) | PVA | Check ingredients on label and use glove appropriate for primary ingredient |

| Chemical | Recommended Glove | Comments |
|---|---|---|
| **PAINT THINNER** (*usually aliphatic hydrocarbons [petroleum distillates]; may also include aromatic hydrocarbons and alcohols*) | Neoprene rubber, Buna-N, nitrile | Check ingredients on label and use glove appropriate for primary ingredient |
| **PETROLEUM DISTILLATES/ALIPHATIC HYDROCARBONS** (*gasoline, kerosene, mineral spirits, naphtha*) | Nitrile, neoprene, Buna-N, and polyethylene | |
| **TERPENES/D-LIMONENE** | Nitrile | |
| **TURPENTINE** | Buna-N, PVA, and nitrile | |
| **PHOTOCHEMICALS:** | | |
| **THYMOL** | Neoprene, latex/neoprene, and butyl rubber | |
| **MISCELLANEOUS** | | |
| **AMYL ACETATE** | Nitrile, PVA | |
| **ANILINE** | Neoprene and natural rubber | |
| **ETHYL ACETATE** | Neoprene, butyl rubber, and PVA | |
| **FORMALDEHYDE** | Natural rubber, neoprene, Buna-N, butyl rubber, PVC, and polyethylene | |
| **MORPHOLINE** | Butyl rubber and PVC | |
| **TETRAHYDROFURAN** | Butyl rubber | |

What length of glove is required? Solids or well-contained materials may require hand protection only. By contrast, liquids that splash or are used in deep vessels may require protection of the forearms as well.

■■

Is lining desirable? Although manual sensitivity may be decreased, linings absorb perspiration, provide a better feel inside the glove, and offer limited protection from hot solutions and irritating textures.

■■

Are gripping qualities important? Many gloves have textured surfaces that facilitate gripping of objects, especially those that are wet. PVC and PVA gloves have the best gripping qualities.

Gloves must be maintained and periodically decontaminated. When chemicals get inside the gloves, promptly remove them. Chemicals inside gloves may be absorbed by the lining, and the glove can then become a co-irritant. Moreover, heat, containment, and perspiration can enhance skin absorption of some chemicals that get inside gloves. Be alert, keep gloves clean, and always keep a spare pair in the studio. Discard damaged or significantly contaminated gloves.

**Barrier hand creams,** applied before work, can decrease chemical interaction with skin. Water-soluble barrier creams may decrease absorption of organic solvents, oils, paints, lacquers, and varnishes. Water-resistant barrier creams may protect against water-containing art materials such as acrylics, mild acids, and dyes. When tactile contact is essential for one's art work, barrier hand creams offer a reasonable, though less desirable, alternative to gloves. Creams do not, however, provide sufficient protection against corrosive substances and will wear off with friction. Creams must be used with awareness of their limitations; they are not a substitute for protective gloves.

**Protective Clothing** There are procedural, economic, and health reasons to wear protective clothing in the studio. First, it is easier to focus on artmaking if one is not distracted by the possibility of damage to clothing worn outside the studio. Second, one can save money by designating clothing to be worn only in the studio. Third, and most important, protective clothing can prevent illness.

Corrosive acids and alkalis, used especially in photography and printmaking, are known to splash and burn holes through clothing as well as the skin. Thus, in addition to gloves, wear an apron, smock, or lab coat made from impervious plastics or synthetic nonwoven materials that are waterproof and acid chemical resistant. Inexpensive, disposable barrier garments, often available at art and photo supply outlets, can reduce the possibility of contamination. If acid or alkali solutions splash onto permeable clothing, remove the affected garments immediately, wash affected skin areas, and promptly soak clothing in water.

∎∎

Dusts and powders cling to clothing and can be transported from one environment to another. Therefore, if you work with dusts and powders such as pastels, charcoal, dyes, or pigments, wear special garments kept in the studio. Upon arrival, put them on; remove them before leaving. Launder these garments frequently and separately from other clothing.

∎∎

Inks and paints splatter onto clothing. When dry, they may become dust particles. Again, we suggest use of impervious garments kept in the studio and laundered frequently.

## Safe Handling of Toxic
## Substances

**E**ffective ventilation, appropriate personal hygiene, and protective attire will not necessarily prevent all routes of exposure to hazardous art materials. Maintenance of good health also requires an ability to handle art materials carefully. Every material has distinctive qualities that determine its use by artists. Many art materials, however, have chemical qualities that permit classification by composition and mechanism of action. These common characteristics permit common approaches for safe handling.

Acids and alkalis are among the most underrated toxic chemicals used by artists. To economize on shipping and selling, acids and alkalis are frequently distributed in concentrated forms. The higher the concentration, the greater the danger, therefore artists may have to dilute them for their work. It is essential to avoid skin contact with acids and alkalis. When mixing them, use a glove box or wear protective gloves, goggles, an acid-proof apron, and a face shield. To prevent splashing, always add liquid acid and alkali concentrates slowly to water. **Never** pour water into acid or alkali concentrates because an exothermic, or heat-generating, boiling, splashing, and potentially explosive reaction may occur. If solutions splash on skin, rinse with water immediately. For splashes into the eye, immediately irrigate the eye with running cool water for 20 minutes and seek medical attention promptly.

∎∎

Adhesives and binders are manufactured for specific uses. Many are considered safe a drop at a time but may give off irritant solvent vapors when applied over broad areas. Others, such as cyanoacrylates (Super Glue), tightly bond to skin within seconds. Thus, appropriate (impervious to binding) gloves should be worn when using such glues. The solvent content of some adhesives may warrant use of effective local exhaust ventilation. In the absence of ventilation, the organic vapor cartridge mask is a less desirable alternative.

∎∎

Dyes and pigments in both powdered and liquid forms may pose significant hazards, depending upon their chemical composition. Observe all dust, powder, and liquid precautions. As a safer alternative, use liquids whenever possible.

∎∎

Fumes, gases, and vapors that arise from molten metals, liquids, or solvents present some of the most challenging hazards that artists face. Avoidance of ingestion, inhalation, and skin contact with these agents requires effective ventilation and protective attire, such as gloves and apron. Again, in the absence of adequate ventilation, an acid gas/organic vapor mask is an important, though less desirable, alternative means of protection.

∎∎

Liquids should be handled with care proportionate to the product's toxicity. Always pour organic solvents, acids, alkalis, and other chemical solutions slowly and attentively to prevent spilling and splashing.

∎∎

Mists and sprays are liquid particles suspended in air. Precautions should reflect the toxicity of the liquid being sprayed. For paints and fixatives, use a spray booth or an organic vapor cartridge mask.

■ ■

Powders should be ladled gently, slowly, and in small quantities, using a small spoon or scoop, in a ventilated (but nonbreezy) environment. We recommend local exhaust systems or glove boxes. Otherwise, wear a toxic dust respirator. Mix powders into solutions only in amounts you and your studio can handle safely. Dilute the concentrates as needed, following the procedures outlined in this section for acids, alkalis, and liquids. If powder concentrates splash on the skin, rinse the area immediately with copious cool water. If powders get into the eyes, irrigate (rinse continuously) with cool water for 20 minutes and seek medical attention.

■ ■

Preservatives include common household bleach, boric acid, formaldehyde, naphthol, mercuric chloride, phenol (carbolic acid), and sodium fluoride. These components are found in very small amounts in most art products. Their presence does not call for extraordinary precautions beyond those required by the overall nature of the product.

■ ■

Solvents, due to their volatility and potential for skin absorption, offer particular challenges to safe handling. In general, use them only in environments that are well ventilated with local exhaust. Wear appropriate impervious gloves and, if necessary, an organic vapor cartridge respirator.

## Art Hazards and Our Health

**A**s new knowledge about hazardous art materials emerges, responsible members of the art community must strike a balance between ignorance and indifference to art hazards, on one hand, and fear and hypochondria on the other. Some writings about dangers in the arts sound alarmist, while major art textbooks neglect or ignore art hazards altogether. We seek to provide options for informed awareness, where artists, art educators, and art employers are alert, without hysteria, to unusual reactions or suspicious ailments.

## EMERGENCY FIRST AID

If the precautions we suggest in this book are taken, we believe there will seldom be a time or situation that requires emergency attention. Accidents, however, do happen, and it is important to be prepared. First aid, by definition, should be quick, simple, and effective. Your knowledge, and that of your colleagues, should be appropriate to the circumstances in which you work. This requires awareness of the kinds of injuries associated with materials and equipment in your particular workplace.

Many artists work alone and therefore must be prepared for self-administered care. First aid equipment, products, and running water with a spray attachment should be readily available in the studio. The names and telephone numbers of your physician, poison information center, and emergency medic or ambulance service should be posted near your telephone and carried in your wallet.

When disaster strikes, try to stay calm and think clearly. When you call for medical assistance, state the type of injury, your location, and the number of persons injured. While awaiting help, stay on the telephone to describe symptoms and signs as well as to obtain further advice.

### First aid supplies for the studio

- One-half pound of activated charcoal for accidental ingestions
- An irritative inhalant (e.g., aromatic spirits of ammonia) for fainting or unconsciousness
- Aspirin tablets (5-grain) for headache
- Ipecac syrup to induce vomiting in children (call poison control center for proper indications for the use of ipecac); do *not* induce vomiting following ingestions of acids, alkalis or solvents; this will prevent additional irritation of the esophagus and aspiration of stomach contents into the lungs (see emergency treatments later in the chapter)
- An antiseptic (any over-the-counter antiseptic)
- Clean eight-ounce paper cups
- A selection of Band-aids up to two to three inches with nonstick pads
- Three-inch sterile gauze pads/compresses
- One-inch adhesive tape
- Sterile adhesive eye pads
- Two- and four-inch bandage compresses with nonstick pad and ties
- A 40-inch triangular bandage for fashioning a sling
- Scissors and tweezers

## Treatment priority guidelines

♦ Stay calm

♦ Flush acids or alkalis from eyes or skin with water

♦ Call for medical assistance

♦ Stop bleeding by applying pressure with a sterile gauze pad

♦ Treat for shock by laying patient down with head low to assist blood flow to the brain

♦ Restore breathing with artificial respiration

♦ If victim has head injury or broken bones, do not move unless threatened by fire or noxious vapors

♦ Remove victim from hazardous conditions (but protect yourself in the process)

♦ Provide clean air

Remember that, unless you know what to do, it may be better to do nothing than to treat yourself or others inappropriately.

## Emergency treatment for common accidents

(listed alphabetically):
### Acid Spills, Splashes, and Ingestion:
**1.** Without delay, rinse affected area and/or eyes with cool running water. While rinsing, remove contaminated clothing.

**2.** If no eye wash station is nearby, get immediately into a shower and rinse eyes and affected skin areas with flowing water for a minimum of 20 minutes. Acid injury to the eye is an ophthalmologic emergency and requires consultation with an eye doctor as soon as possible after the eye irrigation.

**3.** If possible, have a responsible person call the local poison control center or your doctor. Identify the irritant, explain the situation, and get further instructions. If no one is available to call, place the call yourself only after the immediate 20-minute rinsing period.

**4.** If acid is ingested, drink two glasses of water or milk (no more) to dilute it as quickly as possible.

**5.** Do *not* induce vomiting, for that risks additional damage to the gastrointestinal tract, especially the esophagus.

### Alkali Spills, Splashes, and Ingestion can be especially devastating:

**1.** Rinse eyes and affected skin areas with flowing water for a minimum of 20 minutes. Alkali injury to the eye is an ophthalmologic emergency and requires consultation with an eye doctor as soon as possible after the eye irrigation.

**2.** If possible, have a responsible person call the local poison control center or your doctor. Identify the irritant, explain the situation, and get further instructions. If no one is available to call, place the call yourself only after the immediate 20-minute rinsing period.

**3.** If an alkali is ingested, drink two glasses of milk or water (no more) to dilute it as quickly as possible.

**4.** Do *not* induce vomiting to prevent additional damage to the gastrointestinal tract.

**Bone Fractures:** Seek immediate medical attention.

### Burns:
**1.** Cool first-degree burns (skin reddens) and second-degree burns (blisters form) by immersing affected area in cold water. When pain decreases, apply a sterile, protective dressing. Avoid breaking blisters.

**2.** With second-degree burns over a large area and third-degree burns (skin and underlying growth cells are destroyed), seek immediate medical attention. Do not try to remove clothing from burned area or use any ointments. Cover burned area with sterile dressings that have been soaked in a solution of five teaspoons of bicarbonate of soda (baking soda) in a quart of water (five cc per liter). Watch for symptoms of shock (pallor, clammy skin, sweaty palms and forehead, trembling, nausea, shallow breathing), and treat if present (see treatment list for **shock**).

### Cuts, Wounds, and Bleeding:
**1.** Allow a small cut or wound to bleed for a few minutes to flush out foreign material; then wash thoroughly with soap and water. Cover with a sterile dressing.

**2.** If the laceration is deep, seek medical attention for possible suturing and tetanus inoculation.

**3.** In the event of heavy bleeding, apply pressure to the wound with a sterile gauze pad, adding additional ones to the first as needed. Elevate the wound above the patient's heart. In cases of blood loss or emotional distress, watch for symptoms of shock (pallor, clammy skin, sweaty palms and forehead, trembling, nausea, shallow breathing), and treat if present (see treatment list for **shock**).

**Dermatitis** (skin inflammation and irritation): Mild corticosteroids are available without prescription, but if irritation persists beyond a few days of treatment, seek medical attention. Avoid skin treatments that contain benzocaine, which may aggravate the condition.

**Dusts and Powder Spills or Clouds:**
**1.** Leave area, seek fresh air, breathe deeply and spit out (do not swallow) dust and phlegm that is coughed up.

**2.** Seek medical attention for any persistent cough or shortness of breath.

**Electric Shock:**
**1.** Turn off electric current.

**2.** If the victim is unconscious, apply artificial respiration or cardiopulmonary resuscitation (CPR). Continue resuscitation efforts, even if there are no signs of life, until medical help arrives.

**Fume Inhalation** (fumes caused by heated metals in a suspended state):
**1.** Leave area immediately; seek fresh air, breathe deeply, and expectorate (do not swallow) phlegm that is coughed up.

**2.** Fumes (especially cadmium, zinc, copper, magnesium, nickel, chromium, manganese, and aluminum) can cause metal fume fever. Symptoms (acute fever, chills, cough, sweating, headache, weakness, and shortness of breath) typically occur suddenly four to eight hours after exposure and last 12 to 48 hours. It is usually a short-term, nonemergency condition, but if symptoms persist beyond 48 hours, seek medical attention.

**Gas Inhalation** generated by printmaking and photography; gases can include ammonia, chlorine, nitrogen dioxide, carbon monoxide, sulfur dioxide, and hydrogen cyanide. In the event of excessive inhalational exposure:
**1.** Leave area, seek fresh air, breathe deeply, and expectorate (do not swallow) phlegm that is coughed up.

**2.** Seek medical attention for any persistent cough, shortness of breath, headache, or dizziness.

**Heat Stress** may lead to fatigue, dizziness, mental confusion, nausea, and in extreme cases dehydration, unconsciousness, and even death.
**1.** Replace body fluids by drinking salted water (one-quarter teaspoon per quart or four cc per liter).

**2.** Cool down by applying moisture (cool water) to the skin.

**3.** If the victim is unconscious, apply cool water, consider the need for CPR, and seek medical assistance.

**Mists and Spray Exposure** Follow emergency directions for acids and alkalis if the mist or spray is from those sources. If an organic solvent is involved, as in paint and fixative sprays, see the directions for solvent inhalation.
**Shock** is caused by a failure of the circulatory system in response to trauma, blood loss, burns, and/ or extreme emotional stress. Symptoms include pallor, clammy skin, sweaty palms and forehead, trembling, shallow breathing and nausea.
**1.** Obtain medical assistance.

**2.** Until access to medical aid is gained, have the patient lie down with head low to maximize blood supply to the brain. If there is head bleeding or injury, keep the head above the heart and elevate the legs.

**3.** Stop any bleeding.

**4.** If patient is chilled, wrap with blankets and keep warm, applying heat if available. Shivering and trembling, while symptomatic of shock or hypothermia (severe chill), is nature's way of warming and should not be repressed.

**5.** After bleeding is controlled, if the victim is conscious, call your doctor, local poison center, or nearest hospital emergency room for further advice.

**6.** If the victim is unconscious, place spirits of ammonia (commercially available in premeasured vials) or dilute ammonia or acetic acid (vinegar) applied to a gauze pad or cloth, near the victim's nose and mouth.

**7.** If breathing has ceased or pulse seems to have stopped, and you know artificial respiration or cardiopulmonary resuscitation (CPR), apply these techniques appropriately.

**Solvent Exposure** includes petroleum distillates such as mineral spirits and kerosene; turpentine; aromatic hydrocarbons such as benzene, toluene, and xylene; chlorinated hydrocarbons; and alcohols. They are present in all non-water-soluble paints, inks, lacquers, varnishes, paint removers, etc.

♦ Inhalation: Leave area immediately, seek fresh air, breathe deeply, and expectorate phlegm that is coughed up.

♦ Ingestion: Do not induce vomiting, as the solvent can be aspirated into the lungs causing inflammation, pulmonary edema, or even death. Call your local poison information center; seek emergency medical attention.

♦ Skin contact and irritation: Wash thoroughly with mild soap and lots of water, and apply a soothing cream. If irritation is painful or persistent, seek medical attention.

**Vapors** are the airborne, respirable form of volatile liquids:

**1.** Inhalation: Leave area immediately, seek fresh air, breathe deeply, and expectorate phlegm that is coughed up.

**2.** Seek medical attention for any persistent cough, shortness of breath, headache, or dizziness.

# WHEN TO SEEK MEDICAL ATTENTION

A serious accident absolutely requires medical attention, and a few examples were reviewed in the preceding section. By contrast, it is not always clear when medical attention is appropriate for an undramatic but suspicious condition. Tune in to your body. Any abnormal condition that persists over a period of time, is unusual, or lacks clear attribution and treatment warrants medical investigation and perhaps intervention. We suggest that if you experience illness that may be attributable to art materials, get medical or toxicological advice. Do not wait for suspicious symptoms to go away, no matter how subtle or indistinct (see Table 4.4).

**TABLE 4.4**    SYMPTOMS ASSOCIATED WITH ART MATERIAL USE

| Symptoms | Organ or System Affected | Possible Causes |
|---|---|---|
| SKIN DRYNESS, ITCHING, INFLAMMATION | Skin | Solvents, resins, cutting oils, Fiberglas; photochemicals |
| EYE INFLAMMATION, IRRITATION, TEARING | Eyes | Acid and alkali vapors, dusts, irritating gases, smoke, sprays |
| TINNITUS (EAR RINGING), TEMPORARY DEAFNESS | Ears | Excessive noise, caffeine, quinine, hydroquinones (in photochemicals) |
| HAYFEVERLIKE: SNEEZING, RUNNY NOSE, COUGHING, SORE THROAT | Nose, throat | Dusts, fumes, gases (ozone, ammonia, etc.); vapors from solvents, printmaking, and photochemicals |
| ASTHMALIKE: WHEEZING, COUGHING, SHORTNESS OF BREATH | Bronchial tubes, lungs, throat | Dusts and powders: rosin, silica; alkali, photochemical, and solvent vapors |
| FLULIKE | General body aches and pains | Metal fumes |
| DIZZINESS, DROWSINESS, HEADACHE, TINGLING IN ARMS, HANDS, LEGS, AND/OR FEET | Nervous system | Solvent vapor inhalation, asphyxiant gases (carbon dioxide, acetylene, etc.), carbon monoxide, cyanide |
| ABDOMINAL DISCOMFORT | Gastrointestinal system, liver, pancreas, reproductive system, spleen, etc. | Photo and printmaking chemicals; solvents |

When you visit a health professional, describe in detail your art-making activities, especially those we have suggested may be associated with health problems. Keep in mind that few physicians have specific knowledge of the possible hazards of art materials. If your health professional lacks this knowledge, get a second opinion from a toxicologist, occupational health physician, or industrial hygienist. These specialists are trained to evaluate toxicity and adverse health effects of chemicals in the workplace. Ask your physician or the local poison information center to recommend an appropriate consultant.

Finally, remember that the dose makes the poison. An occasional drop of turpentine on skin is not likely to cause peripheral nerve damage. By contrast, prolonged daily inhalation for many hours over many years may contribute to health problems. The prudent artist's key to good health is to minimize exposure to toxic chemicals wherever and whenever possible. It is our belief that many artists have not yet acquired the knowledge they need to protect themselves properly yet reasonably.

## ART HAZARDS AND HUMAN REPRODUCTION

Little is known about the effect of chemical exposure on reproductive functions, including pregnancy outcome. In the absence of scientific evidence, some artists may wish to presume that any materials potentially harmful to humans or animals are potentially harmful to the reproductive organs of both men and women, to the embryo and fetus of a pregnant woman, and to a nursing and young child.

Reproductive effects associated with toxic chemical exposure are varied. Men may experience loss of sex drive, impotence, decreased fertility, sterility, or testicular changes or damage. Women may experience loss of sex drive, decreased fertility, sterility, genetic damage to ova, or menstrual changes or disorders. Toxic chemical exposure during pregnancy is associated with spontaneous abortion, fetal death, retarded growth, premature birth, low birth weight, birth defects, and other complications.

If one is contemplating having a child, the most cautious course, for mother and father alike, is to reduce exposure to potentially harmful chemicals for at least six months prior to conception. Specifically, however, we recommend cessation of all work with solvents and other volatile agents during pregnancy and breast feeding.

Before moving on to the following chapters, which focus on specific materials and methods for drawing, painting, printmaking, graphic design, illustration and photography, it seems appropriate to summarize the foregoing chapters with a few guidelines for making art safely:

Select a studio suitable to your needs that will not contaminate other individuals. Keep it clean and well ventilated. Store materials safely, and eliminate physical and mechanical hazards.

■■

Know your materials. Read labels, and use the least toxic materials possible. Wear and use personal protective attire and equipment whenever necessary. Minimize food consumption in the studio; wash hands and relevant surface areas before eating.

■■

Be prepared (have a game plan) for fire and medical emergencies.

■■

Above all, recognize that "making art safely" is the key to being a healthy artist.

# 5

Making marks on a surface to express a feeling, experience or vision is part of the magic of drawing. To enhance this expression, one must have an understanding of and sensitivity to the tools and materials of drawing— their potentials as well as their limitations and hazards.

# Drawing

Drawing media include dry implements such as charcoal, colored pencils, crayons, graphite, and pastels, and liquid mediums such as inks, markers, and paints. Many are relatively safe to use (color plates 5.1–5.4).

As with all the chapters that follow, we presume that readers have read chapters 1 through 4 and that terms such as *sensitizer, acute* and *chronic illness,* and *general* and

♦ ♦ ♦

*specific irritant* are understood; that principles of ventilation, safe storage, and disposal of art materials and chemicals are known; and that personal protective attire, emergency procedures, and when to seek medical attention, are recognized and observed. Remember, wherever children are present, hazardous materials should be stored in locked cabinets.

## Potential Hazards of Drawing and Appropriate Precautions

The hazards in drawing are few yet significant:

- Dusts and powders from dry media such as graphite, charcoal, and especially pastel pigments
- Fixative sprays used to adhere dry media to paper
- Solvent-based markers and inks
- Solvents used to blend both dry and oil media

### DUSTS AND POWDERS

To best avoid inhalation of dusts and powders, work in an area with a flameproof, variable-speed, lateral slot exhaust system mounted on a wall that vents directly to the outside. Situate the work area, table, or easel one to two feet away from the slots so that, as you work facing the exhaust slots, dust is drawn away from you toward the exhaust system. A source of fresh air intake should be located across the room, to your back. While drawing, use the exhaust fan on a low speed; when spraying drawings with fixative, use high speed. In both cases, make sure the intake vent provides the appropriate amount of fresh replacement air.

Do not use a hooded or ceiling exhaust system that may pull dust upward into your personal intake area. Some dust, such as that from graphite, is relatively heavy and may fall to the easel tray or floor, thereby escaping the exhaust system. Daily cleanup with a damp mop or microfilter vacuum cleaner keeps dust from accumulating. Ideally, turn off the exhaust fan and leave the area for an hour to allow dust to settle, then mop or vacuum all surfaces around and under your work area. If additional dust has settled when you return to the studio the next day, wipe it up or vacuum again before turning on the exhaust system and getting to work.

When the ascent of dusts to personal breathing space is unavoidable, wear a dust mask: either a NIOSH-approved toxic dust mask or the more pro-tective half-face cartridge respirator with an NIOSH-approved spray paint/particulate filter. Use the mask as recommended by the manufacturer and as indicated in chapter 4, "Eye, Face, and Respiratory Protection." In addition, follow the recommendations for daily wet-mopping and vacuuming.

### FIXATIVE

Workable and permanent **fixative** sprays represent a noteworthy potential hazard of drawing. They contain about 2 percent resins or plastics suspended in 98 percent solvents, most of which are flammable and toxic. If the contents are not designated on the label, obtain the Material Safety Data Sheet to ascertain ingredients, particularly solvents. (Also refer to Table 6.3.) Spraying action via aerosol, pump, or mouth-blowing produces respirable mists, vapors, and gases. A mouth-blower should be used only if the descending tube is inserted into a cork that fits in the top of the fixative container and is valved so that no fluid may back up into the user's mouth. Few pump sprayers provide sufficient propulsion for spray to mist evenly onto the drawing. Due to their convenience and variety—ranging from matte to glossy, workable and permanent—aerosol sprays are most often used.

Note exactly which solvents are included in your fixative sprays; those with alcohols are safest. Avoid fixative that contains n-hexane, which is associated with peripheral and central nervous system damage. Also avoid excessive exposure to those that contain aromatic hydrocarbons such as toluene, which may initiate nervous system, kidney, and liver damage from excessive exposure.

A spray booth that exhausts to the outside is best for containing and safely venting mist from fixatives because a filter removes particulates before discharge of solvent vapors into the environment. Select an explosion-proof booth with interior measurements to accommodate the size of your work. Use the booth only as recommended and, for safety and efficiency, regularly clean and replace filters as advised by the manufacturer.

Due to flammability of solvent sprays, use exhaust ventilation with explosion-proof wiring and motor. In the vertical method, with the work propped on an easel one to two feet in front of the outtake, the fan should run at high speed to exhaust noxious vapors and stray mist particles quickly and efficiently. In the horizontal method, a low-speed fan is preferred; if the air draw is too strong, less fixative will fall onto the drawing and more will be pulled away by the fan.

When spraying lacquers and fixatives, the best protection is a NIOSH-approved organic vapor respirator with a spray prefilter, as well as exhaust ventilation in the studio. In the absence of an exhaust fan, take your work outdoors to spray on a relatively calm, clear day. With a smoke tube or child's bubble blowing kit, test the direction of the wind, set up your work (fastened to a backing board) with your back to the wind, and spray. Leave the finished work outside until it is thoroughly dry and odors have dissipated.

## SOLVENTS

Solvents for mixing dry and oil-based drawing materials should be used only with appropriate exhaust ventilation and protective attire, including gloves. Use the least toxic solvents for the job. Do not use turpentine. Substitute odorless mineral spirits, Turpenoid, Stoddard's solvent, or Varsol. Even better, use isopropyl alcohol or acetone if they will achieve the results you desire. Despite fewer hazards, however, effective ventilation and skin protection are still recommended.

## Dry Drawing Mediums

The primary format for drawings is **paper,** although almost any surface imaginable has been the recipient of markings, both artistic and otherwise. There are essentially no hazards associated with the tremendous variety of papers available today. (In chapter 7, however, we address hazards which may arise during preparation of handmade papers.) Also, wetting paper occasionally releases preservative vapors that irritate some sensitized individuals.

## GRAPHITE

The ordinary "lead" pencil no longer contains lead. Today's pencil contains graphite, bound with varying degrees of clay. Graphite is a naturally occurring, slightly oily carbon with a dark gray, metallic sheen. Graphite in pencils is virtually hazard-free.

Graphite is available in fine sticks for mechanical pencils, in wood-encased pencils, as compressed bars and pencils, and as a powder. Depending upon the proportion of clay, graphite varies in degrees of hardness designated by H increasing to 8H, and softness (less clay) by B to 6B. The softer the grade, the less

clay is used to bind it together, so the graphite is more likely to generate dust and powder. The common No. 2 pencil falls in the middle of this range, indicated in drawing pencils as F or HB. Precision drawings made with hard graphite on vellum create negligible dust. At the other extreme, working with 6B bars on large, rough-textured papers can create quite a mess. Fortunately, graphite dust is heavy, so most of it falls to the easel tray or floor. While drawing with graphite, do not use fans to blow dust around. Keep the studio clean with daily damp mopping or vacuuming of settled dust. If you work with graphite powder—dusting it on, into, and around your drawing—or do a lot of rubbing likely to raise dust squalls, we suggest wearing a NIOSH-approved dust mask. Blend tonalities with paper stumps, tor-

**Figure 5.1.—Edgar Degas:** *Dancer Adjusting Slipper,* **pencil and charcoal on faded pink paper, heightened with white chalk, 12⅞″ × 9⅝″. (Courtesy Metropolitan Museum of Art, New York, bequest of Mrs. H. O. Havemeyer, 1929, the H. O. Havemeyer Collection.)**

Figure 5.2.—Winslow Homer: *Flower Garden and Bungalow, Bermuda*, 1899, pencil and watercolor, 13⅝″ × 20½″. (Courtesy Metropolitan Museum of Art, New York, Amelia B. Lazurus Fund, 1910.)

Figure 5.3.—Betz Green: *Oriental Still-Life*, 1989, pencil, 19″ × 14″. The materials I use for my artwork consist of B pencils, water-soluble color pencils, and watercolor. These non-toxic materials are easily transported and cleaned, leaving no dust or vapors to be inhaled. This is important to me as I have young children whose health must be maintained.

tillons, or chamois cloths rather than your bare hands. Graphite may be burnished to a higher sheen with metal burnishing tools.

## METALPOINT

Silver, but also platinum, gold, copper, and brass, are safe, dust-free drawing materials that lend themselves to refined, delicate drawings. Metalpoints are most effective with coated paper that will receive and hold the metal markings. While a number of industrial papers have such a coating, virtually none have desirable archival qualities: acid-free or 100 percent rag content. Therefore, a smooth rag paper may be coated, using a broad sable or camel hair brush, with Chinese white watercolor or gouache pigment. Silver, copper, and brass deposits will oxidize with time, turning a warm black in the case of silver, and irregular browns and greenish browns for copper and brass. Platinum and gold deposits remain a delicate silver gray. Precious metal wires and a spring vise with adjustable chucks to hold wires in place are available from jewelers or jeweler supply houses. Brass and copper wires are sold by hardware stores; the copper is stripped from heavy-duty three-wire electric cable (a supportive merchant may give you a three-inch piece gratis). Smooth and sharpen metalpoints by rubbing on emery cloth, available at both hardware and jeweler supply stores. Metalpoint drawings are permanent and do not erase well. No fixatives or other potentially toxic substances are required for preservation.

## CHARCOAL

When carbon pigment is obtained by firing twigs in a kiln, the product is known as charcoal, probably the oldest art-making medium, dating to Paleolithic times. Today charcoal is available as charred vine sticks, compressed into bars, as a powder, and in pencil form. As used by most artists, charcoal is not considered a significant health hazard. Some carbon black and lampblack pigments, however, are formed from soot of unknown origin. With long-term exposures, soot of undefined origin may cause cancer in laboratory animals, so carbon black of unknown origin has been banned from use in mascara and other cosmetics. Therefore, pay a little extra and get the finest artist-quality charcoal from a reputable art materials manufacturer.[1] Because it is lighter than graphite, charcoal dust may become airborne for periods of time. We advise use of a NIOSH-approved particulate mask. Do not rub and smudge charcoal with your bare hands, but use paper stumps and chamois leather to blend tones. For heavy or lengthy hands-on activity, wear nitrile gloves.

## PASTELS

No matter how glorious and arresting pastel drawings may be to the eye (colorplate 5.1), or how responsive the medium to immediate tactile expression, pastels present significant hazards because they:

♦ Are made from a variety of pigments, some of which may be toxic

♦ Are dusty and respirable

♦ Are bound by easily ingested gum arabic or gum tragacanth

♦ May contain other agents, such as preservatives

The powders formed from using pastels on textured, toothy paper can be so light and fluffy that they not only may be inhaled but also may cling to skin, hair, and clothing. These powders therefore may be hazardous through all routes of entry long after we have put them away for the day.

To work safely with pastels (color plate 5.2):

Use the least dusty pastels; pencils produce less dust than sticks and bars.

■■

Figure 5.4.—Judith E. Stone: *The Years are Too Short*, 1986, graphite, Conte, 22″ × 30″.

Use pastels designated as nontoxic,[2] with Art and Craft Materials Institute (ACMI) AP, CP, and No Label Required designations. Be wary of imported brands that may not comply with current labeling laws.

■■

Substitute less toxic colors for the more highly toxic colors (see in chapter 6, "Coloring Agents").

■■

Use Material Safety Data Sheets from the manufacturer to learn the pigment, binder, and filler composition.

■■

If you use toxic pigments, wear a NIOSH-approved particulate filter/toxic dust respirator mask. If you do not know whether you are inhaling pastel dust, tie a white handkerchief over your nose and mouth and breathe through it when working. Color deposits on it suggest inhalation and the need for a dust mask.

■■

may accumulate. Because the most dangerous dusts and powders are those that are invisible to the human eye, never dry-sweep.

■■

Do not blow accumulated pastel dust from your drawing; if the drawing is vertical on an easel, tap it lightly so that loose dust will fall to the tray or floor. If it is horizontal, use a minivac held above the paper to draw off excess residue.

■■

Keep children out of pastel work areas. Most pastels are not suitable for use by children.

**Chalk,** used as a common filler in pastels, is available in lightly tinted "colored" chalks that are safe for use by children. If the limited and bland color range befits your vision, those for use on paper (as differentiated from chalkboard) offer a safe alternative to pastels.

Like graphite and charcoal, pastels and chalks must be bound to the paper or drawing surface if they are not to rub, wear off, and eventually disappear. Spray finished drawings with fixative, taking the precautions described previously. Preserve drawings by placing them under glass in a sealed frame.

## COLORED PENCILS

Colored pencils are pigments bound with gum arabic, clay, wax, or a combination thereof, encompassed in a wood sheath. They come in a variety of colors and qualities from inexpensive to choice and elegant. Colored pencils can be sharpened to a very fine point for exquisite detail work, and those that are water-soluble can be blended with a brush dipped in water for a wide variety of effects. Because the coloring agents vary and are seldom specified, colored pencils, even those designated nontoxic, should never be moistened with the tongue or lips. A small, shallow cup or saucer of water with a nearby paper towel will permit flexible, controlled moistening of the pencil.

**Erasers** remove and modify drawn lines and images, and there are no undesirable health effects associated with their use. The kneaded rubber eraser is versatile, leaving virtually no crumbly residue, and can be shaped to a point for selective lifting of the drawn image created with most of the foregoing dry drawing materials. Those that follow are generally not erasable.

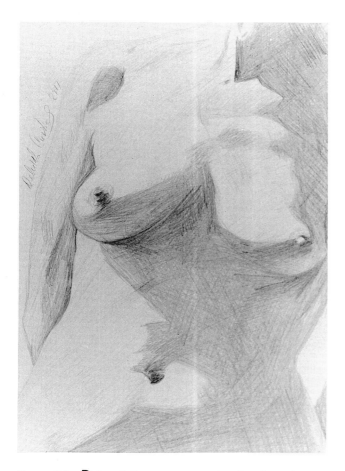

Figure 5.5.—**D**eborah Curtiss: *Torso*, silverpoint, 7" × 5". Working with dust-free precious metal deposits on coated illustration board or paper is, for me, an act of pure visual and tactile love. I use a spring vise holding two gauges of two-inch pieces of silver wire.

Wear separate work clothes that are kept in the studio and laundered regularly. Do not wear pastel-dusted studio clothes into your living area.

■■

Do not eat, drink, smoke, chew gum, or apply cosmetics around pastel dust.

■■

Wash hands and other contaminated areas thoroughly after working.

■■

Keep the studio clean with daily damp mopping or vacuuming with a microfilter vacuum cleaner. If you work regularly with pastels, periodically clean all surfaces throughout the studio where pastel dusts

# CRAYONS

Crayons are pigments and filler bound with wax. Because they are the color drawing medium most used by children, they are today among the safest art materials available. The California State Department of Education list "Acceptable Art and Craft Materials for Grades K–6" includes ten art product manufacturers who make more than 30 kinds of crayons that are considered safe. Because the safest crayons are not necessarily the highest quality, art material manufacturers have produced several different kinds of artist-quality crayons.

**Conte crayons** and pencils are made from high-quality pigments compressed with a combination of clay and wax. Available only in black, white, and a few earth tones, and in several degrees of hardness, they are permanent, produce little dust, and do not erase well. In most cases they adhere well to paper, so fixative is not required.

Nontoxic, artist-quality, **water-soluble crayons** provide an exciting medium for expression. Examples include Caran d'Ache Neocolor II Aquarelle crayons and Reeves Paintstix by Windsor & Newton. Colors may be blended with a brush or other implement dipped in water, so that the appearance may range from graphic, as with crayons, to painterly, as with watercolors. Their multiple use in printmaking techniques is described in chapter 7.

# OIL PASTELS AND STICKS

**Oil pastels** were also developed to address the shortcomings of wax crayons (color plate 5.3). Used alone without solvents, oil pastels offer immediacy and rich color in a safe and permanent drawing medium.[3] The range of graphic, painterly, and textural qualities is wide, and they can be used effectively on primed canvas as well as on a variety of white and colored papers, mylar, and other ground supports. Vapors from oil pastels are negligible, and if any adverse reactions to oil pastels have occurred, they have not come to our attention. Since oil pastels contain small amounts of solvents and solvents may be added for special effects, we suggest wearing nitrile gloves when working with these materials.

A number of artists alter oil pastel drawings by applying solvents with brushes, sponges, rags, crumpled paper, and so forth. Because the potential hazards of solvents are many, use them only with effective exhaust ventilation and while wearing appropriate gloves (see Table 4.3). Do not use turpentine, but substitute odorless mineral spirits, Turpenoid, Stoddard's solvent, or Varsol. If they achieve the results you desire, isopropyl alcohol or acetone may be even safer than the mineral spirits.

In quality of application and presence, **oil sticks** (color plate 5.4) hover somewhere between oil pastels and oil paints. Mixed with linseed oil and wax, they produce, with pressure, the quality of thickly applied oil paints with the convenience of an oil pastel. With lighter pressure, the appearance is that of crayonlike graphic marks. Available in most of the standard oil colors, they become liquid as they are applied and dry completely in 72 hours. They are advertised as nontoxic and hypoallergenic, but considering their pigment and vehicle composition, we would recommend using them in a ventilated space with proper skin protection. Like oil pastels, oil paint sticks may be blended and altered with solvents. Use appropriate precautions.

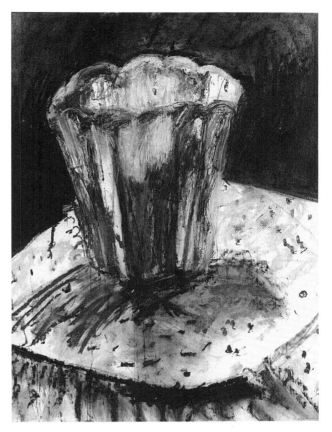

Figure 5.6.–**M**arion Spirn: *Vessel*, 1989, mixed media: charcoal, Caran d'Ache watercolor crayons, oil paintsticks, and acrylic paint on paper, 30″ × 22½″. I use all AP nontoxic materials, avoiding turpentine and mineral spirits.

**D**rawing **inks** are available in a broad range of qualities, colors, and delivery systems. Usually sold premixed in jars, inks for drafting, drawing, and lettering can be nonwaterproof and water-soluble, or waterproof and permanent after drying. Permanent drawing inks are comprised of pigment, shellac, and borax in water or in organic solvent. Water-based drawing inks are the safest alternatives. However, sepia ink, a classic, popular, cool brown color synthetically manufactured or derived from the cuttlefish, is also acceptable. Likewise, India ink, which contains a nonhazardous form of carbon black, is not associated with significant health risk. Look for the certified nontoxic label, or obtain the names of approved waterproof, technical, and nonwaterproof inks from the California list mentioned in note 1.

Colored inks are prepared from pigments described in chapter 6 in "Coloring Agents." Some recently developed organic pigments yield brilliant and permanent colored drawing and lettering inks. Until we learn more about their safety, avoid skin contact or ingestion. Do not point brushes in your mouth. Like all water-based media, inks are subject to spoilage and may contain allergenic preservatives. To avoid contaminating your inks, use only distilled or boiled water for thinning and cleaning pens and brushes.

Colored marking pens, known as **markers,** are highly prized by children, illustrators, and graphic designers. They are so popular that many brands and qualities are produced, both water-soluble and permanent (organic solvent based). Although health problems have not been associated with the coloring agents, some may contain inadequately researched dyes, so they should never be used to decorate the skin unless it is a product specifically formulated and designated as safe for that purpose.

Water-soluble markers are safer than permanent markers and may be used by children and all designers who can achieve acceptable results with them. Drawings with water-soluble markers take longer to dry than those made with organic solvent-based markers. Water-soluble markers may also cause lightweight papers to wrinkle. Yet, in response to concerns about solvent-based markers, a number of manufacturers have produced water-based ink markers from play-school to outstanding professional quality in up to 60 different colors. Colors may be blended with brush and water. A limited number of bright, opaque, and nonfading water-soluble colors and fluorescent colors, suitable for signs and posters on porous and nonporous surfaces, are available.

Figure 5.7.—**R**andle Twaddle: *Bend Sinister*, 1990, charcoal on paper, 60½" × 43½". (Courtesy Damon Brandt Gallery, New York City).

Many permanent markers contain aromatic hydrocarbons such as toluene and xylene, which, in large amounts, can cause narcosis (drowsiness) and damage to the upper respiratory system, kidneys, and liver. Unfortunately, one of the most popular design markers is of this type, as is a colorless solvent marker that releases and blends colors on the page. Markers containing toluene or xylene should be used with effective local exhaust ventilation and with care to prevent contact with skin.

In response to artists' concerns for safety, several manufacturers have produced permanent markers with a less toxic n-propyl alcohol solvent. One manufacturer produces an alcohol marker in more than 150 colors with several differently shaped replaceable nibs, and refillable ink barrels. Like the older permanent markers, they dry quickly and do not run, smear, or cause paper to buckle. Alcohol-based marker advertising frequently makes claims such as "a pure denatured grain alcohol-based ink that emits NO toxic fumes," which is true, but misleading. Another says, "NON-TOXIC FUMES." Fumes are, by definition, metal particles suspended in air caused by heating metal to the melting point and have nothing to do with markers. Alcohol emits, with a high rate of evaporation, intoxicating *vapors* that, in sufficient quantity, cause symptoms of drunkenness, i.e., central nervous system depression. In addition to anesthetic qualities, the vapors may cause eye and mucous membrane irritation. Liquid alcohol is drying to the skin. Denatured alcohol is a combination of grain or ethyl alcohol (from which liquors are made) and methyl alcohol, which (in pure form) is extremely toxic by ingestion (two to three ounces can be fatal). Denatured alcohol markers are safer than aromatic hydrocarbon-based markers, but not totally harmless. Like all markers, alcohol-based products rely heavily upon dyes as coloring agents, so one should avoid skin contact and use them in a ventilated space. These markers are effective not only on paper but also on wood, glass, ceramics, metals, plastic, film and acetate.[4]

Permanent markers are available in metallic, opaque white and in water- and chlorine-resistant colors for fabric. There are also a limited number of oil-based paint markers that will write on any surface. Despite some "nontoxic" labeling provided by manufacturers rather than ACMI, solvents are rarely specified. Use markers prudently with adequate ventilation and appropriate skin protection.

**Ink pads,** popular in the home office and with hobbyists, are available in a broad range of colors in both water-based and solvent-based inks. Water-based ink pads (that do not contain glycol ethers) are safer, with the usual advantages of easy cleanup and reactivation with water.

**M**any of the world's most treasured visions and images have been captured by painting, a form of expression familiar to artists, hobbyists and the general public. By contrast, the potential hazards in materials used in painting are less well understood. There are today many

# Painting

paint mediums and processes available to apply in countless ways. To paint safely, artists should learn the ingredients and safe use of paint products. In this chapter we attempt to update essential information on paints so that artists can achieve a balanced approach to materials and methods of painting. The recommendations are best considered following a review of chapters 1 through 4.

◆ ◆ ◆

The 1988 Labeling of Hazardous Art Materials Act requires manufacturers to list all potentially harmful substances in their paints and paint products. By now most labels should contain a fairly complete list of ingredients, including pigment composition, binder, solvents, and additives. Trade secret confidentiality, however, permits some manufacturers to withhold potentially important information from the consumer. For example, if the amount of formaldehyde (a preservative) in a product is less than 1 percent, a manufacturer does not have to list this ingredient on the label. Unfortunately, the complete contents of some paint products, especially those purchased before November 1990, may be impossible to identify.

Most paint generally consists of:

- Coloring agents: pigments or dyes

- A binder: acrylic, egg, gum arabic or tragacanth, milk/casein, linseed oil, plaster, resins, wax, or a combination thereof

- A liquid vehicle: organic solvent, oil, or water

- Inert ingredients to add bulk: chalk, diatomaceous earth, marble dust, mica, silica, talc, or whiting

Paints may also include additives that affect paint characteristics:

- Preservatives to decrease acidity and mold

- Plasticizers, stabilizers, surfactants, and wetting agents to suspend and disperse color in the binder and vehicle

- Retarders (antioxidants) or driers to slow or accelerate drying time

- Chemicals to prevent discoloration

- Agents that affect sheen: matte to glossy

## COLORING AGENTS

Pigments and dyes provide a broad range of rich, radiant colors. They are ubiquitous in art supplies. As long as artists maintain their voracious appetites for color, new hues will continue to be marketed and used despite absence of information concerning long-term health effects. Careful handling of these products is nonetheless appropriate because: (1) the contents of some pigments and dyes present well-documented health hazards and (2) the long-term health effects of chronic exposure to mixtures of coloring agents remain unknown. We suggest the following:

- Buy premixed paints, inks, and crayons from high-quality manufacturers with informative labels.

- Avoid dry pigments and agitation, spraying, sanding, heating, or other activities that create airborne powders.

- Avoid ingestion of paints by washing hands thoroughly after painting and before eating or smoking, by prohibiting eating or smoking in the studio, and by eliminating unconscious hand-to-mouth activities.

- Wear gloves and/or barrier hand cream to keep paints from contact with skin.

- Learn which chemicals (and potential hazards) are associated with a particular color; check the product label or Material Safety Data Sheet for the Color Index Number.

**Pigments** Pigments are insoluble, naturally colored materials that are crushed into a fine powder and mixed with binders and vehicles to form paints. Sources of pigments include:

- Inorganic minerals and metals, natural or manufactured, ground into a fine powder

- Organic (animal and plant) substances

- Synthetic (chemical) manufacture

Powdered forms of dry pigments are easily inhaled or ingested. Do not use powders around children. Even relatively safe powdered tempera colors should be premixed into solution by responsible, protected adults. If possible, mix all powdered pigments in a glove box. When grinding oils, wear a particle filter mask and use local exhaust ventilation. Observe daily cleaning procedures as outlined for pastels in chapter 5.

As noted previously, few pigments will ever be fully tested for potential human health hazards. As of this writing, the metal-containing pigments raise the most significant health concerns. The use of antimony, arsenic, cadmium, chromium, lead, lithium, manganese, mercury, and nickel has recently been curtailed by a few manufacturers of paint and ink pigments. In the classroom, do not use pigments known to contain these metals; in the studio, avoid inhalation, ingestion, skin contact, and contamination of the environment (Table 6.1).

**TABLE 6.1**   METAL PIGMENTS USED IN PAINTS AND INKS

| Pigment | Toxicity | Health Effects | Uses | Precautions |
|---|---|---|---|---|
| **ANTIMONY** (*antimony sulfate, barium sulfate*) | Moderate to High; Reproductive toxin | All paths of entry; dusts and fumes irritate eyes, upper respiratory tract; known to affect enzymes, heart, lungs; ingestion may cause acute digestive upset, liver, kidney damage and, in extreme exposures, possible respiratory failure, coma, death; can cause ulcers on skin and anemia | Pigment in antimony white 11, Naples yellow 41; usually contaminated with arsenic, a suspected carcinogen | Avoid powder; prevent skin contact; symptoms with ingestion may include: metallic taste, vomiting, diarrhea, irritability, fatigue, muscular pain |
| **ARSENIC** | High; Suspected carcinogen | All paths of entry; corrosive to skin, mucous membranes; peripheral nervous system, kidney damage; possible skin cancer, bone marrow damage, lung cancer | A metal in pigments such as cobalt violet (cobalt arsenate); green 21 and 22: emerald green (copper acetoarsenite), Scheele's green (copper arsenite), English, Paris, Schweinfurt, Veronese greens; and yellow 39 | Do not use; symptoms: numbness in hands and feet; chest pain, headache, vomiting, diarrhea, coma, death |
| **CADMIUM** (*cadmium sulfide, cadmium selenide*) | Moderate to High; Reproductive toxin; Suspected carcinogen | All paths of entry; irritating to eyes, skin, respiratory tract: cough, chest pain, chills, breath shortness, weakness; ingestion: nausea, vomiting, diarrhea, abdominal cramps; kidney, lung damage; anemia; associated with lung and prostate cancer | Ingredient in red, orange, yellow pigments: red 108, 113; orange 20, 23; yellow 37 | Avoid powdered and pastel pigments; wear gloves; avoid ingestion and inhalation |
| **CHROMIUM COMPOUNDS**<br><br>**CHROMIUM III** (*chromic oxide, chromic sulfate*) | Moderate to High | By all routes; dermatitis, corrosive to skin, mucous membranes; respiratory irritation, lung damage; severe enteritis, fluid loss, even shock if ingested | Metal in orange, yellow, green pigments | Avoid if possible; use local exhaust ventilation and wear gloves; avoid powdered pigments |
| **CHROMIUM VI** (*Barium, lead, strontium, and zinc chromate*) | Suspected carcinogen | Associated with cancer | | Avoid |
| **COBALT** (*cobalt arsenate, oxide, phosphate; potassium cobaltinitrite*) | Slight to Moderate | Eye, skin irritant, allergen; chronic inhalation may cause asthma, pneumonia, fibrosis; ingestion may cause vomiting, diarrhea; heart damage | Ingredient in blue, green, yellow, and violet pigments | Avoid powdered pigments or paint sprays; wear gloves |
| **COPPER** | Slight | By skin, may discolor skin, contact with eyes may cause conjunctivitis and corneal ulcers; elemental copper is poorly absorbed orally | A metal in pigments | Avoid powdered pigments; wear gloves |
| | Moderate | Heavy exposure through skin absorption or inhalation can cause nausea, abdominal pain; respiratory allergy and irritation; metal fume fever, possible ulceration, or perforation of nasal septum | | |

| Pigment | Toxicity | Health Effects | Uses | Precautions |
|---|---|---|---|---|
| **LEAD COMPOUNDS** *(lead antimoniate, lead carbonate, lead chromate, others)* | High; Reproductive toxin | Inhalation and ingestion are major routes that cause lead poisoning, symptoms of which are anemia, gastroenteritis, weakness, malaise, headaches, irritability, joint and muscle pain, kidney, liver, and nervous system damage; may affect neurological development in fetuses and children; accumulates in bones and tissues; children and fetuses are more susceptible to lower doses | Pigment, paint ingredient: flake white, Naples yellow, chrome yellow; white 2, 4; red 103, 104, 105; orange 21, 45; yellow 34, 46; green 15; ingredient in ink driers | Avoid |
| **LEAD CARBONATE** *(flake white, lead white)* | High | See lead compounds | A pigment used in painting and primers | Avoid; substitute non-lead gesso paints and inks |
| **LITHIUM** | Slight to Moderate | Mild skin, eye, and mucous membrane irritant; if ingested, may cause fatigue, dizziness, and gastrointestinal upset; systemic absorption can cause kidney damage; tremors, muscular weakness, seizures, coma, death | Found in some pigments | Avoid inhalation of powdered pigments |
| **MANGANESE COMPOUNDS** *(manganese ammonium phosphate, dioxide, silicates; barium manganate)* | Moderate to High | Can irritate eyes, mucous membranes, and respiratory tract; chronic inhalation can produce behavioral disturbances and a degenerative nervous system disorder resembling Parkinsonism | Drier in inks; pigment ingredient in mars brown, raw and burnt umber, manganese blue and violet; red 48, blue 33, violet 16, black 14, 26 | Avoid; substitute cobalt linoleate as drier; wear gloves; do not use powdered pigments |
| **MERCURY** *(mercuric sulfide)* | High | Acute inhalation of large amounts of elemental mercury vapor may cause respiratory irritation and pulmonary edema; skin contact leads to irritation and/or sensitization; mercury salts, if acutely ingested, can cause intestinal, liver, kidney, and nervous system damage; chronic inhalation of elemental vapor, or chronic ingestion of mercury salts, can cause gum disorders, kidney damage, and permanent impairment of nervous system | Pigment ingredient: vermillions, cinnabar, mercadium colors; red 106 | Avoid all mercury compounds; vapor has no odor warning |

| Pigment | Toxicity | Health Effects | Uses | Precautions |
|---------|----------|----------------|------|-------------|
| **NICKEL** | Moderate; Suspected Carcinogen | Common cause of severe skin allergy, chronic eczema; ingestion of salts can cause giddiness and nausea; fumes irritate respiratory tract and may cause pulmonary edema; inhalation of some nickel compounds associated with nasal and lung tumors | Pigment ingredient: yellow 53, 57, green 10 | Avoid powdered pigments and spray paints |
| **TUNGSTEN** | Slight; Flammable | May irritate respiratory tract; some salts may release acid on contact with moisture; chronic exposure to tungsten carbide has been associated with lung scarring | Metal used in pigments | Avoid powdered pigments and spray paints |

Although it has been banned from consumer paints for years, lead is still found in automobile, boat, and artist paints. The luminous and pearlescent qualities of lead white have been prized by some artists. Most health professionals, however, strongly discourage use of any lead-containing products by artists of any age. Currently, there is unequivocal consensus among health experts that lead in the forms used by artists presents a substantial health hazard. We therefore strongly recommend the elimination of all lead-containing products from the studio. Artists may wish to experiment with micaceous pigments to replace use of lead white.

Some reds, oranges and yellows, highly prized by many artists, may contain cadmium, a suspected reproductive toxin and suspected carcinogen. As of this writing, artist advocacy groups are endeavoring to retain production of cadmium colors for artist use. Realistically, we know that some artists will persist in the belief that the esthetic merits of lead white and cadmium colors outweigh any health risks. For those artists, we have three suggestions: (1) minimize dermal, oral, and inhalational exposure to any material known to contain these pigments; (2) inform everyone who enters the studio that toxic pigments are present; and (3) label any work that contains lead, cadmium, or any of the previously mentioned less common toxic metal pigments, so that purchasers and conservators have notice of the presence of a toxic metal.

Pigments that contain metals such as iron oxides (Indian, Mars, Tuscan, bole, ochre, Venetian, Spanish oxide colors), silver, strontium, tin, titanium, tungsten, zinc, and zirconium have no known significant hazards.

**Metallic,** premixed paints with a sheen formed by combinations of aluminum, bronze, copper, gold, silver, stainless steel, tin, titanium, and zinc with other colors are relatively nonhazardous. Metallics containing mercury as well as lead should be avoided. Premixed **iridescent** and interference colors given a sheen by inclusion of mica and micaceous oxide (hematite) powders have not been associated with adverse health effects.

**Organic Pigments and Dyes** Organic pigments offer an increasingly diverse array of new, alternative colors. For many years, alizarin crimson and phthalocyanine blue and green were the only organic pigments of satisfactory quality for artist oil paints. Some coloring agents incompatible with oil, however, proved to be miscible, brilliant, permanent, and colorfast in an acrylic base.

Organic pigments and dyes attach or bind to materials (including skin) in the absence of a binding agent. Most of the organic coloring agents listed in Table 6.2 have no significant short-term health effects. Long-term effects remain unknown or are under study. Therefore, responsible, prudent artists avoid inhalational, oral, and dermal exposure when using organic pigments.[1]

For many years, phthalocyanine colors were manufactured with polychlorinated biphenyls (PCBs). PCBs are probable animal carcinogens that may also cause skin problems (chloracne) and liver dysfunction in humans. PCB contamination of dry colors was strictly regulated by the Environmental Protection Agency in the early 1980s. If you have phthalocyanine paints produced prior to 1982, discard them. Dry pigments produced after 1982 may still contain trace amounts of PCBs at levels less than 50 parts per million (Rossol 1988).

Many commercial dyes are marketed as dry powders that must be dissolved in water to color fabrics and other materials. There are fiber-reactive dyes,

**T A B L E   6.2**    ORGANIC PIGMENTS USED IN PAINTS AND INKS

| Pigment | Toxicity | Health Effects | Uses | Precautions |
|---|---|---|---|---|
| **ANILINE** | Moderate; Suspected carcinogen | May cause drowsiness, eye irritation, methemoglobinemia | A class of dyes derived from petroleum, used to tint photographs; not colorfast | Use with exhaust ventilation; avoid if possible |
| **ANTHRAQUINONE** | NKT[1] | No acute effects; under study for long-term health hazards | A dye used in paints and inks; red 83: alizarin crimson, madder lake, rose madder; indanthrene blues | |
| **AZO** (hansa, many other names) | Slight | Under study for long-term health hazards | An organic pigment/dye in: reds: acid scarlet, acid red, red 3, arylamide, arylide, azo, Bon, hansa, helio, lake, lithol, napthol, para, permanent, pigment scarlet, pyrazolone, scarlet lake, segnale, toluidine; oranges: dinitro-analine, hansa, napthol, orthoanaline, Persian; yellows: hansa, toluidine yellow 1, 3, 5, 10; browns: arylide, Bon | |
| **DIARYLIDE, DIARYLANILIDE** | Carcinogn | | Contains benzidine | Do not use |
| **DIOXAZINE** (carbazole dioxazine) | NKT | | Purple pigment: dioxazine violet | |
| **EOSIN** | Slight | May cause dermatitis when used in cosmetics | Red dye | Avoid skin contact; wear gloves |
| **HANSA (SEE AZO)** | | | | |
| **LAKE** | Variable | Check Material Safety Data Sheet; by ingestion when precipitated by barium carbonate | A barium precipitate of an organic chemical dye in inks and paints; red 53 | Avoid powdered pigments |
| **MOLYBDATE** | High | If contains lead | Orange and red pigments, red 104 | Avoid |
| **NAPHTHOL** | Slight | Skin irritant; if absorbed through the skin in large amounts, may cause gastrointestinal and systemic symptoms; long-term effects under study | Reds, acid green, naphthol green 12 | Avoid; use only with local exhaust ventilation; wear nitrile gloves |
| **NIGROSINE** | Slight | May be a skin and respiratory allergen | Acid black, nigrosine black | |
| **PHTHALOCYANINE** (Monastral, manufacturer's name blue or green) | Slight to Moderate; possible carcinogen; possible reproductive toxin | If contaminated by polychlorinated biphenyls (PCBs) | Blue and green pigment ingredient, green 7, blue 15 | Use recently prepared, adequately labeled products; avoid powdered pigments and paint sprays; discard products purchased prior to 1982 |

| Pigment | Toxicity | Health Effects | Uses | Precautions |
|---------|----------|----------------|------|-------------|
| **QUINACRIDINE** | Unknown | | A dye; coloring agent in yellow, red, and orange paints and inks | |
| **RHODAMINE** | Suspected carcinogen | No known short-term hazards | A dye in Day-Glo, fluorescent colors | Wear gloves; avoid powdered pigments, spray paints |

[1] NKT, no known toxicity.

direct dyes, natural dyes, and mordant dyes containing agents that enhance permanent coloring. Natural dyes are prepared from plants, insects, and algae, while synthetic dyes are made primarily from five aromatic hydrocarbons: anthracene, benzene, naphthalene, toluene, and xylene.

Benzidine dyes are known to cause bladder tumors. Prior to 1979, more than 200 benzidine-containing products were used as direct dyes by craftspersons and textile workers. Benzidine dyes are no longer made or sold in the United States; if you have any in your studio, call your local health agency for advice on proper disposal.

Cold-water (fiber-reactive) dyes, commonly used for batik and tie-dyeing, have occasionally been associated with respiratory and skin allergies. Dyes are not listed in *Art & Craft Materials Acceptable for Kindergarten and Grades 1–6*, as prepared by the California State Department of Education.[2] Young children should not handle dye solutions. Mature children should work with dyes only under the supervision of adults who recognize the value of handwashing, protective gloves, and/or barrier creams. Children also need frequent reminders to keep hands away from faces and mouths.

Because of natural binding qualities and unknown health hazards, powdered or liquid dyes should be handled with care.

Avoid inhalation of dye powders:

Use commercially mixed concentrated solutions whenever possible. Do not buy dyes by the ounce or pound; buy only in premeasured packets that will be used in their entirety. Do not leave partially used packages lying haphazardly around the studio. Submerge dye packets in water before opening to discharge powder directly into the water; some packets are made from materials that will dissolve.

Alternatively, mix dry dyes into concentrated solution inside a glove box or while wearing a NIOSH-approved toxic dust mask.

Avoid ingestion of dye powders:

◆ Do not eat, drink, or smoke when dye powders and solutions are present.

◆ Keep them in well-labeled containers out of reach of children.

Avoid skin contact with dyes (liquid or dry) to prevent dermatitis or skin absorption; wear rubber gloves and an impermeable apron.

Most dyes are stable when exposed to light, air, and moisture; they also tend to resist biodegradation. There is no evidence, however, that dyes present a health threat when disposed of through sewage treatment systems or septic tanks. Thus, the artist can dispose of dye solutions down the drain with running cold water. However, dyes that contain concentrated acids may require neutralization prior to disposal.

**Mordants** are chemicals (usually acids or alkalis) that make dye permanent. Alum, ammonium alum, cream of tartar, formic (methanoic) acid, Glauber's salt (sodium sulfate), urea, and vinegar are the safest mordants. Check Appendix I, Art Material Chemicals, for additional information on hazards and precautions.

## SOLVENTS

Nonaqueous painting mediums (alkyd, encaustic, magna, oils, and resins) require and/or contain **organic solvents** for mixture and use. Organic solvents are the primary components of:

◆ Aerosol sprays of adhesives, fixatives, and paints

◆ Brush cleaners

◆ Lacquers and lacquer thinners

◆ Mineral spirits and turpentine

- Paint removers and strippers
- Varnishes and varnish thinners

Many solvents are highly volatile (evaporate quickly in air); most are flammable or combustible. Prolonged inhalation, ingestion, or absorption of large quantities can lead to drowsiness, headaches, dizziness, feelings of intoxication, fatigue, mental confusion, increased heart rate, nausea, and loss of coordination. Inhalation of vapors from most organic solvents can also cause eye, nose, and throat irritation. High-level exposure may also cause breathing difficulty, convulsions, kidney damage, and even death.

Solvents such as the aromatic hydrocarbons (benzene, toluene, xylene), chlorinated hydrocarbons (carbon tetrachloride, trichloroethylene), aliphatic hydrocarbons (petroleum distillates, kerosene, mineral spirits), and turpentine are all skin irritants that may be absorbed into the body. These solvents have been associated with the potential for both short- and long-term adverse effects on the nervous system. Although the mechanisms of solvent neurotoxicity are poorly understood, some authorities believe the adverse effects can be cumulative.

The presence or absence of odor is not correlated with, or an indicator of, toxicity or potential hazard. For example, although deodorized mineral spirits may be less toxic because some of the aromatic hydrocarbons have been removed, adverse effects, such as headaches and drowsiness, may still occur in some people. In another example, highly undesirable solvents such as benzene and toluene have pleasant odors, while acetone, a preferred solvent, has a strong, disturbing odor. Constant, prolonged inhalation of solvent vapors can decrease odor (and hazard) awareness. Therefore, we recommend use of as little solvent as possible. Keep all solvent containers tightly closed when not in use. Contact lens wearers should avoid exposure to solvent vapors, which may shorten the life of lenses by slowly dissolving the material from which they are made.

Considering the array of potential problems associated with solvents, we encourage you to:

**1. Avoid** as many organic solvent exposures as possible by substituting water-soluble paints.

**2. Adopt** the safest available handling procedures for the toxic substances that you must use:

**a.** Prevent inhalation, ingestion, and skin contact with effective ventilation, careful work habits, and protective attire.

**b.** Keep only small amounts of organic solvents (a one-day supply) in the work area.

**c.** Store solvents and solvent-soaked rags in appropriate, labeled, and tightly capped safety cans.

**3. Alert** yourself to any symptoms that may be associated with use of potentially hazardous art materials. If you believe your workplace may be contributing to a health problem, do not procrastinate: seek medical attention. Early detection and treatment promotes health and prolongs lives.

Many solvents can and should be totally avoided. We suggest you rid your studio of the following solvents:

- Benzene (benzol)
- Carbon tetrachloride
- Methyl-n-butyl ketone
- n-Hexane
- Styrene
- Chloroform
- Dimethylformamide
- Trichloroethylene
- Trichloroethane

Dispose of these solvents promptly and safely (see in chapter 3, "Safe Disposal of Dangerous Materials"). Then consider safer substitutes for the solvents you feel are necessary for your work (see Table 6.3).

Finally, change work habits to reduce the use of nonaqueous solvents. (See our suggestions in the "nonaqueous mediums" section later in this chapter.) When you must work with nonaqueous solvents and solvent-containing materials, observe the following safeguards:

Use effective, specific exhaust ventilation to carry vapors away from you and out of the studio.

■■

Avoid skin contact by wearing protective attire and gloves. Many solvents penetrate gloves or dissolve glove material. Consult Tables 4.2 and 4.3 to determine which glove is best. Use gloves according to manufacturer's specifications, and have spare gloves available to replace those that have dissolved or been penetrated. In addition, wear an appropriate hand cream whenever possible.

**T A B L E   6.3**     SOLVENT HAZARDS, PRECAUTIONS, AND SUBSTITUTIONS

| Solvent | Toxicity | Health Effects | Uses | Precautions |
|---|---|---|---|---|
| **ACETONE** | Slight to Moderate; Extremely flammable | Associated with eye, mucous membrane, skin irritation; central nervous system depressant, may cause headache, dizziness, and unconsciousness if used with inadequate ventilation; no known long-term health effects | A ketone solvent for plastics, lacquers, inks, paints and varnishes; also used to clean brushes, rollers, and tools | Use local exhaust ventilation; wear neoprene gloves |
| **ALCOHOLS** (ethyl, isopropyl, methyl) | Variable; Anesthetic, Flammable | All routes; irritant to eyes and mucous membranes; drying to skin; central nervous system depressant | Solvent for lacquer, paint, rosin, shellac, varnish | Use exhaust ventilation; wear nitrile, neoprene, or natural rubber gloves; ethyl and isopropyl among the safer solvents |
| **ALIPHATIC HYDROCARBONS** (petroleum, distillates) | Variable; Anesthetic; All flammable, some explosive | All routes of entry; irritating to mucous membranes and skin; central nervous system depressants, narcotic effects by absorption and inhalation | Multiuse, fast-drying solvents (benzine, gasoline, hexane, kerosene, mineral spirits, paint thinner, petroleum ether, Stoddard solvent, VM&P naphtha); see also N-hexane | Avoid; substitute effective isopropanol or acetone; use only with local exhaust ventilation away from flame; wear nitrile gloves |
| **AROMATIC HYDROCARBONS** | Moderate to High | All routes of entry; especially irritating to skin; dissolve waxy layers to increase skin absorption; cause local irritation | Solvents: benzene, styrene, toluene, xylene | Avoid; substitute isopropyl alcohol or acetone; use with local exhaust ventilation; wear gloves and protective attire |
| **BENZENE** (benzol) | High; Suspected carcinogen; Extremely flammable | All paths of entry; associated with aplastic anemia; may cause some forms of leukemia; known to damage bone marrow where blood cells are formed | An aromatic hydrocarbon solvent | Do not use; substitute isopropyl alcohol or acetone |
| **BENZINE** (VM&P Naphtha) | Slight to moderate; flammable | Less toxic than benzene to which it is related, but also depresses nervous system; may be an eye, skin, and respiratory irritant | An aliphatic hydrocarbon solvent, one of the least toxic in this class | Use with local exhaust ventilation; wear nitrile gloves, protective attire; a preferred substitute for turpentine |
| **BUTYL CELLOSOLVE** | Moderate | All paths of entry: irritating to eyes, skin, upper respiratory tract; absorbable through skin to cause blood, kidney, and liver damage | A solvent, less toxic than methyl Cellosolve | Use less toxic solvents such as acetone |
| **CARBON TETRACHLORIDE** | High; Suspected carcinogen; Anesthetic | Skin absorption can cause dermatitis; kidney and liver damage; central nervous system depressant | Chlorinated hydrocarbon solvent | Do not use |
| **CELLOSOLVE** (a glycol ether, 2-ethoxyethanol) | Moderate; reproductive toxin | Mildly irritating on direct skin contact; inhalation, ingestion, and skin absorption may cause kidney damage, anemia, narcosis, and behavioral changes; testicular atrophy and low sperm count observed in animals | Once common in color processes such as Cibachrome | Avoid; if used, wear goggles and gloves; use exhaust ventilation or an organic vapor respirator |

| Solvent | Toxicity | Health Effects | Uses | Precautions |
|---|---|---|---|---|
| **CELLOSOLVE ACETATE** | Moderate; Combustible; Reproductive toxin | Significant skin absorption; eye, nose, and throat irritant; associated with central nervous system depression; kidney and lung damage | Solvent for resins, lacquers, inks, paints, varnishes, and dyes | Avoid |
| **CHLORINATED HYDROCARBONS** | Moderate to High Anesthetic; Suspected carcinogen | Depress central nervous system; cause dizziness and loss of consciousness; with massive exposure, death; liver and kidney damage; defat skin; some decompose in heat or ultraviolet light to emit highly toxic phosgene | Solvents: carbon tetrachloride, ethylene dichloride, methyl chloroform, chloroform, trichloroethylene, methylene chloride, and others | Poor odor warning; do not use |
| **D-LIMONENE** | Slight | Long-term effects under study; may cause eye, mucous membrane, or gastrointestinal irritation | A paint solvent in waterless hand cleaners and paint thinners; a terpene extracted from citrus peels | Do not use around children, as citrus odor renders it attractive to ingestion; reasonable alternative to turpentine |
| **ETHER** (diethyl ether) | Moderate; Anesthetic; Extremely flammable, potentially explosive | Irritant to eyes, skin, and respiratory tract; overexposure produces nausea, headache, dizziness, and respiratory arrest | For thinning etching grounds | Avoid; use ethanol, isopropanol (isopropyl alcohol) or acetone |
| **ETHYL ALCOHOL** (ethanol, grain alcohol) | Slight; reproductive toxin | Irritant to skin, mucous membranes; central nervous system depressant via inhalation and ingestion; acute ingestion causes intoxication; chronic ingestion may cause liver damage, or adverse effects on fetal development | Solvent | Safest solvent; use with exhaust ventilation and gloves; denatured alcohol may contain small amounts of methyl alcohol, which must not be ingested |
| **GASOLINE** | Moderate; extremely flammable | May cause dermatitis; vapors irritate eyes, respiratory tract; high doses by inhalation may cause incoordination, dizziness, headaches, and nausea; may contain highly toxic benzene and/or organic lead | Solvent | Do not use; substitute deodorized mineral spirits, acetone, isopropyl alcohol or ethyl alcohol |
| **GLYCOL ETHERS** (cellosolves) | Slight to Moderate; Combustible | Mild skin irritants; can be absorbed through skin; vapors may irritate mucous membranes, upper respiratory tract; may reduce sperm counts in men; under study for reproductive effects | Used to disperse pigments in solution; solvents in photo resists, offset lithography, and many paint products and varnishes; contained in dyes, lacquers, resins, epoxy products, and cleaners | Avoid Cellosolves; propylene glycol ethers are preferred over ethylene glycol ethers; use with local exhaust ventilation, and wear neoprene or natural rubber gloves |
| **HEPTANE** | Slight; Flammable | Mild skin, eye, and respiratory irritant; may predispose heart to abnormal rhythms | Rubber cement solvent | One of the least toxic aliphatic hydrocarbons; preferred over hexane |

| Solvent | Toxicity | Health Effects | Uses | Precautions |
|---|---|---|---|---|
| **HEXANE** (N-hexane) | High; Extremely flammable | Mild skin, eye, and respiratory irritant; high-dose exposure may produce headache, dizziness, gastrointestinal upset; major concern is onset of peripheral nerve damage; early symptoms include cramping, numbness, or weakness in hands, legs, and feet; loss of appetite and weight; fatigue | An aliphatic hydrocarbon solvent of rubber cement, contact glues; ingredient in multipurpose solvents | Do not use except with effective local exhaust ventilation and protective attire; substitute safer adhesives such as those containing heptane; methyl ethyl ketone enhances neurotoxicity of n-hexane |
| **ISOPROPYL ALCOHOL** (household rubbing alcohol, isopropanol) | Slight to Moderate; Flammable | In high concentrations, vapors produce mild eye and respiratory irritation; ingestion of 50–250 ml of rubbing alcohol may cause drunkenness, abdominal pains, vomiting, low blood sugar, and central nervous system depression | Solvent | Good odor warning; one of least toxic solvents; no evidence of long-term hazards |
| **KEROSENE** | Slight to Moderate; Flammable | A central nervous system depressant; can be absorbed through skin; skin and mucous membrane irritant | Aliphatic hydrocarbon, petroleum distillate, relatively inexpensive solvent | Avoid; use local exhaust ventilation and nitrile or neoprene gloves; if ingested, do not induce vomiting, which could cause fatal pulmonary edema |
| **KETONES** (Acetone, cyclohexanone, isophorone ketone, methyl butyl ketone, methyl ethyl ketone, methyl isobutyl ketone) | Variable; Flammable; Anesthetic | Methyl butyl ketone associated with neuropathy; all cause mucousal irritation, dermatitis; high exposures associated with central nervous system depression, narcosis | Organic solvents for lacquers, oils, plastics, vinyl inks, and waxes; paint removers and photo supplies | Acetone least toxic, but extremely flammable; wear gloves and use local exhaust ventilation |
| **LACQUER THINNER/ SOLVENT** (lacquer film adherents) | Variable; Anesthetic; Combustible | See ingredients for hazards and precautions; check Material Safety Data Sheets | A solvent mixture that may contain various alcohols, aliphatic, aromatic, or chlorinated hydrocarbons, ketones, and acetates | Avoid; substitute ethyl or isopropyl alcohol if effective; use only with local exhaust ventilation or an approved organic vapor respiratory mask; observe fire precautions |
| **LIMONENE** (d-limonene) | Slight | Long-term effects under study; may cause eye, mucous membrane, or gastrointestinal irritation | A paint solvent in waterless hand cleaners and paint thinners; a terpene extracted from citrus peels | Do not use around children; citrus odor renders it attractive for ingestion; reasonable alternative to turpentine |
| **LITHOTINE** | Slight to Moderate | Eye, mucous membrane, and skin irritant | Contains mineral spirits and pine oil; a cleaner and paint thinner | Use local exhaust ventilation; nitrile gloves; if ingested, do not induce vomiting as accidental entry of lithotine into lungs can cause fatal pulmonary edema |

| Solvent | Toxicity | Health Effects | Uses | Precautions |
|---|---|---|---|---|
| **METHYL ALCOHOL** (*methanol, denatured, wood alcohol*) | High | By ingestion, as little as 3 tablespoons has led to blindness; larger doses have resulted in death; inhalation, skin absorption can contribute to injury of internal organs and the nervous system; in high concentrations, vapors are irritating to eyes and skin | Solvent | Avoid; use isopropyl alcohol |
| **METHYL BROMIDE** | High | Vapors irritate eyes, skin, and respiratory tract; inhalation of high concentrations may lead to lung edema or nervous system impairment | A preservative and fungicide | Avoid; substitute O-phenyl phenol |
| **METHYL BUTYL KETONE** (*MBK, methyl n-butyl ketone, 2-hexanone*) | High; Flammable | All paths of entry; skin, eye, mucous membrane irritant; inhalation: headache, lightheadedness, nausea, vomiting, dizziness, unconsciousness; peripheral nerve damage occurs with repeated exposures | A ketone solvent | Do not use; substitute acetone |
| **METHYL CELLOSOLVE** (*ethylene glycol monomethyl ether, or 2-methoxy ethanol*) | Moderate; Reproductive toxin | All paths of entry: demonstrated in animal tests to cause testicular atrophy, miscarriages, and birth defects; may cause bone marrow and central nervous system depression, heart, lung, and kidney problems | A glycol ether in photochemicals and photoetching (in KPR developer, KPR photo resist and KPR blue dye) | Avoid; permeates most glove materials in minutes; can be absorbed through skin; use presensitized zinc plates to eliminate using KPR photo resist and KPR blue dye |
| **METHYL CELLOSOLVE ACETATE** (*2-methoxy ethyl acetate, ethylene glycol monomethyl ether, or 2-methoxy ethanol*) | Moderate; Reproductive toxin | Absorbed through skin; eye, skin, lung irritant; based on animal studies, may cause bone marrow and nervous system depression, kidney damage, decreased sperm counts, and birth defects | A glycol ether in photochemicals and photoetching (See above) | Avoid; use gloves and local exhaust ventilation; see above precautions |
| **METHYL CHLOROFORM** (*1,1,1-trichloroethane*) | Slight to Moderate; Flammable if heated<br><br><br>High | Absorbed through skin; vapors irritate eyes, skin, respiratory tract; chronic overexposure may cause dermatitis; acute inhalation or ingestion of large amounts may lead to nausea, euphoria, incoordination, dizziness, drowsiness; lung, liver, and kidney injury; abnormal heart rhythms; and nervous system impairment; decomposes in heat to produce phosgene gas and hydrochloric acid | Chlorinated hydrocarbon in photo color retouchers and correction fluids; as a printmaking solvent; component in aerosol sprays and film cleaners | Do not use; substitute water-soluble inks, correction film or tapes; if necessary to use, wear gloves; use local exhaust ventilation |

| Solvent | Toxicity | Health Effects | Uses | Precautions |
|---------|----------|----------------|------|-------------|
| **METHYL ETHYL KETONE** (*MEK, 2-butanone*) | Anesthetic; Flammable<br><br>High | May cause dizziness, headache, lightheadedness, nausea, vomiting, and unconsciousness if inhaled at high concentrations; vapors irritate eyes, skin, mucous membranes, and respiratory tract; enhances nervous system damage caused by methyl butyl ketone (MBK) or by n-hexane | Organic solvent for lacquers, oils, plastics, vinyl inks, and waxes | Use with local exhaust ventilation; natural rubber gloves best, but will permeate in 10 minutes; substitute acetone; do not use with n-hexane |
| **METHYLENE CHLORIDE** (*dichloromethane, methylene dichloride*) | Moderate; Asphyxiant; Suspected carcinogen | Eye, skin, and upper respiratory tract irritant; metabolized (transformed) to carbon monoxide; inhalation of large amounts has been associated with heart attacks; causes lung, liver, breast cancer in laboratory animals | Chlorinated hydrocarbon used in printmaking, lithography; spray paints | Avoid |
| **MINERAL SPIRITS** (*lithotine, naphtha, Stoddard solvent, Varsol*) | Slight to Moderate; Combustible | Absorbed through skin; vapors irritate eyes, skin, and respiratory tract; chronic overexposure associated with fatigue, headache, bone marrow depression | Petroleum distillate solvents | Use with local exhaust ventilation; wear nitrile gloves; do not induce vomiting if ingested |
| **NAPHTHA** | Variable | See mineral spirits | A solvent that can be a petroleum distillate (aliphatic) or coal tar derivative (aromatic hydrocarbon) | Use with local exhaust ventilation; wear nitrile gloves; different from VM&P naphtha |
| **N-HEXANE** (*Normal hexane*) | High; Extremely flammable | Mild skin, eye, and respiratory irritant; high-dose exposure may produce headache, dizziness, gasrointestinal upset; major concern is onset of peripheral nerve damage; early symptoms include cramping, numbness, and/or weakness in hands, legs, and feet; loss of appetite and weight; fatigue | An aliphatic hydrocarbon solvent of rubber cement, contact glues; ingredient in multipurpose solvents | Do not use except with effective local exhaust ventilation and protective attire; substitute safer adhesives such as waxers or those containing heptane; methyl ethyl ketone enhances neurotoxicity of n-hexane |
| **NITROBENZENE** | Moderate to High<br><br>Reproductive toxin | Well absorbed by all routes; irritating to eyes, skin, mucous membranes; may cause headache, cyanosis, weakness, intestinal upset, liver injury, and methemoglobinemia; chronic exposure associated with anemia, bladder irritation; adverse reproductive effects | Solvent | Avoid; use only with local exhaust ventilation; wear supported polyvinyl alcohol (PVA) gloves; odor of shoe polish is a good warning property |
| **PAINT THINNER** | Variable; Combustible to Flammable | Check label and Material Safety Data Sheets for ingredient health effects and precautions | Most are mineral spirits, but some may contain aromatic and chlorinated hydrocarbons or ketones | "Odorless" paint thinners have removed most aromatic hydrocarbons; remain alert to health effects; avoid any product containing benzene |

**TABLE 6.3**   CONTINUED

| Solvent | Toxicity | Health Effects | Uses | Precautions |
|---|---|---|---|---|
| **PETROLEUM DISTILLATES** (*gasoline, hexane, kerosene, mineral spirits, VM&P naphtha*) | Variable; Combustible to Flammable | Vapors irritating to eyes, skin, respiratory tract; check Material Safety Data Sheets on specific agents; central nervous system depressants; narcotic effects by absorbtion and inhalation | Common solvents, paint thinners, cleaning fluids | Avoid; use only with effective local exhaust ventilation and nitrile gloves; if accidentally ingested, do not induce vomiting |
| **PETROLEUM ETHER** (*petroleum spirits, local spirits*) | Variable; Flammable | Vapors irritating to eyes, skin, respiratory tract; check MSDS on specific agents; central nervous system depressant | Common solvent in paint thinners, cleaning fluids | Avoid; use only with effective local exhaust ventilation and nitrile gloves |
| **STYRENE** (*vinyl benzene*) | Moderate; Anesthetic; Suspected carcinogen | Via all routes; can cause neurological impairment (narcosis); eye, skin, and respiratory system irritant; associated wih liver and nerve damage | Aromatic hydrocarbon solvent | Do not use; more toxic than toluene and xylene; attractive, sweet odor at low concentrations; penetrating in larger doses; substitute isopropyl alcohol or acetone (flammable) |
| **TERPENES** (*limonene*) | Slight | May cause eye, mucous membrane, skin, or gastrointestinal irritation; undergoing further study for health effects | Hydrocarbons extracted from plants, limonene a common example; used in waterless hand cleaners, paint thinners, cleaning products | Reasonable alternative to turpentine; keep away from children |
| **TOLUENE** (*methyl benzene, toluol*) | Moderate; Flammable; Anesthetic | Absorbed through skin, may cause dermatitis; vapors irritate eyes, skin, and respiratory tract; chronic overexposures associated with headache, dizziness, nausea, drowsiness; acute inhalation of large amounts may cause lung, liver, and kidney injury; abnormal heart rhythms, nervous system impairment, and muscle damage | Common aromatic hydrocarbon solvent used to thin and dissolve acrylics, lacquers, varnishes, paints. Ingredient in photo-resist materials and retouching fluids | Avoid; substitute isopropyl alcohol or acetone |
| **1,1,1-TRICHLOROETHANE** (*methyl chloroform, 1,1,2-trichloroethane*) | (Slight to Moderate; Flammable if heated; Suspected carcinogen | Absorbed through skin; vapors irritate eyes, skin, respiratory tract; chronic overexposure may cause dermatitis; acute inhalation or ingestion of large amounts may lead to nausea, euphoria, incoordination, dizziness, drowsiness, lung, liver and kidney injury, abnormal heart rhythms, and nervous system impairment; decomposes in heat to produce phosgene gas and hydrochloric acid | Chlorinated hydrocarbon in photo color retouchers and correction fluids; a printmaking solvent; component in aerosol sprays and film cleaners | Do not use; substitute water-soluble inks and resists; correction film or correction fluids; if necessary to use, wear gloves and use local exhaust ventilation |
| **TRICHLORO-ETHYLENE** | Suspected carcinogen | See health effects and precautions for trichloroethane | Chlorinated hydrocarbon solvent in paints, varnishes, and correction fluids | Do not use; substitute less toxic solvent such as acetone or alcohol |

| Solvent | Toxicity | Health Effects | Uses | Precautions |
|---|---|---|---|---|
| **TURPENTINE** | Moderate; Flammable | Via all routes; vapors irritate eyes, skin, mucous membranes; repeated exposure associated with chronic bronchitis, pulmonary edema, and kidney disease; absorbed through skin and may cause dermatitis; ingestion of one-half ounce can be fatal to children | Oleoresin solvent distilled from coniferous trees | Avoid; substitute deodorized mineral spirits such as Turpenoid; use with local exhaust ventilation and hand and skin protection with nitrile gloves; gum turpentine is less irritating than wood or steam-distilled |
| **VARSOL** | Slight to Moderate; Combustible | See mineral spirits | | One of the least toxic solvents |
| **VM&P NAPHTHA** (*varnish makers' and painters' naphtha, ligroin*) | Slight to Moderate; Flammable | Irritating to eyes, skin, and respiratory tract; depresses central nervous system at high doses | Aliphatic hydrocarbon solvent, also known as benzine, ligroin, high-boiling petroleum ether and paint thinner | Local exhaust ventilation; eye, skin, and hand protection; nitrile gloves; avoid any product containing benzene |
| **XYLENE** (*dimethyl benzene, xylol*) | Moderate; Combustible; Anesthetic | Absorbed through skin, may cause dermatitis; vapors irritate eyes, skin, and respiratory tract; chronic overexposure associated with headache, nausea, dizziness, drowsiness; acute inhalation of large amounts associated with lung, liver, and kidney injury, abnormal heart rhythms, nervous system impairment, and muscle damage | Aromatic hydrocarbon solvent; lacquer, paint, varnish and solvent; metal cleaner | Avoid; substitute isopropyl alcohol, acetone, or deodorized mineral spirits |

Avoid use of nonaqueous solvents to clean skin or hands. Instead, apply baby or mineral oil to the area to be cleaned, followed by soap and warm water. If needed, use a waterless hand cleaner (Art Gel, Art Wipes, Artguard), followed by soap and water and a soothing cream.

## Painting Procedures

In the pages that follow, we separate aqueous mediums, which may be thinned and cleaned with water, from nonaqueous mediums, which require organic solvents for mixture and cleanup. We identify as many ingredients as possible, their potential adverse health effects, appropriate precautions, and, when relevant or possible, safer material and procedural alternatives (color plates 6.1–6.21).

## PREPARING THE PAINTING SURFACE

Some painting processes require that the ground—whether canvas, paper, or panel—be sized, sealed, or primed to achieve a desirable painting surface. Most handmade or commercial papers for watercolor painting are presized as part of the manufacturing process, but many artists choose to modify them with coatings of Chinese white or acrylic gesso.

**Gesso** originally referred to a concoction specifically designed to isolate wood panels so their resins would not bleed into egg tempera. Later, gesso also meant a procedure designed to size and isolate canvas from oil paints. Gesso began with two or more coats of rabbit-skin glue (dissolved in water in a double boiler) to seal the wood or size the canvas. This was followed by one or more coats of whiting (calcium carbonate) and zinc or titanium white bound with rabbit-skin glue. With the advent of oil painting on canvas, the procedure was often completed with one or two coats of white lead paint to provide a luminous undercoat. Each step involved hours of drying time, such that several days may be required to properly prepare wood panels or canvas by this method.

With the development of acrylic paints in the late 1940s, acrylic gessos were formulated to mimic the sizing and undercoating properties of traditional gesso. Essentially nontoxic (some individuals may have allergic reactions to preservatives), premixed acrylic gesso consists of acrylic polymer mediums, titanium white, whiting (marble dust or calcium carbonate), and small amounts of ammonia and preservatives. Acrylic gesso simultaneously sizes, isolates, and primes a canvas for use with either water-based or oil-based paints. Equally effective with canvas, paper, wood, and synthetic panels, the ground support can be prepared for painting in half a day.

While easier, more convenient, more versatile, faster, and safer than traditional gesso, there is at least one hazard associated with acrylic gesso. To avoid accumulation of ammonia and preservative vapors, which irritate the eyes and mucous membranes, apply acrylic gesso in a well-ventilated space. Cross-ventilation with open windows and doors is usually sufficient.

**Figure 6.1.—C**harles Burchfield: *Dandelion, Seed Balls and Trees*, 1917, watercolor on paper, 22¼″ × 18¼″. (Courtesy Metropolitan Museum of Art, New York City, Arthur Happock Hearn Fund, 1940.)

## AQUEOUS MEDIUMS

Water-soluble paints offer the safest painting options. These paints, however, may develop mold or increased acidity, so preservatives must be added. Formaldehyde or formalin, mercury (phenylmercuric acetate), ammonia, bleach (sodium hypochlorite), and/or phenol (carbolic acid) can be found in paint formulations. Mercury was regulated in the manufacture of paints in 1990 but still may be present in paints you have on hand. Water-soluble paints containing mercury should be discarded. In most products, the amount of formaldehyde is very small; adverse health effects are not expected. However, if you use water-soluble mediums regularly (including drawing media and printing inks) and have breathing difficulty or skin rash typical of allergic reaction, seek medical evaluation.

**Watercolor** paints contain finely ground pigments bound with gum arabic or gum tragacanth. When thinned with water, these paints deposit a light, translucent film of color onto paper (color plates 6.1–6.2). By definition, they are not waterproof;

they may be altered by moisture. A wide range of quality and colors is available, in both tube and dried form. Tube colors may also contain nontoxic glucose or glycerin. The opacity of **gouache** colors is due to high concentrations of pigment and chalk, whiting compound, or talc opacifiers suspended in gum binder (color plate 6.3). Professional-quality gouache can be brilliant in its coloring power. Like tempera, casein, and acrylic colors, gouache may be used as underpainting to reduce time spent with nonaqueous solvent-based oil paints.

Binding power may be enhanced in watercolors or gouache by increasing the proportions of sugar syrup, gum arabic, or gum tragacanth. Always add gum powders in a glove box or while wearing a particle mask. Except for rare allergic responses to preservatives, watercolor and gouache are not associated with adverse health effects.

**Tempera** is a term with many meanings (color plate 6.4). It may refer to pigments sold in powdered form or suspended in a variety of binders and vehicles. The classic form of tempera is **egg tempera,** which uses egg yolk as the binding agent and water

Plate 5.1.—Roberta Schofield: *Massive Glass:* 1983, pastel on paper, 18″ × 20″. Brushing and rubbing with charcoal and pastels creates considerable dust. Thus I wear a Wilson respirator #R685 and R21 cartridges, approved #TC-23C-50, and R10 dust filters, TC-21C-140.

Plate 5.2.—John Slivjak: *House in a Woodland Setting:* 1989, pastel on prepared Masonite, 18″ × 24″. I use an exhaust fan whenever possible to decrease inhalation of pastel powders. Prolonged contact with dust dries my skin and causes itching. I alternate media, so this is not a big problem.

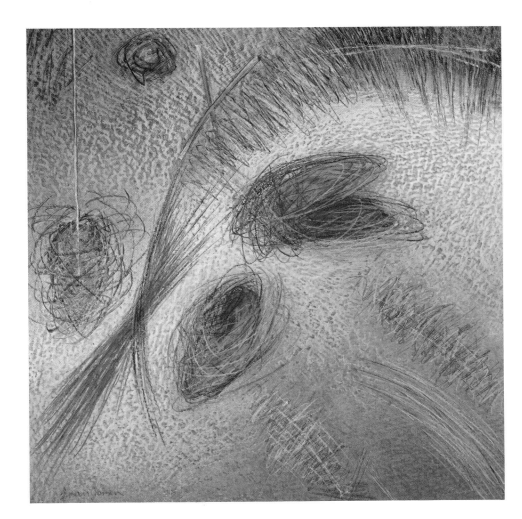

Plate 5.3.—Susan Gorsen: *Still 2 of a Kind*, 1989, oil pastel on paper, 31″ × 30″. Around 1980, my four-year-old daughter experienced allergies and bronchitis. I too had respiratory problems, as well as frequent headaches. I decided that oil painting was not in our best interest. After trying a number of nontoxic drawing and painting media, I developed a process of working in layers from hard colored pencils to the creamiest oil pastels, building up a pebbled texture luminous color. I now work exclusively with oil pastel crayons. Making art no longer makes us sick.

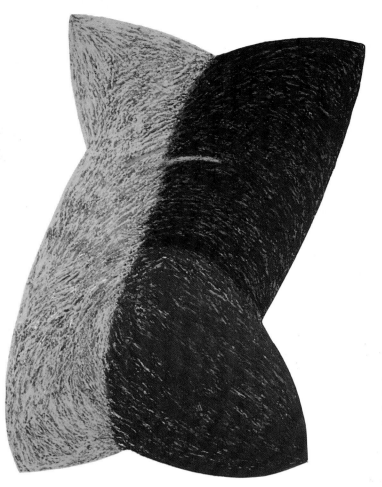

Plate 5.4.—**J**oy Walker: *Wig Walk:* 1989, oil stick on steel sheet, 33″ × 20″. In my early work with acrylics, I used airbrush and spray guns without ventilation. I started coughing. My doctors were not sure what caused the problem but I felt I had to change my way of working. I switched to oil stick on cutout galvanized iron sheets and now enjoy good health.

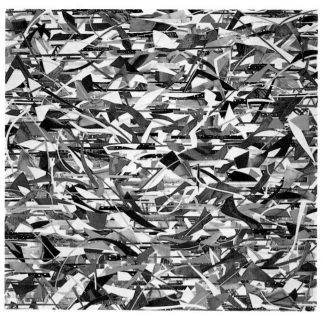

Plate 6.1.—**R**ichard Hamwi: *Chime*, 1983, watercolor and ink on paper collage using acrylic medium as adhesive, 36″ × 36″. My choice of watercolor has to do with preference for a paint which requires a nontoxic solvent and the qualities of transparency. I also use acrylic medium as an adhesive because it is nontoxic and has long-lasting conservation properties.

Plate 6.2.—**J**ames Green, Philadelphia: *Cawing Crow*, watercolor, 14½″ × 19½″. I choose to work in watercolor because of its spontaneity and detail; also I can work in any room of my home without concern for toxic solvents or vapors.

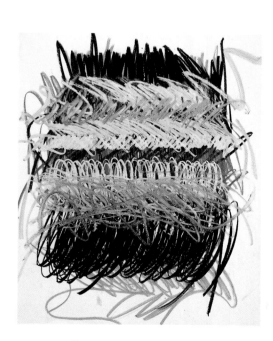

Plate 6.3.—**D**orothy Powers: *3BPE*, gouache, inks, charcoal, colored pencils on paper, 50″ × 38″. In 1988 I developed health problems and was advised by my doctor to avoid art materials and work elsewhere. I now spend weekends experimenting with nontoxic materials: watercolor, charcoal, and inks.

Plate 6.4.—**F**rancesco Clemente: *Sun*, 1980, tempera on paper mounted on cloth, 91" × 95". (Courtesy Philadelphia Museum of Art, Edward and Althea Budd Fund, Katherine Levin Farrell Fund, 20th Century Art Restricted Fund, and Funds contributed by Mrs. H. Gates Lloyd.)

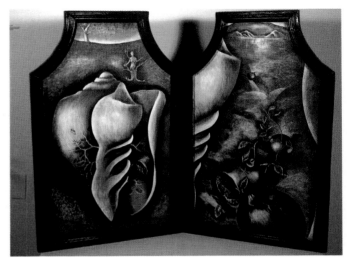

Plate 6.5.—**M**yra Herr: *The Water Near Where We Lay Down to Sleep*, 1987, egg tempera on wood, 31" × 32" × 2". After working in theaters, using toxic materials, and in art restoration with lacquer, lacquer thinner, lead paint, benzine, and other toxic materials in a poorly ventilated work environment, I developed problems with my health. In 1986 I began working with egg tempera: egg yolk mixed with gouache and distilled water. Wood panels are sealed with rabbit-skin glue and gesso, and so far I haven't had any reactions to these materials. I learned a painful lesson. (Photo: Ron Testa.)

Plate 6.6.—**V**incent Falsetta: 88-4, 1988, acrylic on canvas, 65" × 80". I use acrylics because oils and oil paint solvents make me 'uncomfortable.' I like the color intensity, fast drying, and the way color can be built with acrylic paints. I use a speckling technique that is achieved by flexing a toothbrush. This method is safer than an airbrush (less airborne paint) and provides a more direct approach. I wear a mask during this procedure. Having had asthma as a child, I am aware of the need for safety measures in painting. (Courtesy O.K. Harris Works of Art, New York City)

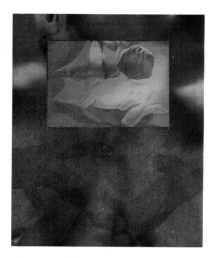

Plate 6.7.—**D**eborah Curtiss: *APSE* (Synapse Series), 1986, acrylic on two canvases: 24″ × 33″ inset in 66″ × 54″. I began painting exclusively with acrylic polymer paints in 1968 to collage fabrics onto paintings and to work on unprimed canvas. Tube acrylics, diluted to stain into rinsed canvas resulted in muted, washed-out colors until, in 1983, I discovered Golden's Fluid Acrylics (many ACMI AP). The brilliance and versatility of the colors continues to amaze and delight me.

Plate 6.8.—**E**dgar Buonagurio: *Fantail*, 1980–81, acrylic on canvas, 40′ × 18′. Since 1974 I have made abstract paintings with an unusual, personally developed process. Using industrial sanders, I systematically grind into the painting's surface. To contain the acrylic dust generated, I work in a specially outfitted room, isolated from the larger studio space, that is equipped with a filtered, negative-pressure blower system. I also use an industrial-quality collection unit with a specially designed pickup to extract as much dust as possible close to the source. For direct protection I wear a Bullard air-supplied positive-pressure respirator hood designed for sandblasting and a disposable Tyvek jumpsuit. (Collection, IBM, Inc., Stamford, Conn.; Art Options, Inc., New York City.)

Plate 6.9.—**M**erle Spandorfer: *Winter Jungle*, 1990, acrylic paint, water-soluble printing ink, collage using methyl cellulose on Arches Cover paper, 30″ × 40″. Due to respiratory irritation, I avoid using oil-based paints and inks as well as thinners. The only solvent required for *Winter Jungle*, a combination of painting, collage and collagraph, was water.

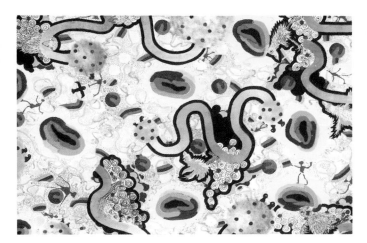

Plate 6.10.—Victoria + Dodd: *Tone Pome*, 1987, acrylic fabric paint on unprimed, washed stretched silk, 44″ × 86″ detail. Because we both experienced health problems during use of oil-based and latex paints, we searched for a nontoxic, brilliant, and permanent painting medium. We discovered DEKA-Permanent, an acrylic polymer–based paint developed primarily for the decoration of clothing.

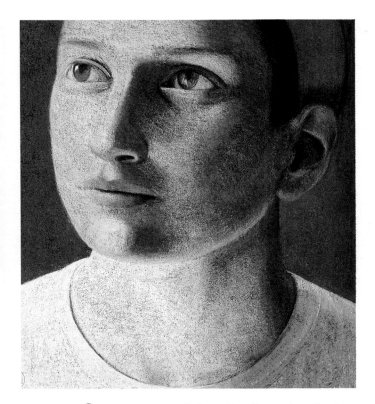

Plate 6.11.—Susan Moore: *Untitled*, acrylic, oil pastel, and paint sticks on paper, 80″ × 70″. I worked with only these media during both my pregnancies, and also wore gloves. Now that I am oil painting again, I begin with an acrylic underpainting to cut down on exposure to vapors. (Photo: Joseph Painter.)

Plate 6.12.—Bruce Pollock: *Fil et Auguile*, 1989, acrylic on canvas, 48″ × 40″. In 1977, I began using enamel paints because I liked their flow, easy mixing, and smooth, glassy surface. Over the years I noticed I was irritable, restless, and groggy when leaving the studio. I would have difficulty falling asleep and felt heavy and depressed. I read about toxic solvents used as media and driers in enamels and slowly realized that the materials which gave me so much pleasure were also poisoning me. I sent for MSDSs from the manufacturers to learn what mask to wear when painting. It made a difference, but the symptoms did not completely disappear. I didn't like wearing the cumbersome mask and continued to fear the toxins penetrating my skin. At first there was little effect, but after seven years, just opening a can of paint triggered symptoms. I began to experiment with acrylics, casein, and oils and found acrylics to be best suited to my work. In fact, acrylics opened a whole new world of possibility in painting for me. (Courtesy Janet Fleisher Gallery, Philadelphia.)

Plate 6.13.—**G**eorge Kozmon: *Nimes #64*, 1989, acrylic on canvas, 66" × 119". Several years ago I experimented extensively with pastels and other materials. I was disturbingly conscious, however, of the fine powder I was constantly inhaling. This led me to work with acrylics. As I construct a new studio, I am providing good ventilation and an immediate source of clean water.

Plate 6.14.—**M**ark N. McBeth: *Sometimes They're There*, oil and pastel on handmade paper, 57" × 38". During my final days of school, my artistic hygiene was less than impeccable. I often allowed paint to remain on my hands and arrived in the cafeteria like a Van Gogh wannabe. I was creating works on paper that included the use of oil paints, pastels, fixative, and solvents in which I would apply, scrape, and reapply until a satisfying result of layered pigment was achieved. I inhaled pastel dust, solvents, and paints, and also allowed these materials to contact my skin. During the course of these activities, I developed a cough and tightness in my chest. Medical tests revealed the cadmium content of my urine to be higher than expected. I was told that cadmium can cause respiratory problems. (In addition, cadmium is now a suspected carcinogen—Authors.) Now I work with better ventilation, wear a protective hand cream and a mask when using pastels, wash my hands immediately after working, and avoid cadmium colors.

Plate 6.15.—**L**aura Bockno: *Late Afternoon*, 1988, oil on canvas triptych, 83" × 25½". For years I would work all night, breathing in turpentine vapors and mistaking my coffee cup for painting medium. Then my husband became seriously ill from inhaling arsenic fumes during glassblowing, and my consciousness was instantly raised. I now work in a high-ceilinged studio with an exhaust fan. I paint wearing latex gloves, use Turpenoid instead of turpentine or paint thinner, and keep food and beverages away from my working area. (Photo: Stan Sadowski.)

Plate 6.16.—**S**hao Chang Yi: *Country Road*, oil, 18″ × 20″. As a professor at Sichuan Fine Arts Institute (Chongqing, People's Republic of China), I take my students outdoors to paint from nature. I seldom paint at home but go to the countryside where the fresh air is an improvement over that of the city or studio.

Plate 6.17.—**R**oberta Schofield: *Jan's Dance*, 1989, oil on board, 60″ × 144″. I experienced skin inflammation and peeling when using turpentine, so I switched to mineral spirits and saw improvement. Because of publicity about possible harmful long-term effects of many pigments, I apply Artguard barrier cream. (Photo: Burton Gilmore.)

Plate 6.18.—**N**athan Oliveira, Stanford University studio. There is a vacuum fan directly above the painting area to exhaust vapors. Artists, like those in other professions, must become aware of what they use in creating their work. Dangerous chemicals and solvents are no different for artists to use than anyone else. It is my belief that it is the responsibility of learning institutions to educate young artists about the hazards that do exist. It then is the artist's responsibility to use, or not use, safeguards as they see fit. I further believe that it is not in the interests of artists to ban hazardous art materials from the marketplace.

Plate 6.19.—**M**ary Barton: *Focus on the Changing*, 1989, oil on canvas, 45″ × 35″. About a year ago I became highly sensitive to vapors from turpentine and other cleaning solvents— even the ones claiming to be odorless irritated me. I substituted baby oil as a solvent for cleaning brushes with good results.

Plate 6.20.—**T**ina Spiro: *Hermitage*, 1987, acrylic, casein, and oil, 5′ × 12′. My method of working included airbrushing oil paints, some of which contained lead, diluted with thinner. I worked in a poorly ventilated studio (in Kingston, Jamaica) and occasionally wore a dust mask. I developed several disorders and began to suspect art materials. I now use acrylics and minimize airbrushing. Pigments are selected for low toxicity. I use a mask with an acid/organic filter when spraying and always wear rubber gloves. Further, I now have a large exhaust fan in the studio. As a result of these precautions, I believe my health has improved.

Plate 6.21.—**M**att Schwede: *Study XXXII—Everyman in Noplace Series*, 1989, mixed media on paper, 10″ × 8″. I live in one of this country's most chemically saturated areas, Baton Rouge, La. My images, made under hazardous conditions, are essentially a comment on these conditions. The *Everyman in Noplace Series* depicts human forms in a vitriolic environment, and, like my images, the materials and processes I use are hazardous: white lead, fluorescent acrylic and enamel spray paints, asphalt, charcoal, and acrylic paints. Some areas are blackened by burning, and the final product is sealed with a matte acrylic spray. I wear protective gloves when handling the paints, use a NIOSH-approved respirator and work under an exhaust hood when spraying and burning, and even wear the respirator while cleaning my studio. I have had no ill effects to date.

Figure 6.2.—Francis Picabia: *Negro Song (Chanson Negre)*, watercolor on paper, 26" × 22". (Courtesy Metropolitan Museum of Art, New York City, The Alfred Stieglitz Collection, 1949.)

Figure 6.3.—John Marin: *Pertaining to Deer Isle— The Harbor no. 1*, watercolor, 12¾" × 16½". (Courtesy Metropolitan Museum of Art, New York City, gift of an American, 1936.)

as the vehicle (color plate 6.5). Since egg serves as an emulsifier that allows mixing of oil and water, egg tempera can be combined with linseed oil, resin, wax, or gums. Each combination has unique qualities for the artist. Another common tempera mixture, known as *distemper*, contains dry pigments bound with rabbit-skin glue. Due to their brilliant flat color, distemper mixtures are frequently used to paint stage scenery. Tempera may also refer to cheaper forms of gouache sold in tubes. **Poster paints** consist of dry pigments and reducers (inert fillers such as whiting) in a medium of gum, water, and preservatives. **Casein** paints contain pigments mixed with a powerful binder derived from dried skim milk curd. While they may be thinned with water, they become insoluble in time with exposure to air. Few pigments mix effectively with casein binder, but as it mixes well with gouache and tempera, bringing its binding power to them, a wide range of colors and effects is possible. **Casein tempera** is an emulsion of casein solution and oil. All of these paints tend to be brittle and should be used on a firm support to avoid future cracking and flaking.

Figure 6.4.—William Blake: *The Wise and the Foolish Virgins*, watercolor with pen and ink, 14⅛" × 13¹/₁₆". (Courtesy Metropolitan Museum of Art, New York City, Rogers Fund, 1914.)

Figure 6.5.—**M**ary Cassatt: *Portrait of the Artist*, gouache on paper, 23½″ × 17½″. (Courtesy Metropoli- tan Museum of Art, New York City, bequest of Edith H. Proskauer, 1975.)

Figure 6.6.—**R**alph Goings: *Still Life Threesome*, 1989, watercolor and gouache on paper, 4″ × 6″. (Courtesy O.K Harris Works of Art, New York City.)

Since all temperas contain water, they may be thinned and cleaned with water, preferably distilled or boiled to prevent contamination. Adverse health effects are not expected with use of tempera as paint or as a method of painting.

**Finger paints** and other water-soluble paints for use by children are considered safe if they have the AP or CP seal of the Art and Craft Materials Insti- tute (ACMI) and/or listing in the latest edition of *Art & Craft Materials Acceptable for Kindergarten and Grades 1–6.*

**Acrylic polymer** paints are water-based paints bound with synthetic acrylic resins. Acrylic polymer paints must be distinguished from organic solvent- based acrylics such as alkyd or magna paints. Acryl- ics use many of the same pigments as oil paints and, depending upon application, may be indistinguish- able from oil paintings in texture, sheen, and overall appearance (color plates 6.6–6.13). Acrylics dry quickly to form a permanent film that is impervious to water. Although hot water will soften them, acrylics are remarkably stable; extremely high tem- peratures (over 500°F) are required for decomposi- tion. Moreover, they are less likely than oil paints to be altered by changes of temperature or humidity.

Acrylics are exceptionally flexible in use and application:

- They may be used on unprimed as well as primed paper, canvas and other fabrics, wood and Mason- ite panels, acetate, Plexiglas, or any oil-free sur- face.

- Pigments are easily mixed with a vehicle to pro- vide great richness of color.

- They resemble watercolors when thinned with water, without loss of permanence.

- A flat, opaque, even color can been achieved.

- They may be applied with palette and painting knives for thick impasto.

- Addition of retarding medium slows drying time and enhances color blending on the canvas.

- Supplemental mediums, known as gels and mod- eling pastes, readily take on and extend colors to significantly increase the variety of consistencies and textures.

- The gluelike binding qualities of acrylic mediums allow incorporation of fabrics, papers, fibers, me- tallic particles, sand and other materials.

- Depending on the vehicle or medium, acrylics can have a matte, semiglossy, or glossy finish.

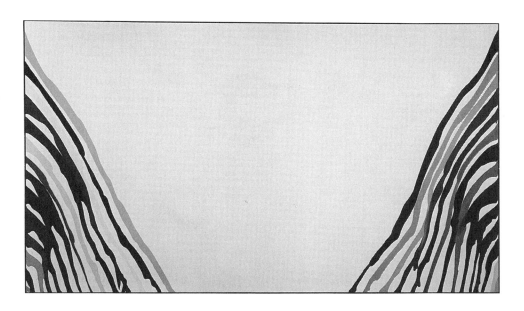

Figure 6.7.—Morris Louis: *Alpha Pi*, 1961, acrylic on canvas, 102½″ × 177″. (Courtesy Metropolitan Museum of Art, New York City, Arthur H. Hearn Fund, 1967.)

◆ When dry, acrylics are essentially inert and are not significantly altered by changes in temperature and humidity.

Acrylic polymer paints are safer than oil paints, primarily due to the absence of toxic solvents (color plate 6.6–6.13). No reproducible adverse health effects have been associated with use of acrylic polymer mediums. Like all water-soluble paints, however, acrylics contain preservatives (usually formaldehyde in small amounts). Thus some individuals may experience occasional allergic reactions. These individuals may benefit from use of exhaust ventilation or a formaldehyde mask offered by 3M. As of this writing, however, there is no NIOSH-approved formaldehyde respirator.

A faint odor of ammonia can be detected in some acrylics, leading to eye and mucous membrane irritation when used extensively. General dilution or window exhaust ventilation should eliminate this problem.

Miscellaneous components of acrylic paints may include:

◆ Surfactants (chemical agents that prevent pigment separation or polymer coalescence)

◆ Extenders and fillers: diatomaceous earth (Celite), silica, mica, or whiting (nontoxic)

◆ Antifoaming agents

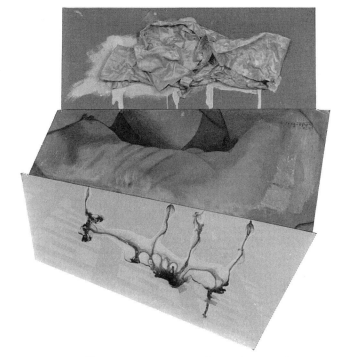

Figure 6.8.—Deborah Curtiss: *Threnody III*, 1990, acrylic on three canvases: 12″ × 28″ unprimed linen base and collage affixed with acrylic gel; 8″ × 30″ x 14″ × 32″ trapezial unprimed cotton duck wetted and stained with fluid acrylic colors, and overlaid with graphite acrylic paint; and 10″ x 33″ x 15″ x 26″ trapezial lightly primed (with thinned fluid acrylic medium) natural silk pongee with Golden interference and graphite acrylic paints, 36″ × 34″ overall.

Figure 6.9.—Merle Spandorfer: *Road to Hana*, 1989, acrylic on canvas, 42″ × 38″. The entire canvas was first covered with washes of blue and green acrylic paints diluted with water. After the canvas was thoroughly dried, different shades of white acrylic paint were added and the surface was drawn with a palette knife, scratching the surface to reveal underlying colors. I prefer acrylics because they thin, mix, and clean with water. Colors remain permanent and brilliant, dry rapidly, do not yellow, and are water-insoluble when dried. I use gloves and exhaust ventilation when painting.

♦ Antifreeze agents (usually small amounts that present no health hazard)

Most manufacturers will not disclose the precise contents of their acrylic mediums. Thus, the artist must rely upon the ACMI seals of approval to determine safety.

To keep acrylics workable, keep them moist by misting with water or by using a moisture-retaining palette. Use distilled or boiled water to avoid contamination. Covering paints with a lid or plastic wrap can retain moisture for days. Some artists use acrylic retarders to slow drying and assist blending.

**Brushes** used with acrylic paints most commonly employ nylon bristles, and are available in various sizes and styles. Synthetic bristles, unlike natural bristles of oil paintbrushes, are not significantly al-tered by water. Acrylic paintbrushes, however, must be kept moist to prevent paint from drying to a tough, gummy film that is difficult to remove. We discourage use of brush cleaners and baths that contain toluene or xylene. As an alternative, soak brushes overnight in an acrylic medium to soften the paint. Afterwards, the brushes can be cleaned with strong laundry soap (Lava, Fels Naphtha), warm water, and a good scrubbing. If solvents must be used, try acetone or isopropanol with effective ventilation and protective gloves. Nylon brushes with wayward bristles may be reshaped as follows: suspend the bristles for one minute in boiling water; comb the bristles straight; press flat brushes with a weight or wrap round brushes with a rubber band; and allow brushes to dry.

The tough film provided by acrylics rarely needs varnish. If acrylic varnish is desired, apply with adequate general ventilation. Some acrylic varnishes are solvent based, and require more stringent precautions.

Acrylic paints are not biodegradable; they may leave a residue in plumbing. To discard acrylics safely, let them dry in disposable containers. Recycle or discard as appropriate for solid waste.

Commercial **latex** house paints have been used by some artists who are attracted by their relative low cost. We vigorously discourage this practice. Quality-control standards for housepaints have not been as rigorous as those for artist products, and prior to its regulation in 1990, mercury was routinely used as a preservative in **latex** paints. Also, the inclusion of various inexpensive fillers and extenders precludes precise determination of chemical composition of many of these products.

**Fresco** is a highly specialized process. Water-soluble colors are applied to walls covered with wet lime plaster. The colors blend with the plaster and dry to a durable film of crystalline carbonate of lime, which contains varying amounts of corrosive calcium hydroxide. Fresco artists should wear protective gloves and goggles to avoid eye and skin irritation. In the twentieth century, fresco has largely been replaced by murals made from ceramic tiles or extremely tough epoxy resin paints (see next section).

## NONAQUEOUS MEDIUMS

**Oil Painting**   For over 500 years oil painting has remained a highly esteemed painting process. Brilliant colors are achieved by grinding pure pigments into linseed oil and thinning with turpentine. Oils may be applied to a flexible ground of primed canvas on stretchers, wood or Masonite panels. Over the years the process of oil painting has been enhanced with drying oils, complex emulsions, lacquers, varnishes, and a variety of thinners (color plates 6.14–6.18).

There are no adverse health effects associated with linseed oil itself. Potential hazards may arise, however, with the inclusion of pigments, turpentine, or solvents (color plate 6.7). Many volatile solvent-based products can be inhaled or absorbed through the skin. Thus, oil painting requires effective ventilation and appropriate skin and respiratory protection. A variety of oil painting **mediums** improve the consistency of oil paints. **Stand oil,** which is linseed oil in a specially heated form, and resins are most commonly used. Resin dusts may provoke allergies in some individuals. Read labels and Material Safety Data Sheets, and check Appendix I, "Art Material Chemicals," for hazards and precautions.

**Driers** are commonly found in printing inks, oil paints, and painting mediums. Check labels and MSDSs. Cobalt linoleate is the preferred drier. Products containing cobalt naphthenate, lead, or manganese should be avoided.

Turpentine is the solvent most commonly used by oil painters. Turpentine vapors irritate the eyes and respiratory tract. Massive inhalation of turpentine may lead to headache, dizziness, nausea, confusion, and rapid pulse. Massive ingestions of turpentine have been associated with kidney damage, confusion, and even death. We strongly recommend that all artists replace turpentine with safer deodorized turpentine substitutes. (See in chapter 1, "Evil Spirits," for additional information on solvent hazards.)

Remember that rags and papers (those indispensable companions when painting) may become saturated with linseed oil and solvents. They are flammable and may ignite by spontaneous combustion. Store them only in safety containers especially designed for the purpose. Dispose of them frequently in a closed metal container.

For protection, finished oil paintings may be coated with a transparent film of **varnish** or lacquer. Most varnishes contain natural or synthetic resins suspended in multisolvent mixtures. They should be used only with appropriate exhaust ventilation and protective attire. Spray varnish and lacquer products contain a high percentage of organic solvents and propellants relative to resin content, and spraying increases the possibility of inhalation of these solvents. Small works may be varnished in a spray booth or outdoors. If your paintings are large, find an industrial spray paint room where you can spray varnish safely. The preferred method of brushing on

Figure 6.10.—**E**laine Cohen Nettis: *Forging Ahead No. 86,* 1989, mixed media on paper, 30″ × 22″. I wear gloves when I work with acrylics and drawing media. I keep my acrylic jars loosely covered to simply lift and lower as paint is needed.

varnish should be done in a well-ventilated indoor environment. In the absence of effective ventilation, wear a NIOSH-approved organic vapor cartridge mask.

During **cleanup** of solvent-based paints, wear protective gloves and/or barrier creams on hands and wrists. Wash inadvertent paint splashes from the skin as soon as possible. First, soften with baby oil; then scrub with soap and water and follow with a protective skin cream (color plate 6.19). To clean oil paintbrushes, soak them overnight in mineral or baby oil to loosen accumulated paint. Wipe them as clean as possible. Wash with a strong laundry soap (e.g., Fels Naphtha, Lava) in water. Wear protective gloves, and use kitchen scrubbers to remove softened but stubborn paint. Let natural-bristle

**Figure 6.11.—Joan Shrager:** *Once & Future*, 1990, acrylic, mixed media, 10″ × 12″. For many months I worked intensely with crayons, bearing down very hard as I worked. I developed joint inflammation in the forefinger of my right hand; the pain radiated into my palm and other fingers. My doctor prescribed rest from all art work, but then allowed me to resume it slowly. He suggested that a variation of crayon and brush thickness would be helpful, so I began using wraps to change the diameters of the crayons and brushes. To further avoid the problem, I take more breaks and concentrate on varying both techniques and materials.

brushes dry thoroughly before reuse to prevent trapping of water. If you need to soak brushes in an organic solvent, place brushes in a capped container under a fume hood, or wrap aluminum foil around the brush handles and over the container to reduce solvent vapors in the workspace. Observe the safe studio practices recommended in chapter 3 (color plates 6.20–6.21).

If your preference is to work with oil paints and washes, exposure to solvent hazards may be reduced by using acrylic gesso (rather than an oil prime) and developing the imagery with water-soluble mediums such as casein, gouache, tempera, and acrylics.

**Other Solvent-Based Paints** **Alkyd** paints are oil paints modified with synthetic resins and solvents to accelerate drying. They offer no safety advantage; precautions for oil paints and their solvents should be observed at all times.

**Waterbased oil** paints reduce solvent use because they can be thinned and cleaned up with water. These products, however, still contain solvents such as alcohol or turpentine. Check the label or Material Safety Data Sheet before purchase and use.

**Encaustic** is a suspension of dry pigments in molten, white, refined wax. Encaustic may also contain drying oils, Venice turpentine, and natural resins. It is a potentially hazardous medium. Wear a particulate filter mask to prevent inhalation of powdered pigments. There is no mask cartridge that will protect against hot wax vapors (Rossol 1982) that arise from warming the palette to keep the wax workable. Heat accelerates evaporation and release of potentially harmful vapors and pigment fumes from the various ingredients. Thus, effective local exhaust ventilation is essential with encaustics.

Avoid overheating since hot wax may catch fire. Melt wax safely in an electric slow cooker (available in kitchenware stores) that maintains a dependable temperature below boiling. Alternatively, a double boiler on a hot plate may be used with caution. Never place a melted wax pot directly on the hot plate, and never use a flame heating element (torch) to heat wax applied to a surface. A heat-generating flood lamp in a ceramic receptacle with a bracket handle is safer. For cleanup, use Turpenoid with appropriate precautions. Avoid the more toxic solvents such as carbon tetrachloride.

**Solvent-based acrylic** paints, also known as **Magna** colors, are permanent pigments ground in an acrylic resin with solvents and a plasticizer. These paints contain more solvent than many oil paints; therefore, their use requires all the precautions of oil painting. Use odorless mineral spirits or Turpenoid for thinning and clean-up.

**Epoxy** paints are pigments suspended in an epoxy resin to which an amine hardener is added immediately before use. These extremely tough paints are most often used for murals, especially on exterior walls. Some alkaline amines may irritate skin. Epoxy vapors may irritate the upper respiratory system, while epoxy mists irritate eyes and may damage the cornea. The hardener may contain toluene or another hydrocarbon solvent. Thus, work with epoxy paints requires effective ventilation. Both indoors and outside, a respirator with an organic vapor cartridge will enhance protection. In addition, we suggest heavyweight nitrile gloves and protective clothing to keep paint off skin. Have replacement gloves available since epoxy paint ingredients may occasionally penetrate some gloves.

**Enamels, stains,** and **lacquers** commercially produced for house paints and surface coatings have been employed by some artists. Many contain complex combinations of solvents, driers, and extenders. Lingering emissions from commercial paints may contribute to respiratory irritation in poorly ventilated work areas. If and when you do use these products, do so only with effective ventilation, while wearing protective attire to prevent inhalation, skin contact and ingestion. Trade paints also have questionable archival quality, and works by artists who have employed them as recently as the 1950s have presented significant conservatorial challenge.

The magic of painting will continue to seduce some individuals to devote their lives to it, regardless of risk or price. The cost of succumbing to the charms of painting, however, does not have to include poor health. This chapter has suggested how the artist might lower the price paid in terms of adverse health effects. Improved understanding of the painting environment **can** significantly decrease its contribution to health problems. Therefore, young people, students, and colleagues must learn how to use paint products and procedures in the safest way possible.

**P**rints are works of art made by transferring a layer of ink from a printing surface, or matrix, such as woodblock, screen, plate, or stone, onto paper, fabric, or plastic. Printmaking possibilities range from methods so simple that they are accessible to children, to the

# Printmaking

technical and connoisseurial demands of fine art. They include intaglio, relief, screenprinting, lithography, monoprints, and collagraphs. In this chapter, we identify environmental and health hazards associated with these procedures; we also suggest appropriate precautions and feasible, safe alternatives.

◆ ◆ ◆

To better understand this chapter, review chapters 1 through 4, so that you know:

- How chemicals enter the body

- The difference between acute and chronic illness and the difference between irritant and sensitizer

- The principles of effective ventilation, safe storage, and disposal of art materials

- Use of personal protective attire, emergency procedures, and appropriate medical attention

## Potential Hazards of
## Printmaking

Traditional printmaking techniques involve potentially hazardous **physical** (mechanical) acts such as cutting, drilling, filing, and grinding plates; carving wood and linoleum blocks; and printing with heavy-pressure presses.

**Chemical** challenges in printmaking include:

- **Acids** used to etch zinc, copper, or other metal plates for intaglio printing; to etch the surfaces of lithostones and plates; and to process photosensitive emulsions

- **Alkalis** used for cleaning, conditioning, and photosensitizing

- **Dusts** and **powders** found in photochemicals, rosin, talc, and whiting

- **Solvents** used to thin and dissolve inks, to clean and prepare plates for printing, and to adhere silkscreen film

- **Photochemicals** used to sensitize and develop images on intaglio, silkscreen, and lithography matrices

All of these processes and materials present significant challenges to making prints safely. Changes of habit as well as changes in materials, procedures, and equipment may be warranted.

## Safety and Precautions

The most significant and practical step toward safer printmaking is to replace solvent-based and oil-based inks with water-soluble inks. In addition, remove as many organic solvents and solvent-containing materials as possible from use and your studio. Finally, assure appropriate ventilation and efficient use of time, space, and equipment.

## THE PRINTMAKING STUDIO

A printmaking studio must be properly located. The home is not a good place for acids, alkalis, solvents, dusts, powders, and photochemicals. The artist and other family members should not risk daily, 24-hour exposure to potentially hazardous materials and vapors.

A safe printmaking studio must have a source of running water for printing, first aid, and cleanup procedures.

Effective ventilation is essential. Therefore, avoid basement studios that lack windows. Find an upper floor where vapors can be vented to the roof. Beware, however, of air pollution codes or building requirements that prohibit release of airborne substances that may reenter homes, offices, or studios of others.

Additional safeguards for the studio:

- Adequate lighting decreases eye strain and accidents.

- Avoid gas heaters if using flammable solvents. Designate at least two exits in case of fire.

- Keep a list of emergency phone numbers (doctor, poison control center, police, fire department, and ambulance service).

- Place permanent labels on all containers. Identify contents and special hazards.

- Store hazardous materials in unbreakable containers.

- Do not store large containers on high shelves. Prevent the possibility of falling and breaking.

- Keep all containers closed, even when working, to prevent escape of vapors, dusts, etc.

- Do not eat, drink, or smoke in the studio.

- Wear gloves when working with solvents, acids, alkalis, and inks.

Observe basic safety precautions in the use of power tools, workshop equipment, and printing presses:

- Keep area clear of debris.

- Maintain smoke alarms, fire extinguishers (that you know how to operate), and a first aid kit in the studio/print shop.

- Use plate choppers and paper cutters carefully and only according to directions; do not use a guillotine cutter without a guard.

- Use properly wired and grounded electric equipment away from water (ground fault circuit interrupter is preferred).

- Secure plates while working on them.

- Be alert. Accidents usually result from carelessness.

Printing presses, which exert high pressure, have crushed many hands and fingers. Select a press designed for optimum safety with guardrails or shields that keep hands away from the roller. Get one that is geared to allow movement without undue strain. Install the press properly so that it is held firmly in place. Set presses apart in the studio where they will not interfere with other activities. A motorized press should have a safety stop switch. Presses must be used with utmost caution and care:

- Tie back long hair, neckties, ribbons, and scarves.

- Do not wear loose clothing near presses; roll up sleeves to the elbow or above; tuck in shirttails.

- Never put your hands on the press bed when it is moving.

Effective ventilation is the most important safeguard for printmakers using any of the materials cited here in Potential Hazards of Printmaking. We recommend that you work in studios with local exhaust ventilation (preferably lateral slot) and general dilution ventilation, especially if you use acids, inks or solvents.

Careful handling of materials and consistent use of protective attire and equipment are also essential to making prints safely.

## PRINTMAKING INKS

Inks contain a mixture of pigments and dyes suspended in a vehicle and are made workable with materials that thin, retard, extend, and dissolve the ink. The ingredients and their proportions vary according to the kind of printing for which they are

Figure 7.1.—Evan Summer, using his new Takach Garfield electric-powered etching press featuring a Lexan safety shield to keep fingers and hands away from the upper roller and press bed. A microswitch on the shield provides an additional safety feature. When the shield is deflected toward the roller, the microswitch is tripped; this disconnects power to the drive motor, thereby stopping the press. (Photo: Jerry Goffe.)

intended. Information about pigments and dyes used in inks and other art materials may be found in chapter 6, "Coloring Agents."

Water-soluble inks are almost odorless, do not emit unwanted vapors, are mostly nonflammable and nonpolluting, and thus do not require extensive ventilation. Exclusive use of water-based inks in educational institutions can save thousands of dollars in reduced ventilation and safety requirements. They are the only inks permitted in the absence of effective local exhaust and general dilution ventilation, and they are the only inks that should be used by children.

Formerly the domain of children's printing inks in eight or nine basic colors, water-soluble inks are now produced in an increasing range of colors and qualities. Options in water-soluble inks will expand as long as (1) more artists, teachers, and printmaking studios integrate water-based printing into their procedures and curricula; and (2) ink manufacturers

**Figure 7.2.—**Multitiered lateral slot exhaust system located by the drying racks at Ontario College of Art, Toronto.

are increasingly restricted by stringent waste disposal regulations for toxic materials. Some manufacturers continue the search for water-based products; they predict further improvements in the stability, printability, and opacity of various formulations (Duccilli 1989).

Depending upon the purpose for which they are intended, water-soluble printing inks include several shades and tints of over a dozen hues. Appearances and effects are quite extensive; from transparent or translucent to opaque, from light to heavy, from flat to glossy, and from fluorescent or iridescent to metallic. Some manufacturers who produce professional-quality water-soluble printing inks include Advance, Createx by Color Craft, Hunt Speedball, Naz-Dar, and Shiva. Other water-soluble painting and drawing materials are compatible with the inks and are worthy of experimentation both on the plate and postprinting. For example, watercolor crayons, colored pencils, tempera, gouache, and watercolor paints can be incorporated into water-based printing to yield a wide range of interesting effects and expressions. Patient experimentation with these versatile inks soon leads to control, competence, craft, and vision. We have discovered that water-soluble printmaking inks have expanded creative options in every method except lithography (which intrinsically requires oil inks).

The vehicle/binder in water-soluble printing inks may be gum arabic or acrylic polymers. Bases vary according to degrees along the transparency-opacity spectrum. The thinner-solvent is usually water, although inks with an acrylic base may require a compatible acrylic thinner, retarder, or extender (see in chapter 6, "Aqueous Media," acrylics).

A review of Material Safety Data Sheets will confirm that hazards associated with water-soluble inks are significantly fewer than those faced with oil-based inks. Many pigments, which can be inhaled in their powdered state or absorbed through the skin in the presence of organic solvents, are relatively inaccessible, except through ingestion, when suspended in a water-soluble medium. However, because preservatives and dispersing agents can be found in water-based inks, we recommend that individuals with allergic tendencies use water-repellent barrier hand creams and/or gloves (natural rubber, neoprene, polyvinyl alcohol [PVA], Buna-N, butyl rubber, polyvinyl chloride [PVC], or polyethylene). If any of these gloves feels wet inside, consider the possibility of glycol ether permeation. Wear PVA gloves for full protection. Most people, however, are not adversely affected by the ingredients in water-soluble inks.

Oil-based inks suspend pigments and dyes in a vehicle that usually consists of linseed oil and various organic solvents and additives. Solvents are used to mix and blend inks to appropriate consistency, to thin to a lighter consistency, and to clean the ink from plates and other surfaces. Additives, some of which are solvents or solvent-based, affect the performance of inks by:

◆ Increasing bulk and reducing intensity (extenders)

◆ Increasing drying time so that ink remains workable (retarders)

◆ Increasing transparency

◆ Adding gloss (varnishes)

◆ Thinning inks (thinners)

◆ Decreasing drying time (driers)

Driers include cobalt linoleate (slightly toxic), cobalt naphthenate (a suspected carcinogen), lead (highly toxic), and manganese (highly toxic). Consult labels or Material Safety Data Sheets for the presence of driers. Discard inks that contain lead, manganese, or cobalt naphthenate driers. Use inks with cobalt linoleate or no drying agent.

Some pigments and dyes are hazardous, particularly in powdered forms. It is safer to use inks formulated by dependable manufacturers and labeled with all ingredients. If you must use powdered pigments, use a glove box or wear gloves and a dust mask. Undesirable pigments (moderately toxic to highly toxic) include the cadmiums, chromates, cobalts, leads, and talc (if it contains asbestos). Dyes to avoid include diarylide, molybdate, and red lake. More detailed information on pigments and dyes is found in chapter 6, "Coloring Agents."

Since they enter the body so easily, solvents (see "Solvents" later in this chapter) used in oil-based printing inks present significant challenges to health. Therefore, when working with oil-based printmaking inks, observe the following precautions:

Assure effective exhaust ventilation to carry vapors away from you and your work area and to consistently replace the work environment with clean air.

■■
Check the ingredients of your inks (on labels and Material Safety Data Sheets) to learn what solvents are used. Check the glove charts in chapter 4 and wear gloves that provide the most protection.

■■
Do not wipe plates with bare hands. Rubbing (friction) damages skin and its protective barrier, enhancing the risk of solvent absorption. Wear appropriate gloves; use rags for wiping.

■■
Wear an impermeable apron or garments that will not allow solvents to get on clothing and skin.

Children should not be allowed to use oil-based inks (Qualley 1986). Moreover, we encourage all artists, if possible, to change to water-soluble inks. Some have speculated that environmental concerns will hasten the demise of solvent-based inks. The printing ink industry has already virtually eliminated the use of heavy-metal-based pigments in most printing inks.[1] "Most graphics ink manufacturers agree that current environmental regulations are making the future of solvent-based inks look increasingly doubtful . . . they are becoming more and more closely regulated." (Moret 1989). Specific procedures for using water-soluble inks, which have in fact greatly expanded the spectrum of printmaking, will be discussed later in "Printmaking Processes."

In metal plate etching, lithographs, and some photoprocesses, it is not possible to replace all potentially harmful printmaking materials with less toxic substitutes. Control and protection, therefore, must be incorporated into these printmaking activities.

## ACIDS

Acids are used to etch zinc, copper, or other metal plates for intaglio printing; to etch the surfaces of lithostones and plates; and to process photosensitive emulsions. Those used in printmaking that are highly toxic include: **carbolic** (phenol), **chromic** (also a suspected carcinogen); **Dutch mordant** which is a solution of hydrochloric acid, potassium chlorate (an oxidizer), and water; **nitric, phosphoric, sulfuric/oleum,** and concentrated **acetic** acids. In general, acids are corrosive. Depending upon the concentration and duration of exposure, acids cause skin irritation and burns and gastrointestinal damage if ingested. Gases and vapors from these acids irritate the respiratory system and may aggravate chronic bronchitis.

**Nitrogen dioxide,** the gas that evaporates from nitric acid used in etching and lithography, is highly toxic if inhaled in large quantities. Symptoms include fever, chills, cough, shortness of breath, headache, vomiting, and rapid heart beat. Chronic exposure may cause pulmonary dysfunction and symptoms of emphysema.

Safe procedures in handling the acids of printmaking include constant vigilance to avoid inhalation of vapors and contact of acid with skin and eyes. When preparing and handling acid solutions, wear goggles, gloves, impermeable clothing, and, in the absence of local exhaust ventilation, a mask with an acid gas cartridge. Because no cartridge protects against inhalation of nitric acid (Rossol 1982) required for etching and lithography, nitric acid should be used only with effective local exhaust ventilation. Install an eye wash fountain or a flexible spray in your sink to use in case of accidental splashing.

Precautions for working with concentrated acids include:

◆ **Full-face mask:** a (NIOSH/OSHA-approved) combination organic vapor and acid gas cartridge (in a single piece of equipment, it provides the safety of shielded, nonventilated safety goggles, an air-purifying respirator, and a face shield). The combination cartridge is effective for handling both acids and solvents

◆ **Gloves:** gauntlet-cuffed neoprene, buna-N, or NBR rubber

◆ **Attire:** rubber apron or impermeable clothing with long sleeves to protect forearms

◆ **Handling:** when diluting, add acid slowly to water, never water to acid as the concentrate may cause an exothermic (heat-generating), explosive reaction. Use of a fume hood offers additional safeguards; do not agitate or lean over acid trays during etching, and avoid inhaling acid vapors

- **Storage:** return acid solutions to an accurately labeled and appropriate holding container, or keep acid baths tightly covered when not in use; store acid chemicals out of reach of children and away from flammables and alkalis (e.g., sodium hydroxide) in containers and cabinets approved for this purpose

- **Dispose** of dilute acid solutions by neutralizing them with sodium bicarbonate (baking soda); test neutrality with litmus paper and then, with copious running water, slowly pour the neutralized solution down the drain; do not neutralize concentrated acid solutions, do not dispose of acids simultaneously with other chemicals, and always wear protective gloves, apron, and face shield; clean up spills with cold water

- **Emergency** care for acid splashes and burns: flush eyes and skin immediately and continuously with water for 20 minutes or more, remove contaminated clothing, and call for emergency aid

- **Substitute** the "weakest" (most dilute) solution that will accomplish the task; in addition, substitute ferric chloride for Dutch mordant

## ALKALIS

Alkalis that are highly toxic by all routes of entry include **ammonia** compounds used for cleaning and photosensitizing and **sodium hydroxide** (caustic soda) used with lithoplates and sometimes to etch linoleum. Alkalis are corrosive and will burn eyes and skin. Inhalation of dust or solution vapors can cause pulmonary edema, and even dilute solutions are irritating. Many household and heavy-duty cleaning agents are also alkaline. Caustic **alkalis** are used in strong concentration in screen haze removers and in metal plate cleaners. Haze removers should be avoided, if possible, by replacing the screen mesh. If you must use a haze remover or any other alkaline substances, do so only with effective local exhaust ventilation or when wearing an approved mask.

Precautions for working with alkalis include:

- Hooded, flexible, nonvented **goggles** or a **full-face mask** (NIOSH/OSHA-approved for ammonia or ammonia/methylamine use)

- **Gloves:** gauntlet-cuffed natural rubber, neoprene, Buna-N, butyl rubber, or polyethylene

- **Attire:** rubber apron or impermeable clothing with long sleeves to protect forearms

- **Handling:** a glove box or fume hood to avoid inhaling caustic vapors

- **Storage:** return alkali solutions to an accurately labeled and appropriate container; store out of reach of children and away from acids in containers and cabinets approved for this purpose

- **Dispose** of alkali solutions by slowly pouring them down the drain; do not dispose of them with any other chemicals, and wear protective gloves, apron, and face shield; clean up spills with cold water

- **Emergency** care for alkali splashes and burns: flush eyes and skin immediately and continuously with water for 20 minutes or more, remove contaminated clothing, and call for emergency aid

## DUSTS AND POWDERS

Various dusts and powders are encountered in printmaking. Dyes, pigments, photochemicals, rosin, talc, and whiting are some of those most commonly used. Whenever possible, obtain coloring agents and photochemicals premixed into solution.

Precautions for working safely with dusts and powders include:

- A fume hood local exhaust system

- A **glove box** or NIOSH-approved respirator with a particulate filter

- Cleanup of spills with a damp sponge or microfilter vacuum cleaner

## SOLVENTS

Many authorities believe the nervous system can be adversely affected by substantial ingestion, skin absorption, or inhalation of solvent vapors. Headaches, dizziness, feelings of intoxication or "being high," mental confusion, increased heart rate, nausea, and loss of coordination suggest the possibility of significant solvent exposure. Inhalation of vapors from many organic solvents may also cause eye, nose, and throat irritation, and high level exposure has been known to cause breathing difficulty, convulsions, kidney and bladder damage, and even death.

Solvents (refer also to Table 6.3) used in printmaking include:

**Alcohols** (ethyl, isopropyl, methyl), whether ingested or inhaled, depress the nervous system. They irritate the eyes and mucous membranes, and dry the skin, making it vulnerable to cracking and permeation. Ethyl alcohol (denatured alcohol), or ethanol, is considered a safer solvent and is recommended as a substitute for lacquer thinner. Isopropyl alcohol (rubbing alcohol), or isopropanol, is an acceptable substitute for ethanol. Methyl alcohol (wood alcohol), or methanol, should be avoided.

■■

**Aliphatic hydrocarbons/petroleum distillates** (benzine/VM&P naphtha, gasoline, hexane, kerosene, mineral spirits, naphtha, and petroleum ether) may be especially irritating to skin. **N-Hexane,** a highly toxic solvent in rubber cement and other glues, can cause severe axonopathy, which may be manifested as short-term and usually reversible damage to the peripheral nerves (arms and legs). **Benzine,** also known as **VM&P Naphtha** (less toxic than the aromatic hydrocarbon benzene), **mineral spirits, Turpenoid,** and **Varsol** are preferable, less toxic aliphatic hydrocarbons.

■■

**Aromatic hydrocarbons** (benzene, styrene, toluene, xylene) are easily absorbed and irritating to skin. **Benzene** (benzol) is highly toxic, believed to cause some forms of leukemia, and is known to damage bone marrow. Massive exposure to **toluene** (methyl benzene, toluol) may cause dizziness, unconsciousness, and even death. Toluene is easily absorbed and irritating to the skin and may also cause mild liver and kidney damage with substantial chronic use. **Xylene** (dimethyl benzene, xylol) is absorbed through unbroken skin and may cause nervous system damage with substantial long-term use. **Styrene,** which is more toxic than toluene and xylene, should be avoided. Substitute less toxic acetone or mineral spirits whenever possible.

■■

**Chlorinated hydrocarbons** (carbon tetrachloride, ethylene dichloride, methyl chloroform, methylene chloride) have anesthetic properties that can depress the central nervous system. In addition, inhalational exposure has been associated with abnormal heart rhythm. Effects range from mild dizziness to unconsciousness and, with massive exposure, death. Chlorinated hydrocarbons can alter the fatty layer of skin to increase absorption and cause local irritation. Most are not flammable, but some decompose in heat or ultraviolet light, releasing a poisonous gas, phosgene. Avoid them.

■■

**Glycols** and **glycol ethers** (butyl Cellosolve, Cellosolve, methyl Cellosolve, diethylene glycol, ethylene glycol) represent several classes of solvents commonly used to disperse pigments in solution. They are mild irritants that are variably absorbed through skin. Their vapors can irritate mucous membranes and the upper respiratory tract. If ingested, **methyl Cellosolve** (used in photoetching) may cause central nervous system depression and cardiac, lung, and kidney problems. Methyl Cellosolve has been linked to birth defects and testicular damage in animals. The combination of glycol ethers and organic solvents in offset lithographic printing has been associated with bone marrow damage. Glycol ethers are used as solvents for resins, paints, and cellulose inks; as an ingredient in coatings such as wood stains, epoxies, and varnish; in photochemicals; and in cleaning compounds. Most glycol ethers, however, are relatively safe except in massive quantities, and some are approved as ingredients in cosmetics.

■■

**Ketones** (acetone, cyclohexanone, isophorone, methyl butyl ketone, methyl ethyl ketone) are organic solvents that dissolve lacquers, oils, plastics, vinyl inks, and waxes. Methyl butyl ketone has been associated with neuropathy; the others may be mucous membrane irritants. Acetone is the safest.

■■

**Lacquer thinner** is a catchall term for mixtures of alcohols, ketones, and other solvents such as aliphatic and aromatic hydrocarbons (especially toluene). Repeated use has been associated with dermatitis, central nervous system depression, and symptoms associated with exposure to toluene. Because alcohols, ketones, and petroleum distillates are also commonly part of lacquer thinner, health problems may be intensified due to additive or synergistic effects. Lacquer thinners should be avoided; substitute ethyl alcohol to dissolve lacquers, acetone to dissolve plastics, and mineral spirits to dissolve paint.

■■

**Terpenes** such as limonene are hydrocarbons extracted from plants. Terpenes are used in synthetic flavorings, waterless hand cleaners, paint thinners, and cleaning products. A citrus or orange smell and taste make limonene products attractive to children, who occasionally develop gastrointestinal irritation following ingestion of terpenes.

■■

**Turpentine,** an especially irritating solvent frequently employed by artists, is an oleoresin distilled from coniferous trees. Vapors from turpentine are irritating to the eyes and mucous membranes, and repeated exposure has been associated with kidney disease, pulmonary edema, and chronic respiratory diseases such as bronchitis and emphysema. "Allergies" to turpentine may be related to the type of pine tree from which the substance is derived. Turpentine may be life-threatening if aspirated following ingestion. It can penetrate unbroken skin. Turpentine substitutes, such as mineral spirits, Stoddard's Solvent, Turpenoid, and Varsol, are safer, but if any of the common symptoms of solvent inhalation are present—such as headache, dizziness, feelings of intoxication or "being high," mental confusion, increased heart rate, nausea, or loss of coordination—one of these less odorous solvents could be the culprit.

Solvents are hazardous. Avoid accidental or careless ingestion by keeping them away from food, eating areas, and children. In cases of significant ingestion, induction of vomiting is not recommended. Safe procedures for handling solvents used in printmaking include constant vigilance to avoid both inhalation of vapors and skin contact. Because most solvents are volatile, they require effective local exhaust and general dilution ventilation for safe use.

To reduce potential hazards, use the least harmful solvent that will do the job:

- Ethyl or isopropyl alcohol instead of lacquer thinner concoctions

- Acetone instead of toluene or xylene

- Deodorized turpentine such as Turpenoid, paint thinners, or mineral spirits instead of turpentine

Further precautions for working with solvents include:

**1. Respiratory protection:** A NIOSH/OSHA-approved, organic vapor cartridge respirator should be worn in the absence of adequate ventilation. It is effective for solvents and products that contain solvents, including inks, lacquers, paints, aerosol spray fixatives, and adhesives, and for plastic cements, resins, and particles. Wear it as long as solvent vapors are present in the air, until they have been cleared by opening windows and doors.

**2. Gloves:** The correct choice of gloves remains a challenge. Various types are available. Unfortu-

nately, no one kind is resistant to all solvents. For example, butyl rubber is effective against many solvents but not aromatic and chlorinated hydrocarbons. For a given solvent, appropriate gloves are as follows:

**a.** Alcohols (ethyl, isopropyl, methyl): natural rubber, neoprene, butyl, polyvinyl chloride (PVC), polyethylene, and nitrile

**b.** Aliphatic hydrocarbons/petroleum distillates (kerosene, mineral spirits, naphtha, paint thinner): nitrile, natural rubber, neoprene, Buna-N, and polyethylene

**c.** Aromatic hydrocarbons (benzene, toluene, xylene) are most likely to permeate or degrade materials; nitrile and polyethylene gloves are most resistant to toluene and polyvinyl alcohol (PVA) is best for benzene and xylene

**d.** Chlorinated hydrocarbons (perchlorethylene, trichloroethylene): will degrade all but PVA, nitrile, and polyethylene

**e.** Glycol ethers (Cellosolve): polyvinyl alcohol (PVA)

**f.** Ketones (acetone, methyl ethyl ketone): butyl, natural rubber, neoprene, and polyethylene

**g.** Lacquer thinner (ingredients may include aromatic hydrocarbons, alcohols, petroleum distillates): polyvinyl alcohol (PVA)

**h.** Turpentine: Buna-N, PVA, and nitrile

**3. Attire:** Wear clothing with long sleeves to protect forearms; use impermeable aprons.

**4. Handling:** Use the smallest amount of the least toxic solvents required for the job. Limit the amount of solvent used by applying it to a rag rather than splashing it over a plate or screen. Keep containers capped when not in use; pour solvents slowly from a narrow-nozzle dispenser to avoid splashing. Promptly remove clothing contaminated by solvents. A fume hood exhaust is an additional safeguard. Do not use solvents as hand and skin cleaners. Rub baby oil into hands to soften contaminants prior to soap-and-water scrub. Follow up with a soothing cream. If necessary, a waterless nonsolvent-based cleaner (such as Boraxo, Dif, or Gresolvent) can be used, followed with soap and water, and then cream.

**5. Storage:** Cap solvent containers tightly when not in use. Store solvents and solvent-soaked rags in secure, clearly labeled, spring-top containers, away from other chemicals and locked from children. If the solvent is flammable, store large amounts in an approved flammable-storage cabinet.

**6. Dispose** of solvents and solvent-soaked newspapers or rags in tightly closed, labeled, flame-proof containers through regular or, if required, toxic waste disposal pickups. Keep them separate from other chemicals.

**7. Emergency** care for solvents: Flush eyes and skin immediately and continuously with water for 20 minutes or more. Remove contaminated clothing. Obtain medical aid for solvents in the eyes. If ingested, drink two glasses of milk or water and call your poison information center for further recommendations. Wash skin with a mild soap and apply a soothing cream. If skin or respiratory irritation persist, or if symptoms of nervous system damage, such as tingling in arms or legs, dizziness, or headaches persist, seek medical attention.

## PHOTOCHEMICALS

**Photochemicals** used to sensitize and develop images on intaglio, silkscreen, and lithography matrices can present significant challenges to working safely. Emulsions range from relatively safer diazo compounds to ammonium dichromate (suspected carcinogen, flammable) and KPR (flammable). Developers range from simple water rinses to the complex Kodalith solutions, which contain hydroquinone and sodium formaldehyde bisulfite in Part A and sodium carbonate in Part B. These agents can irritate skin, eyes, and mucous membranes, so avoid skin, eye, and clothing contact as well as inhalation of either powders or solutions.

**Photoetching** can be particularly hazardous because it utilizes a number of solvent-based materials (see Appendix I, KPR). For example, working with KPR (Kodak photo resist) products requires exhaust ventilation at the immediate site of use. You may reduce use and exposure by purchasing presensitized plates (which often have a more even emulsion than plates prepared manually) or eliminate exposure altogether by having a professional photography lab expose and develop a plate from your acetate drawing or Kodalith image.

## PROFESSIONAL STUDIOS

Artists may wish to broaden their methods of expression without the expense of setting up their own printmaking shop. To facilitate printmaking with the benefit of master printer expertise, professional studios, or *ateliers*, have been established. These studios generally operate with effective ventilation and safeguards to protect all who work there. A printing atelier is a safer alternative for individuals who employ lithography and etching with oil-based inks but cannot undertake the installation of a suitable ventilation system in their own studios.

## Printmaking Processes

**M**aking plates, inking, printing, and cleaning up are the basic, general procedures shared by all methods of printmaking. In this section we address the ways in which techniques can be employed and prints made as safely as possible (color plates 7.1– 7.22). Included are:

- Relief: woodcuts, linocuts, and wood engravings
- Intaglio: engravings, etchings, and drypoint
- Collagraphs
- Screen printing and stencil
- Lithography
- Unique prints: monoprints, monotypes, transfer prints, and chine collé
- Photo-printmaking processes

Whenever possible, the emphasis is on techniques that expand the use of water-soluble inks to make exciting, creative, and totally professional prints.

Figure 7.3.—**S**hao Chang Yi: *Tree in Wind*, 1985, woodcut printed with water-based inks, 20″ × 24″. (Courtesy Sichuan Fine Arts Institute, Chongqing, People's Republic of China.)

Figure 7.4.—**X**u Zhong-ou: *Birthplace*, 1986, woodcut using oil-based inks on mulberry paper, 36" × 113". Working about 12 hours each day, I make my prints with woodblocks. Sometimes I feel discomfort in the joints of my hands, so I learned 'Chi Kung,' which helps me a lot in my work [Chi Kung is an ancient Chinese system of working with the internal energies of the body]. I can use my mind to control the strength and operation of my hands. I have used this technique for many years to prevent injuries. *Birthplace* took seven months to create and takes 48 hours to make one print. (Translation: Lily Yeh; courtesy Sichuan Fine Arts Institute, Chongqing, People's Republic of China.)

## RELIEF

To make printing plates from blocks of wood or from linoleum, one must use sharp gouges, gravers and knives, and sometimes power tools, to carve out areas of the plate. For safe block cutting:

- Direct cutting tools away from the body and noncarving hand.

- Keep tools sharpened; dull ones are difficult to control.

- Use tools with a large handle or grip that distributes pressure over a large area of the palm (to prevent tenosynovitis or carpal tunnel syndrome).

- Use a solid, steady table with a bench hook to hold the block firmly.

- Position yourself or your light source so that you do not carve in shadow.

- Rest and exercise (shake out, flex, and stretch) the cutting hand and arm frequently. Periodic breaks will prevent strain and injury from repetitive motions.

- Store sharpened tools safely in a container with sharp ends covered or inserted into corks.

- Use only power tools and saws that are properly grounded.

- Use all tools with appropriate knowledge and precautions.

An increasingly common complaint among block printers (and others using repetitive motion) is tenosynovitis, an inflammation of tendons and synovial fluid in joints. Carpal tunnel syndrome, a form of tenosynovitis in the wrist, is associated with repetitive motion and stress that affects the hand, wrist, and arm with pressure, pain, and debilitation in the affected nerves. It is commonly associated with carving tools, but is also known to arise from inking and wiping etching plates. Early symptoms of carpal tunnel syndrome include numbness, tingling, and occasional pain in the fingers and hand.

The condition can be avoided or reduced in several ways. Choose tools with wide handles to spread pressure over a large surface area of the palm. Avoid overgripping because extraneous tension in the hand aggravates the condition. Frequently perform complementary tasks or exercises. For example, while carving, periodically pause to loosely shake the hand. Follow the loosening by flexing and clenching and then stretching and tensing the fingers, hand, and arm. These exercises break the repetition that is the key to developing carpal tunnel syndrome. At reasonable intervals, switch to an alternate activity that requires different, complementary movements. Medical evaluation is useful, and intervention may be required to prevent permanent damage. Total rest of the affected hand and use of anti-inflammatory medication may be appropriate.

**Woodcuts** are made from blocks of wood that lend themselves to clean, straightforward cutting. Clear pine is soft and easy to cut, but pine also

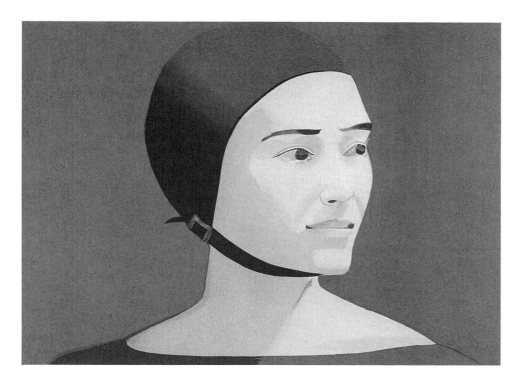

Figure 7.5.—**A**lex Katz: *The Green Cap*, 1985, color woodblock, 12⅛″ × 17⅞″. Using traditional Ukiyo-e woodblock printing, the artist provides a watercolor drawing to be traced. The tracing is pasted on a smooth hardwood slab and a woodcarver cuts away the areas not to be printed. The printer applies a thin coat of transparent, water-based pigment to the block and overlays a sheet of handmade paper. To transfer the pigment to paper, the printer vigorously rubs the back of the paper. The artist and printer work together to alter colors and adjust the number of blocks. (Courtesy Crown Point Press, San Francisco and New York.)

dents, bruises easily, and does not hold up for rigorous printing. Poplar (slightly harder, sturdier, and fine grained) and fruitwoods fare better. For sizes larger than 18 inches, plywood with a fine-grained veneer of bass, birch, cedar, fir, or spruce may be used.

Watercolor woodblock printing was invented and practiced in China as early as 600 A.D. This printing method was the precursor to the Ukiyo-e style that evolved in Japan. The technique remains virtually unchanged since China's golden age of printing during the Sung dynasty (960–1279).

**Woodblock printing** is a relief process in which areas of the image not to be printed are cut away on the blocks. In China, many small blocks, each corresponding to different areas of color, are used to create a single print. (By contrast, in the Japanese method, large, full-sized blocks are used.) In the Chinese technique,[2] blocks of various sizes are cut; the number used to make a single image varies from one to one hundred. Blocks of varying hardness are carved from pearwood or boxwood, with finely grained, smooth, even-textured surfaces. Inks comprised of earth pigments, hide glue, tree resin, and water are prepared by grinding the pigment in an inkstone prior to transfer to porcelain bowls. The inks are permanent and luminous on silk or paper. Blocks are inked with round brushes made from bear or deer hair; to achieve special effects, the pigment

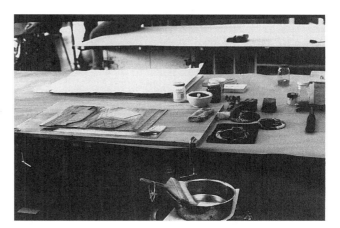

Figure 7.6.—**P**reparing to print in the Japanese tradition. On the table is the carved woodblock. The nori is in a pot in the foreground, and behind it are barens, brushes, and inks.

is often allowed to run, is wiped away, or is built up to achieve gradations of color. Sheets of silk, premounted on paper made from bamboo or mulberry bark, are clamped at one end for registration and placed individually over inked blocks. The printer transfers ink by applying pressure to the back of the silk with a flat, palm-fiber brush. Certain colors are printed and dried, while others are kept moist during printing (color plates 7.1–7.4).

A **wood engraving** consists of carved or incised lines, made with a sharp **V** gouge into the end grain of relatively hard wood such as boxwood or maple. The lines of a wood engraving tend to be fine and clear in contrast with large surface areas to be inked.

**Linocuts,** made from linoleum backed with burlap or mounted on wood blocks, are easier to cut than wood since they require less pressure and have no grain with its own intentions (color plate 7.5). Linoleum cutting is an alternative for persons suffering from tenosynovitis or carpal tunnel syndrome. Flexible Printing Plate, another material available for linocuts, is about an eighth of an inch thick and can be cut with scissors and carved with X-Acto knives into intricate shapes and textures. The material can be glued to a plywood block and cut as a linocut, or separate shapes, cut from the material, can be glued to a block or plate, as with a collagraph, and printed by either relief or intaglio methods.

Informal and inexpensive relief printing plates can be carved from potatoes and erasers.

The primary method of **inking** relief plates is with one color at a time rolled on with a rubber brayer over the entire uncut, raised surface of the block. Water-soluble printing inks may be so applied, or blocks may be inked by brushing water-soluble printing ink colors over the surface to provide a multicolor print with a single block. Let the ink dry; this normally takes 30 minutes. You can expedite drying with a blow-dryer. The Japanese method for printing woodblocks, which is applicable to all relief printing, results in special translucency of color. Increasingly employed by printmakers in the West, it calls for inking with a brush dipped in watercolor paints and a sponge dipped in a rice paste called *nori*. The color and paste are blended over the surface of the block (Johnson and Hilton 1980). The paste delays drying and binds the colors to the paper. It is especially effective on lightweight rice or kozo papers that yield immediately to the inked surface.

Relief **prints** do not require a press. Depending upon the wetness of the ink, dry or premoistened paper is placed over the inked surface. A lightweight plastic or vinyl sheet is placed on top to protect the wet paper from damage, and the surface is rubbed with the back of a spoon or other flat implement, such as the Japanese-made bamboo and rope *baren*. To check the quality of the print, lift a corner to assess whether it has received sufficient rubbing before pulling the whole print away from the plate.

To **clean** relief plates, use water-dampened rags or sponges to wipe residual ink from the blocks. Unless marine plywood has been used, do not submerge the blocks in water, as warping and separation of the plies may occur.

## INTAGLIO

Etching, aquatint, engraving, mezzotint, and drypoint comprise the principal intaglio or incised methods of printing (color plate 7.6). Lines and textures to which ink will cling are processed into and/or onto the plate, and prints are made on dampened paper under pressure with an etching press. Metal (zinc, copper, and steel) **plates** are engraved with tools or acids to create fine lines and textures that will hold ink. Hazards exist with:

- Sharp tools used in engraving: possibility of trauma (cuts, lacerations) and tenosynovitis
- Acids, acid resists, rosin, and sprayed enamels: potentially hazardous via all routes of entry
- Solvents: used for removing the acid resists and oil inks from the plates

**Etchings** and **aquatints** require acids, acid resists, rosin, sprayed enamels, and solvents to make plates.

**Figure 7.7.—R**ichard Hricko: *Nearing the Bridge III,* 1988, color intaglio, 17″ × 26″. I use nonacid intaglio and relief processes including mezzotint, drypoint, engraving, wood cut, and wood engraving. Unfortunately I suffered a nerve problem in my index finger, which was related to the extreme pressure required to engrave end-grain wood and to burnish metal for mezzotint. Soft fabric (cheesecloth) and tape wrapped around handle-less burnishers created a cushion that has prevented further damage to my finger. When etching, I use copper plates and ferric chloride, which is much safer than Dutch mordant. Ferric chloride is comparable in quality, retains its etch strength longer, and is less expensive than Dutch mordant.

Figure 7.8.—Evan Summer: *Landscape XX*, etching, engraving, drypoint, 29″ × 41½″. I recently substituted ferric chloride for Dutch mordant. It etches quickly and is less toxic than the mixture of hydrochloric acid and potassium chlorate which can release irritating chlorine gas.

Thus, appropriate precautions are required. Etchings can be made safely **only** with effective exhaust ventilation. Apply acid resists, blockouts, and enamel sprays with the plate under a local exhaust hood. Heat plates under a hooded exhaust, using a grounded hot plate with an indicator light. Keep potentially explosive rosin dust in an explosion-proof rosin box away from spark or flame.

Etching needles and tools vary in sharpness and may cause injury. They should be used attentively. Highly toxic **acid baths** for etching should be used with lateral slot exhaust ventilation or in a compartment exhaust enclosure; these baths should always be stored or covered when not in use.

Individual artists should not photosensitize their own etching plates. It is safer to purchase, at a moderate cost, a presensitized zinc or copper plate from a printmaking supply store. This permits avoidance of KPR-sensitizing chemicals. Moreover, a commercial coating is more uniform than a manually prepared plate and will print more cleanly and accurately. Commercial plates must be stored in a cool place and used within a few months after purchase. The plate is exposed to a prepared image on Kodalith film (emulsion to emulsion) on a vacuum table or pressed with a sheet of glass and exposed to ultraviolet light, sunlight, or photoflood. Do not use carbon arc lamps except with the precautions described in chapter 9, "Studio and Darkroom Lighting Hazards."

To develop the exposed plate safely in KPR de-

Figure 7.9.—Ghislaine Aarsse-Prins/GAP, pulling a drypoint etching on a Brand Press in her Paris studio. All work carries in itself its own pollution, human beings as we are.

I suggest a well-ventilated studio, barrier hand creams and gloves, comfortable clothing, clean hands and face, and special glasses when using electric tools.

veloper, one must use a nonreactive (stainless steel or glass) developing tray and local exhaust ventilation to draw vapors from the bath. Wear heavy gloves while handling the plate in the developer and while pouring KPR dye to make the image visible. Follow directions carefully. The plate must dry before it is heated on a hot plate at 350°F for five minutes to set the resist. Use this procedure only under a fume hood or lateral slot exhaust. The plate is then ready to be etched, inked, and printed in the usual manner.

Intaglio plates that do not require acids include engraving, mezzotint, and drypoint. An **engraving** plate is typically a soft metal, such as copper, or Plexiglas. Safe engraving includes directing the well-sharpened engraving burin away from the body and bracing hand. Tools with a large handle or grip, which distribute pressure over as large an area of the

Figure 7.10.—Richard Die-benkorn: *CJD*, 1990, dry-point etching, 16½" × 13½". (Crown Point Press, San Francisco.)

palm as is comfortable, will reduce the likelihood of developing painful tenosynovitis. When a sharp burin is held properly, it will incise a clean **V** into the plate. By contrast, a dull burin may raise a burr at the edge of a line. The burr may cut and scratch during inking and compromise the quality of the printed line. Burrs are best avoided, but may be corrected with a keenly sharp, three-sided, wedge-shaped, hardened-steel scraper. Removal of the burr, however, is a hazardous procedure even for experts. Because the scraper must be held so that one of the sharpened edges strokes the plate as horizontally as possible, it may be necessary to place your index finger and/or thumb along the blade for optimum control. To prevent self-inflicted wounds, wrap the scraper blade, toward the handle, with several thicknesses of tape. After the burr or error has been scraped from the plate, the area may require burnishing with a burnisher, a process that demands pressure and tension. Take frequent breaks to avoid injury.

Plexiglas is easier to cut than metal. Engravings on a black plate are more visible; black paper taped to the back of clear Plexiglas will also enhance visibility and safer incising. Store sharpened tools safely in a container or with the sharp end inserted into a cork.

> Although the acid-bitten plate offers the most long-lasting and efficient matrix for an edition of prints, some of the most dramatic, velvety blacks in all of printmaking are achieved through the drypoint and mezzotint methods.
>
> (Ross and Romano 1990, p. 82)

**Drypoint** plates are made by scratching directly into a noncoated plate (as contrasted with scratching through the acid resist to make etchings) with a sharp etching needle on Plexiglas or metal plates. The scratched line can be sharp or sketchy, and the burr created by the scratched line can hold more ink than the line itself. Drypoint etchings are most eas-

Figure 7.11.—John Cage: *R 2/2 (Where R = Ryoanji)*, 1983, Drypoint, 9½" × 23½". Drypoint, the most direct of the intaglio methods of printing, avoids the use of acid and ground. (Printed by Lilah Toland assisted by Marcia Bartholme, Crown Point Press, San Francisco.)

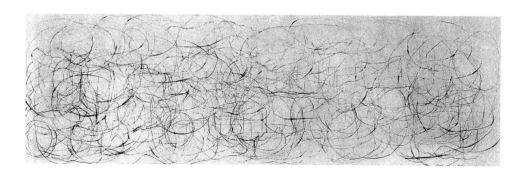

ily made with Plexiglas plates, which are softer than metal and more easily altered by scratching action. Moreover, transparent Plexiglas can be placed over a drawing for accurate replication. Hardened steel, carbide tips and diamond points can be used for making drypoints. Electric engraving tools can also be used for a good drypoint line. When the pointed tool, manual or electric, is held perpendicular to the face of the plate, the line may have a deep incision and make a double burr, one on each side of the line. When the needle is held at a slant, the burr will form only on the point side. These lines will print quite differently. Similarly, lightly scratched lines do not print the same as deeply scratched lines.

For **mezzotint,** the surface of a metal (copper) or Plexiglas plate is abraded and texturized with patterned, serrated, and textured rocker tools that incise many small pits while raising accompanying burrs. In one (negative) method, the entire plate is abraded by rocking in four or more different directions. Then the image is burnished onto the abraded surface. Highly burnished areas print lightly, softer burnishings print in varying shades, and fully abraded areas print the darkest. Devices are available to guide the rockers into regular patterns, or they can be applied freely to develop variable textures.

In the positive method, rockers are applied selectively, to the degree of shading desired, to the areas of the image that will receive ink. A variation for texturizing a printing plate is **criblé,** where punches are hammered into a plate for a dotted texture effect. Engraving, drypoint, mezzotint, and criblé can be used together on one plate or separately in registered, multiplate prints for wide-ranging and effective expression.

With these techniques, mistakes may occur, but corrections can be made on both plastic and metal plates. While the surface may not be restored to pristine perfection, a combination of scraping with an engraving scraper (very delicately on Plexiglass), sanding with a very fine sand or emery paper, burnishing with a burnishing tool, and polishing with fine charcoal and water or jeweler's rouge (rough cloth) will approach a smooth, clean plate surface to be reworked.

**Inking** intaglio plates varies with the composition and method by which the plate was made. Because this book is about making art as safely as possible, we emphasize the use of water-soluble inks in intaglio printing. To ink etching, drypoint, engraving, and mezzotint intaglio plates for printing with water-soluble inks, the viscosity (flow rate) as it comes from the tube is appropriate. Apply a fairly thick coating of ink over the plate, pressing it into the

lines with a cardboard squeegee. Wipe excess ink from the surface with cheesecloth or pages from a telephone book, followed by soft, nonabrasive rags. Vigorous rubbing with the cloth will clean the plate. To prevent skin absorption of ink, do not handwipe for final cleaning. Experimentation and perseverance can lead to colorful and effective inking, wiping, and printing methods.

Artist Naomi Limont, who developed carpal tunnel syndrome from repeated, intense rotating motion while inking and rubbing her plates, suggests using a toothbrush to work stiff etching inks into the

Figure 7.12.—**E**leanor Mink: *Leaves and Moss,* 1991, drypoint etching, 4″ × 6″. To create my drypoint etching, I drew on a Plexiglas plate with a hard carbide scriber. For the darkest areas I held the etching tool straight up and down, and for the lighter areas, I scribbled lightly with the point held like a pencil. The plate was inked with Speedball water-based black ink, wiped and printed on Rives BFK paper.

Figure 7.13.—**N**aomi Limont using a toothbrush to work oil-based ink into the plate. Naomi also uses Easy Wipe in the printing ink to make wiping easier. She keeps solvent in a dispenser with a small opening in the lid that is closed when not in use.

plate. She also notes that Easy Wipe mixed with oil-based printing ink does reduce wiping aggravation.

Because one prints intaglio with dampened paper, the water-based ink remaining on the plate should be dry before running the plate through the press. Printing with wet ink may lead to loss of clarity.

Once an intaglio plate is satisfactorily inked and dried, it is placed face up on the press bed and overlaid with dampened and blotted rag paper suitable for **printing.** To protect the press bed and blankets from "wandering" ink, place newsprint under the plate and over the dampened paper. Felt blankets are placed over the paper, and the entire "sandwich" is run through a variable-pressure etching press. The paper is pressed down into the lines, textures, and burrs, lifting the ink out of them to create prints of superb graphic and textural quality.

The use of water-soluble inks greatly simplifies **cleaning** plates. Run the plate under cool water, and loosen the ink with a soft nylon brush. It will clean quickly and safely.

Printing with water-soluble inks is considerably safer because it avoids the need for solvents, most of which are potentially toxic and flammable or combustible. By experimenting with water-soluble inks, one can significantly reduce potential hazards in the studio.

## COLLAGRAPH

Collagraphs offer an extensive and versatile print-making medium that is especially effective with water-soluble inks and is rewarding for children as well as professional artists. To make collagraphs, the surface of the printing **plate** is created and altered in numerous ways. Raised or recessed surfaces are inked and printed with either intaglio or relief printing methods. Making collagraph plates for inkless or inked intaglio embossments is safer than incising metal plates with tools and acids (color plates 7.7–7.9).

The basic **plate** can be made of readily available material such as inexpensive chipboard or cardboard, illustration board, Masonite, wood, plastic, or metal. A variety of textures, both found and created, are applied with an appropriate adhesive, most often an acrylic polymer glue. The same material can be further textured with a rasp or other implements. Texturing materials include:

- Fibrous materials: fabrics from fine organza and silks to lace, nubby wools, and burlap; papers such as tissue that are textured or crumpled, doilies; straw, hemp ropes, yarn, string, rickrack trimmings; and tapes of various materials and textures.

- Plastics/acrylics: acrylic polymer mediums, gesso, glues; glue gun; modeling paste applied with palette knife, brushes, or pastry tube for a variety of textures and lines; dried impastos that are incised as with wood or linocuts; acetate, mylar, vinyl, and other plastics and films, cut, crumpled, and shaped; linoleum and flexible printing plates cut into shapes.

- Miscellaneous: shaped or crumpled aluminum foil, sandpaper, carborundum, screening, textures from nature, and found objects that apply and serve the artist's vision.

Figure 7.14.—**N**ina Magil Hausner: *Patina*, 1988, collagraph, 14" × 16". Contoured plate made from cardboard with cut-out and glued-cardboard shapes, toothpicks, crushed tinfoil, glitter, sandpaper, and string; gessoed and allowed to dry. Speedball water-based inks applied with brushes, toothbrush, and roller, allowed to dry, and printed with etching press on dampened Arches Cover paper.

In general, one should avoid sharp objects and limit the textural height of materials to be printed to about an eighth of an inch (three millimeters). Thicker textures may tear the paper and, if put through the press, damage the press blankets as well.

Prepared collagraph plates should be sealed with gesso, which avoids the undesirable properties of spray fixatives or shellac, and allowed to dry thoroughly before inking.

Water-soluble **ink** can be mixed and applied to a sealed collagraph plate in several ways: with a rubber roller, brushes, sponges, or rags. Plates may be selectively inked with several colors, layered with different colors, selectively wiped clean or with texture, or otherwise manipulated for various effects. Accurate replication is possible by rolling the ink evenly over raised surfaces and printing as a relief print. A distinguishing feature of collagraphs, however, is the variation of effect achieved by selectively applying color to both raised and recessed surfaces of the plate. After applying the water-soluble ink, let the plate dry thoroughly (expedite with a blow-dryer).

When dampened paper is placed on the collagraph plate and **printed** under pressure with an etching press in the manner of intaglio prints, the dried water-soluble inks are activated and print with great richness. A combination of texture embossment and endless color variety allows several prints to be pulled from each inked collagraph plate. No two prints, however, will be exactly alike. We have pulled as many as twelve prints from one inking, the last print an embossment with hints of colors remaining only in the deepest crevices.

**Cleaning** collagraph plates requires little or no effort because printing with damp paper leaves little ink on the plate. Residual ink traces may leave interesting effects on subsequently re-inked prints. If you wish to clean all ink from the plate, wipe it with a damp soft cloth, use a moistened soft brush to loosen ink in recesses, and blot (running through the press if necessary) with newsprint.

## SCREEN AND STENCIL PRINTING

Screen printing is a versatile and widely used printmaking procedure where stencils are adhered to a fine mesh screen tautly stretched on a wood or metal frame. The frame serves as a container for ink that is drawn across the screen with a squeegee, dispensing a smooth, even application wherever there are open areas in the stencil (color plates 7.10–7.15). The development of permanent, high-

Figure 7.15.—Maria G. Pisano: *Il Palazzo Dipinto*, 1988, collagraph using oil-based ink, 37½" × 27". To avoid etching acids, the plate is made from a ⅛-inch-thick processed board in which I cut the linear design with an X-Acto knife. The plate is covered with a few coats of acrylic matte varnish. It is then inked and printed like an etching using oil-based etching inks. I wear gloves for protection, and my work space is ventilated. I also make my own paper and add collage elements into the pulp. For embossed and textured areas, I use various metals cut to the desired shape and inked with a brayer.

Figure 7.16.—Michael Kolesnikoff (age 16), creating a collagraph plate with string and Elmer's glue.

Figure 7.17.—Claire Cuff (age 15), uses Speedball water-soluble block printing ink on a collagraph plate made from shapes cut from chipboard glued to illustration board and gessoed.

Figure 7.18.—Lori Saunders (age 16): *Les Fleurs des Printemps*, 1990, collagraph using water-based ink, 22″ × 30″. Six collagraph prints pulled from a single inking of the plate, placed consecutively from the first at the bottom of the photo to the sixth at the top.

quality, water-based inks (that yield excellent results on paper) has revolutionized safety in screen printing.[3]

Traditional silkscreen methods using lacquer stencils and oil and vinyl inks are "among the most hazardous activities in the arts and crafts" (McCann, *AHN* 2:7). Cut stencils made from lacquer film and adhered to the screen with lacquer thinner, together with toluene-containing inks, screen washes, and solvents, contributed to health problems such as dermatitis, narcosis, eye irritation, and axonopathy (nerve damage). Frequent complaints, by both artists and students, of illness associated with silkscreen printing gradually forced printmakers and teachers to experiment with water-soluble inks.

Studies of water-soluble inks, occasionally backed by ink manufacturers, have improved and expanded the range of colors. A number of artists have proven that quality water-soluble screenprinting inks can produce excellent prints on paper, fabrics, and other materials. Their enthusiasm is periodically and informatively described in *Screenprinting* magazine and other publications. By abandoning lacquer film stencils and switching to water-soluble inks to avoid health problems associated with solvent use, many screen printers have discovered noteworthy esthetic and expressive advantages. Change has presented technical challenges, but at the same time change has greatly expanded the range of screen printing possibilities and effects.

The use of water-soluble screen-printing inks, however, does not eliminate all hazards associated with screen printing. Some of the emulsions, blockouts, and cleanup materials have potentially hazardous properties. Thus, to make screen prints safely,

Figure 7.19.—Maggie Davis working on acetate for one of the colors for *Big Cypress Swamp*, water-based screen print at Berghoff-Cowden Editions, Tampa, Fla. Our shop uses only water-based, nontoxic Hunt Speedball permanent acrylic inks, which provide excellent color range, strength, viscosity, and subtle transparencies. The inks are fast-drying, easy to clean, and have a good shelf life.

Figure 7.20.—Merle Spandorfer: *The Five Elements*, 1985, water-based screenprinted book, 13″ × 19″, on Okawara paper. My discovery of water-based screenprinting inks was an incredible liberation from the drudgery and headaches associated with the use of oil-based inks and solvents. It also led me to experience the special relationship between Japanese papers and water-based inks.

the artist must stay informed by reading labels and Material Safety Data Sheets and following all safety precautions.

Select a **screen frame** that is three inches larger on all sides than the intended image. This allows proper use of the squeegee with ample reservoir for ink. For printing with water-based inks, use a waterproof screen frame. (Unfortunately, commercially available aluminum frames are expensive.) Wood may be used for the frame if it is free of knots and kiln-dried to reduce warp. Pine, redwood, spruce, and poplar work well. Do not use paper tape and shellac, which are commonly used with oil-based screen-printing inks, to seal the frame and mask the screen. Instead, use vinyl duct tape to cover and seal the wood frame and mask the screen.

The **screen** should be a multifilament or monofilament polyester. Monofilament, while more expensive, lasts longer than multifilament. A 156 to 260 mesh is adequate for normal printing, while 260 to 305 mesh is recommended for halftones and finer detail. Screens may be purchased prestretched to frames in a variety of sizes, or you may stretch your own. After stapling the mesh to the frame, extend duct tape over the wood frame and screen staples to isolate them from water to prevent warpage and rust. Sizing can be removed from a new screen (in a process known as *degreasing*) by brushing or sponging a paste of dishwasher detergent and water to both sides, wiping them off with a clean, rinsed sponge.

The best printing surface or **baseboard** for use with water-soluble inks is a hard, smooth plastic, such as Formica, laminated to plywood. Attach the screen to the baseboard with **retractable hinges**, which permit ready removal of the screen for cleaning and reattachment for subsequent printings.

Figure 7.21.—Anthony Batchelor: *No. 3*, 1983, water-based screen print, 22″ × 30″. As a faculty member at the Art Academy of Cincinnati, I have been a strong advocate of water-based screen printing for many years. In *No. 3*, I used 288 monofilament polyester mesh, photographic and hand drawn positives, Ulano T.Z. direct photo stencil, Hunt Speedball screen filler, permanent acrylic screen printing ink, and transparent base, printed on Arches 88.

The **squeegee** distributes ink over the screen, forcing ink through the mesh onto the paper. Select a squeegee with a plastic or rubber blade about half an inch thick inserted into a plastic, metal, or wood handle that is at least one inch longer than the image to be printed and one inch shorter than the inside of the frame. Squeegee flexibility is measured in durometers; 50 to 60 durometers are best for general use. Preferred for water-soluble inks are blades of gray rubber or translucent polyurethane. The blade must be kept sharp. Metal handles are desirable, but wood handles may be sealed with duct tape to prevent water damage.

There are many safe **stencil options** appropriate for water-based printing. We describe the most common. There are two sides to a silkscreen: the underside, which is placed against the baseboard and printing paper; and the well side, which holds the ink for printing. All of the following stencil materials, unless otherwise specified, are applied to the underside.

Figure 7.22.—**D**onna Neuman, cutting an intricate stencil from freezer paper for a water-based screen print.

**Contact paper,** because of its self-adhesive backing, may be cut into intricate shapes. To retain the same orientation as your concept for drawing, cut through both the backing and contact paper from the backing side. Turn the screen underside up, carefully peel off the backing, and adhere the stencil to the screen, pressing out any air bubbles. Turn the screen over onto the baseboard, place tracing paper in the well side, and gently, through the tracing paper, burnish the screen to the stencil by hand or with a burnisher, bone folder, or other dull implement.

■■

**Freezer wrap paper,** a translucent, plastic-coated paper available at the grocery store, is an inexpensive and readily available stencil material. Place it over the drawing (which might require a light table) with the shiny side up, and cut out the image to be printed. Place a piece of proof paper such as newsprint on the baseboard. Position the stencil on it, shiny side up against the underside of the screen. With the first "dummy" pull of ink, the stencil will adhere to the screen and remain for a small edition of ten to thirty prints. Freezer paper stencils are appropriate for interesting "torn" effects and for introducing screen printing to children.

■■

**Photo mount paper** may be used and adhered as noted for plastic sheeting.

■■

**Plastic sheeting** (0.005 mil, moistureproof Dendril or mylar) can be cleaned and reused. (Acetate is undesirable because it absorbs moisture and stretches when in contact with the ink.) The transparency and reversibility of plastic sheeting make it appropriate for cutting according to a drawing taped underneath. Cut out and remove the areas to be printed. Position the stencil on the baseboard under the screen, tape the stencil to the underside of the screen, and press from the well side to adhere the stencil to the screen. The first pull of ink will also bond the stencil to the screen.

■■

**Tape,** while limited to rectilinear imagery, may be used to create a stencil. Apply it to the underside of the screen and burnish as with contact paper. Thick tape may require extra pressure on the squeegee to provide even coverage of ink. Removal of trace adhesive from the screen may require use of acetone and effective ventilation.

■■

**Drawing Fluid** and **Screen Filler**, made by Hunt/Speedball, can substitute for stencil application. Turn the screen over and draw, spray, sponge, roll, paint, and wipe directly on the screen with the blue-tinted drawing fluid. Any brush, implement, or material that does not damage the screen may be used to achieve a wide range of linear, texturing, and shading variations. Once the drawing fluid has set, dry it thoroughly with a blow-dryer. Then coat the entire screen with preshaken screen filler using a scoop distributor or a cardboard squeegee. Working quickly, spread an even coating in one direction, pressing the filler into the screen. Do not go over coated areas. Allow the filler to dry thoroughly.

When it is dry, run the screen under cold water. The drawing fluid washes off, leaving the filler as a blockout. Any areas that do not rinse away may be loosened with a soft nylon brush. The screen is ready to print when it is thoroughly dry. Any material that will dry to the touch and readily dissolve in warm water can be used as a substitute for the drawing fluid on an open screen. Possibilities include watercolor crayons, a thin syrup of sugar and water, liquid soap such as Murphy's Oil Soap brushed or splattered on the screen, or drawing with a bar of soap shaved to a point.

■■

**Screen filler** may be used alone as an image-making medium to create a negative printing stencil. Shake it well before using and, if desired, thin it for varying degrees of opacity. Apply screen filler with brushes, cardboard, or crumpled paper or cloth to achieve a variety of effects. Allow the filler fluid to dry thoroughly or it may break down during printing. "This type of stencil usually fills in a little for the first three or four prints, so proof the image before starting the edition" (Batchelor 1983).

■■

**Water-soluble stencil** is a water-soluble film attached to a plastic backing that may be used in the traditional manner with oil-based inks. In the method described here, water-soluble stencil is used with screen filler to create a negative image that may be printed with water-soluble inks. This method lends itself to detailed stencil cutting and different creative possibilities. Cut and remove from the backing those areas of the stencil image not to be printed. Position and tack the finished cut stencil (which contains the areas to be printed), plastic backing down, under a 240 to 260 mesh screen. Lift the screen and dampen it thoroughly with water. Lower the wet screen onto the stencil, place newsprint or blotting paper inside the screen, and press out excess moisture against the stencil. When the screen is dry, turn it underside up and carefully peel the plastic backing from the dry, cut stencil. Squeegee a thin coat of a screen filler over the entire underside, and clean excess filler from the edges with a damp sponge or cloth. Allow it to dry. Repeat this process with a second coat in a direction perpendicular to the first. Blow-drying the filler increases its resistance to water and shortens drying time. When the filler is thoroughly dry, take the screen to a sink and wash away the water-soluble stencil with warm water spray. The stencil will wash away, taking the filler coating on it and leaving the filler as a block in the other areas. The printed image will have the form of the temporarily adhered cut stencil.

■■

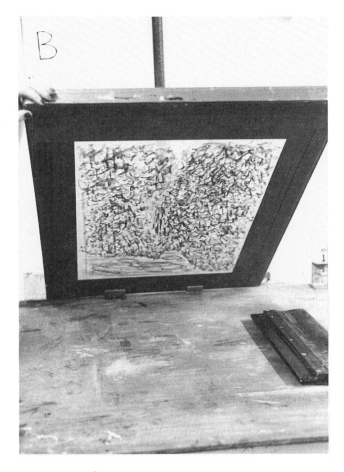

Figure 7.23.—Caran d'Ache Neocolor II Aquarelle crayon drawing on the underside of a screen prepared for water-based monotype screen printing.

**Watercolor crayons** may also be used to draw a positive textural image on the screen. After drawing the image, coat the entire screen with one squeegee pass of screen filler fluid. Allow it to dry thoroughly. When dry, wash out the crayon image with cold water. The resulting print will be of the crayon image drawn on the screen, including its variable shadings. Water-soluble crayons, such as Caran d'Ache aquarelle, may also be used for a small-edition screen monoprint. Make a multicolored drawing, with a generous deposit of the crayons, directly on the underside of a clean screen. Print with ink or with clear or tinted screen printing transparent base. The ink or base releases the crayon colors to print on the paper, but the colors will diminish with each printing, thereby limiting the edition number. Additional crayon colors may be applied between prints, to either the under or well side, for a related monoprint series.

Artist Jane Gregorius has developed a method of **monotype silkscreen** in which one-of-a-kind prints are produced without stencils. The image is created by applying water-soluble screen ink colors (slightly thinned with water or glycerine if necessary) with squeeze bottles, brushes, palette knives, and other implements to the well side of a monofilament screen. Open or white areas are made by covering those areas of the screen with transparent or extender base or with white ink. One squeegee pass makes one multicolored print. The print is removed from the screen, which is washed out and dried for the next print.

**Photo-Screen Printing Options** Although **photographic** applications for screen printing have greatly expanded imaging possibilities, the environmental and health hazards associated with photography cannot be ignored. In this section we spotlight the safest methods pertinent to screen transfer. We encourage those who do not have access to a safe darkroom environment to use a professional photography laboratory or screen printing facility for photo-image transfer to the screen. Many are equipped to develop Kodaliths from your drawings or photographs, apply a photosensitive emulsion to your screen, expose the screen to the drawing or photograph that you have furnished, and develop the screen for printing. To find such a photolab in your area, call those listed in your yellow pages and inquire about their capacity to transfer photo imagery to screens and plates.

Next, we review methods to obtain fine-quality photo images for screen printing. The major manufacturers of **direct photo emulsions** and photo systems produce various products with directions and fact sheets to accompany them. Since products and systems are periodically updated, however, one should request information on water-based photo processes directly from the major companies:

**Ulano Corporation**
**255 Butler Street**
**Brooklyn, NY 11217**
**718-622-5200**

**832 E. Rand Rd., Unit 17**
**Mt. Prospect, IL 60056**
**312-392-1446**

**15335 Morrison St.**
**Sherman Oaks, CA 91403**
**818-986-2690**

Figure 7.24.—**P**hotoProcess Screen Manufacturing Co., Inc.: A, Carmen Di Enno making a Kodalith from artwork. B, Ted Oteri stretching a screen. C, Ted Oteri removing a used screen to reuse frame. We supply the screen industry with custom-made screens of all sizes and meshes. We provide all our customers with Material Safety Data Sheets on the supplies and inks that we sell. Artists who do not have proper ventilation in their studios can bring their artwork to us, and we will custom screen.

Majestech Corporation
P.O.B. 440, Rt. 100, Somers, NY 10589,
800-431-2200

Naz-Dar Company
1087 N. Branch St., Chicago, IL 60622,
312-942-8338

Advance Process Supply Co.
34-06 Skillman Ave., Long Island City, NY 11101
718-937-6400

6480 Corvett St., City of Commerce, CA 90040
213-685-3400

400 N. Noble St., Chicago, IL 60622,
312-829-1400

445 Blvd. Guimond, Longueuil, Montreal,
Quebec J4G 1L8, Canada, 514-879-9000

In a well-ventilated, dimly lit room, or a dark-room with a yellow safelight, coat the screen mesh with the sensitized emulsion according to the manufacturer's directions. When the emulsion is dry, overlay a contact positive image, and expose the screen in an exposure unit or use a No. 2 photoflood bulb. Because photofloods generate a lot of heat, use a blowing fan to cool them. Some artists have been able to achieve good results using a light table for a cooler exposure source. After exposure, wash the screen with warm water. During the washout process, the unpenetrated emulsion in the image areas washes out, leaving them open and ready for printing, while the hardened emulsion in nonimage areas serves as a blockout.

For work with water-soluble inks, diazo salts and ammonium dichromate are the direct photosensitizing emulsions most often used. Because it is flammable, allergenic, corrosive, and a suspected carcinogen, ammonium dichromate should be used with extreme caution. The diazo method is probably less hazardous, it produces sharper images, and treated screens can be stored (in the dark) longer than screens treated with ammonium dichromate. The major suppliers manufacture diazo-sensitized photoemulsions under various names. Some are presensitized emulsions; others are sold with a diazo sensitizer (in powder or syrup form) oxidized with emulsion to sensitize it. Majestech's Majicol T. White photo emulsion yields good results with water-based inks, and its premixed diazo sensitizer may be used directly from the container. Ulano's products 800 PER and 1X-99 are presensitized and ready to use, whereas the 900 series requires mixing of a diazo powder or syrup with the emulsion to sensitize it. Advance makes waterproof, solvent-resistant, and reclaimable Raysol Universal DM-567 Zero Plus.

For best results, use the photo emulsion sensitizing agent on a mesh of 240 to 300 monofilament polyester. Wear a dust mask or use a glove box while measuring and mixing powders into solution, wear gloves while mixing and applying the solution to the screen, and follow the manufacturer's directions carefully. Allow the screen to dry thoroughly before exposure.

The best exposure with your contact positive or drawing on mylar is achieved with a vacuum exposure table that assures tight contact between the image and the sensitized screen. Without a vacuum table, clean glass plates may be used to sandwich the negative and screen together. Use the exposure time recommended with the product or, better yet, time suggested by a previously exposed test screen. After rinsing the exposed screen according to directions, let it dry thoroughly before printing with water-soluble inks of your choice.

Another direct photo method uses a **presensitized photo film.** CDF/TZ by Ulano is a diazo emulsion prefabricated on a plastic backing sheet that will adhere to the screen with water. Effective with water-soluble inks, it is the photo-imaging technique of choice because one does not mix chemicals to make the emulsion. Prepare the screen as described previously, carefully degreasing it, for photoemulsion coating. Because this film is highly photosensitive, it should be handled only in a darkroom with a yellow safelight. Cut a piece of film an inch or two larger than your screen image and re-close the film package. With the emulsion side (tacky to the touch) up on a baseboard padded with several layers of newsprint, thoroughly moisten the screen with water and place it on top of the film. Pass a squeegee with firm pressure to adhere the film, and sponge away excess moisture. Accelerate drying with a low-heat (less than 100°F) blow-dryer or fan. When the screen is thoroughly dry, peel off the backing of the film. Proceed, as described previously for direct photoemulsion, following the product fact sheet for recommended light source and exposure time. To wash out after exposure, use cold or warm water with a hard spray if desired. Wet the printing side first; then wash out the well side until the image opens up. Finish rinsing the printing side, and blot both sides with newsprint paper. After drying, touch up and fill in uncovered surrounding screen area with screen filler and vinyl duct tape.

Whatever the nature or material of the stencil, always check it for pinholes and other problems. Most mistakes can be corrected with screen filler applied with a fine nylon watercolor brush.

**Screen Printing Inks** Water-soluble silkscreen inks for printing on textiles have been available for years. These inks are effective for printing on paper as well as fabrics. Various colors and consistencies of water-soluble silkscreen inks (textile and artist quality) can be mixed and applied with multiple stencils in multiple layers for stunning visual effects. Inks vary. Test different brands and types in small amounts to learn which is best for your work. At the time of purchase, get a color chart and Material Safety Data Sheet for information on pigment composition and vehicle ingredients.

As more artists and schools adopt water-based screen printing methods, manufacturers have responded by producing high-quality inks, retarders, extenders, and transparent bases. However, the properties of each brand vary. Test a few to discover which are best suited to your imagery. The most frequently used are Hunt Speedball permanent acrylic ink and water-soluble inks. The Createx pigments and bases are compatible with both. Their properties and others are described in Table 7.1.

Createx Poster-Fabric inks are opaque and may be mixed with transparent base. When used on fabrics that will later be washed, they must be heat-set either by dry iron on cotton setting (300° to 350°F) applied to the reverse side for three minutes (a protective cloth may be necessary to prevent scorching) or in a home clothes dryer run for 45 minutes at the hottest setting (commercial dryers may require only 30 minutes). Do not wash for a few days following setting.

**Transparent base** (which may contain sensitizing or irritating preservatives and dispersing agents) is a butterlike, water-compatible acrylic medium that may be used to make inks more transparent, to extend them, or to aid in printing fine details such as halftones and sharp lines. To avoid streaking, the base must be thoroughly premixed with ink colors prior to placement in the screen. Proportions depend entirely on the degree of transparency sought.

Many water-based screen printing inks contain **retarders** in their base to keep inks workable for expected printing times. If your inks dry too quickly, retarders are commercially available to mix with ink before putting it on the screen. Gel retarders are especially effective because they reduce evaporation by altering the surface tension of the ink. When the ink is not being used, it pulls together into a heavy jelly, but when exposed to the friction of the squeegee, it breaks down to an almost liquid state, making it printable. After the printstroke, ink retracts (pulls back in), decreasing evaporation. Gel retarders "allow inks to sit in screens for up to three to five hours without significant drying-in problems" (Duccilli 1989). A drop or two of glycerin per four ounces of ink will also slow drying in the screen. By contrast, recall that water alters the consistency of ink and may cause seepage and bleeding on the print. Ink buildup around the edges of a print also indicates excess water. Experimentation, documentation (keeping a record) of the materials and proportions you try, and getting the feel of inks and methods through use and practice are keys to enhancing your vision and creativity.

Prior to printing, mix a small amount of ink to arrive at the color you desire. Test a smear of it on the paper you will use for printing. From 10 to 50 percent transparent base may be added to extend the color and achieve more translucent color. Test after each addition until you get the desired quality. For the best printing quality, the ink should be smooth and buttery. Ink brands vary in viscosity. Some require no additives, while others respond well to extenders, a drop or two of glycerin, water, or retarder to achieve ideal printing consistency. Test again with a test screen and squeegee on the paper you intend to use. Because color often darkens upon drying, wait for the ink to dry before making a visual assessment.

Be sure to mix enough color to complete an edition in one printing. Ink waiting on the screen may dry, clog, and require washout of the screen before resuming printing. Various factors affect the amount of ink required to print satisfactorily: absorbency of the printing surface, mesh size, stencil thickness, squeegee consistency and pressure, and relative humidity in the studio at the time of printing. As a base measurement, a gallon of most screen printing inks may cover up to 1000 square feet of surface; one quart should cover 250 square feet. Multiply the square feet of your print size (height times width in inches divided by 144) by the number of editions to determine how many square feet you need to cover. It is better to mix too much and have color left over than to mix too little and run out prematurely. In general, for each printing, have on hand at least one quart of each color and a gallon of transparent base.

**Printing** from prepared screens with water-soluble inks generally follows basic principles of traditional screen printing. Nevertheless, we offer some additional tips. For example, apply ink just inside the frame, producing a fairly even distribution of ink sufficient to coat the entire screen with one pass of

**TABLE 7.1**  WATER-SOLUBLE SCREEN PRINTING INKS

| Ink | Colors | Additives | Fabric | Clean-up | Remarks |
|---|---|---|---|---|---|
| **ADVANCE EXCELLO WIT SERIES** | 16 transparent, 4 process, and 5 fluorescent colors. | Bodying Agent to thicken, WBC-866; Retarder, WBC-990; Clear Extender base, WIT-800SH; Clear Metallic Base, WIT-780SH. | (Monofilament Polyester) 86—305 | Tap water, DET-34 detergent. | Textile ink, requires heat set on fabrics. |
| **CREATEX PURE PIGMENT (COLOR CRAFT LTD.)** | 30-plus permanent pure colors in plastic dispensing topped bottles. Concentrated colors, use sparingly. | Use Lyntex base for transparency (12 drops pigment to 8 oz. of base), or 2 parts Createx gouache medium to one part Lyntex base for opacity. | (Monofilament Polyester) 240–270 | Tap water. | Do not add water. Small amounts of isopropyl alcohol improve printing. Compatible with Hunt Speedball base and colors. |
| **HUNT SPEEDBALL PERMANENT ACRYLIC INK** | 17 colors; may be expanded with acrylic paints and transparent base. | Transparent Base; Retarder, which contains glycerin to slow drying time to several hours. | (Monofilament Polyester) 150–230 | Hose with warm water, using Wisk or other liquid detergent for difficult areas. | Good quality; work well for opaque and transparent effects; ready to use from container. Prints on paper, wood and textiles. |
| **HUNT SPEEDBALL WATER-SOLUBLE INK** | 10 colors; gouache and watercolors can be added to transparent base to extend color range. | Transparent Base. Add a few drops of water and base to improve consistency. | (Monofilament Polyester) 150–230 | Hose with warm water, using liquid detergent for difficult areas. | Molds may form in warm weather: scrape off, discard mold, and use remainder. |
| **NAZ-DAR/KC AQUA PRINT** | 30 colors, including fluorescent. | Extender base: AQ 51; Thickener: AQ 53; Aqua Lube: AQ 54 | (Monofilament Polyester) 150–230 | Hose with warm water, using liquid detergent for difficult areas. | Consistency usable from container. Prints very well on Japanese papers, heavyweight papers and fabrics. Good shelf life. Requires heat set on textiles. |

the squeegee. Areas of the screen not covered with fresh ink with each pass can dry and clog. Holding the squeegee at a 45° angle, apply firm, steady pressure, and pull it across the screen.

A print may stick to the screen when the screen is lifted, sometimes causing a fuzzing of the image. To avoid this, position the screen about half an inch above the printing baseboard for "off-contact" printing. The screen will contact the paper only when and where the squeegee exerts pressure. By contrast, the screen will return to the raised position at the end of the pull. The paper, now free from the screen, will have a sharp image. Run some proof prints until the image achieves your desired quality and appearance.

If ink begins to dry in the screen, mist it sparingly with water spray, and print several proofs onto newsprint to release the dried ink. Replenish the ink supply and resume printing. It may be necessary to stop printing and clean the screen if it remains clogged with dry ink.

**Cleaning** water-soluble inks from the screen requires immediate attention because inks dry rapidly from evaporation. Promptly after printing, remove excess ink, store it in labeled, lidded jars, and wash the screen before residual ink can dry and clog the screen. If you are not going to save the ink for later use, do not dump it down the drain because it may build up in the plumbing and introduce pigments and resins that your local waste treatment system may not permit. Place unusable ink on several layers of newspaper to absorb moisture and dry by evaporation. Then dispose of it as solid waste.

Remove the screen by detaching it from the retractable hinges. Hose ink from the screen with water. To retain a stencil on the screen, flush with slow-running cold water while rubbing gently with a soft nylon brush to get all the ink out of the screen while leaving the stencil intact. Some stencils, such as freezer paper, should be peeled off before cleaning and placed on a receiving paper (such as freezer wrap) for storage and reuse at another time.

Clean remaining ink from the screen with warm water and a power spray, if available. As inks tend to cling to the mesh, gentle scrubbing with a sponge or brush may help release residue. For stubborn spots, apply a solution of liquid detergent and scrub softly. If that fails, apply isopropyl (rubbing) alcohol to dry screen mesh, and rub both sides of the clogged mesh area vigorously with soft cloths until the area clears. Finally, rinse the screen thoroughly to remove all detergent, alcohol, and ink residue.

After ink is removed, take off the screen filler by applying a strong liquid detergent (e.g., Mr. Clean, Wisk, 409) to both sides of the screen with a sponge

or nylon brush. Let the detergent remain on the screen for 15 to 20 minutes. Scrub gently and flush with strong-flowing warm water, working the filler loose with the sponge or brush. Several applications may be necessary.

Photoemulsions should be removed from screens immediately after printing. The longer an emulsion remains on the screen, the more difficult it is to remove because it hardens with exposure to light. Forceful reclaiming and degreasing techniques may be required.

**Reclaiming** a screen to be used for new imagery may require extra effort. Wear protective gloves and attire, and use the following methods only with effective local exhaust ventilation. Removal of wax crayons often requires acetone or mineral spirits, followed by degreasing with potentially caustic dishwasher detergent paste or a commercial degreaser. Ammonium dichromate photoemulsion may be removed with a one-part household bleach to four-parts water solution or with a commercial screen reclaimer (e.g., MCB Screen Reclaiming Crystals, Ulano #4 Stencil Remover Liquid, or Ulano #5 Stencil Remover Paste). Removal of photoemulsion stencils requires removers that are compatible with the brand used. Carefully observe the manufacturer's directions. Follow with high-pressure water rinsing. Professional and institutional printing studios may have a power washer with a narrow jet stream to be aimed at residues on a screen propped vertically in a sink or drain trough. Hot water effectively dislodges buildup from acrylic materials.

Some inks, such as magenta, brown, and black, may leave a stain that does not wash out with any of these methods. Image residues can be disconcerting when reusing the screen for another image. Residues may also interfere with light sensitivity in photo-imaging. Thus, extremely caustic cleaners known as *haze removers* have been developed to remove such ghosts. We suggest, however, that you restretch a new screen mesh and forget these potentially toxic cleaners. If you must use a haze remover, wear non-vented goggles, impervious clothing, and gloves of natural rubber, neoprene, Buna-N, butyl rubber, or polyethylene. In the absence of appropriate ventilation or opportunity to work outdoors, wear a mask approved for caustic exposure.

**Stencil Printing** Stencil printing is possible without a screen. Known among artists as **stenciling** and/or **pochoir,** the method involves application of ink or paint by daubing with a special brush or spraying with airbrush through cut stencils onto receiving paper, fabric, or object (color plates 7.16–7.17). The stencils may be made from lightweight paper, oiled paper, thin brass or copper, or specially produced plastic-coated stencil paper that has been cut, following a master drawing or design, with scissors and X-Acto knives.

When the ink or paint is water-soluble and applied with a brush, there are, beyond basic safety practices with scissors and knives, no hazards with stencil printing. Precautions for using airbrush are provided in chapter 8.

## LITHOGRAPHY

From its beginning, lithography has attracted artists because of its natural sensitivity to drawing (color plates 7.18–7.19). Recent developments have widened the spectrum of style and techniques. Lithography remains an exciting process with a multitude of possibilities for the artist.

Planographic printing, which means printing and nonprinting areas are at the same level, involves placing paper over the inked stone or plate and applying pressure with a lithography press. The wet area of the stone repels rolled oil-based ink while the grease-based image attracts it.

Fortunately, many of the potential health hazards in lithography can be eliminated by proper safeguards and precautions. Potential hazards include:

◆ Heavy stones

◆ Acids, rosin, and talc used to prepare stones and plates

◆ Vinyl lacquer and lacquer thinner used to set images on metal lithoplates

◆ Oil-based printing inks

◆ Caustic alkalis and solvents used for removing the image and inks from stones and plates

◆ Photochemicals and xylene used in photosensitizing stones and plates

The surface of **lithostones** must be prepared by grinding with carborundum abrasives and water, a process that does not involve significant hazards if dust inhalation is avoided. In addition to cleaning off prior images, the grinding produces a finely grained surface that is responsive to lithographic drawing materials.

**Litho crayons** or pencils contain grease and fatty acids (petroleum and/or animal derived), waxes, soaps, and carbon or lampblack. Litho crayons used as intended pose little hazard.

**Tusche** is a material used for wash effects and is

Figure 7.25.—A lithostone lift can prevent injury from carrying and moving heavy lithography stones.

available in liquid form or in solid sticks. It is dissolved in either water or solvents such as lacquer thinner, turpentine, lithotine, or alcohol. Avoid the use of solvents with tusche. Buy the water-soluble tusche solution, or use solid tusche sticks with water.

Once an image has been applied to the stone by direct or transfer drawing, the surface is dusted with rosin and talc (occasional allergens). Then the surface is etched with a mixture of gum arabic and nitric or phosphoric acid (both highly toxic in concentrated form) or tannic acid (a suspected carcinogen). Counteretches of acetic acid (moderately toxic) or saturated alum solution (slightly toxic) may be employed to resensitize for making additions or to correct the image. A safer alternative is a citric acid counteretch solution. This solution may be used to clean oxidized plates and is also recommended for counteretching to add a new drawing to previously etched plates or stones.

Figure 7.26.—Robert Cumming works on a drawing for a lithograph at Corridor Press. Drawings are made on textured polycarbonate film with litho crayons, stabilo pencil, graphite pencil, and Liquitex paint, scratching from darker to lighter, to make a photopositive plate. Proofs shown in the background. (Courtesy Residency Program, Print Club, Philadelphia.)

Figure 7.27.—Merle Spandorfer: *Maui Waterfall*, 1988, lithograph, 33½″ × 23″. The image was hand-drawn on an aluminum lithoplate with tusche diluted with water. When applying the vinyl lacquer to set the plate, I worked outdoors with a respirator, even though there is a large exhaust fan in the studio. I kept the plate in the fresh air for an hour to allow the lacquer to dry thoroughly and to let vapors dissipate.

A gum etch of weak acid solution that fixes the final image presents no significant health threat.

Fountain solutions, used to wet stones during printing, may contain ammonium or potassium dichromate (flammable, allergenic, corrosive, and suspected carcinogens), phosphoric acid, or phenol (both highly toxic). A safe alternative for fountain solutions is the use of *water* with a small amount of gum arabic. This solution restores and maintains the pH balance of the plate.

Inks in plates and stones are washed out with lithotine. If deep cleaning is required, acetone is recommended. Use effective local exhaust ventilation with impermeable attire and wear nitrile, neoprene, or Buna-N gloves.

*Aluminum lithoplates* are lighter-weight alternatives to lithostones. These may be purchased already grained, counteretched, and ready to use. Lithoplates are an excellent, safer alternative for lithographers.

The image drawn on metal plates sometimes requires fortification with vinyl lacquer, which must be used with effective local exhaust ventilation as it contains aromatic hydrocarbons, usually xylene and ketones. Substitute acetone for lacquer thinners for it is less toxic and just as effective. To delete imagery on the plate or stone, use acetone followed by a gum arabic etch.

Metal plates are inked, printed, and cleaned in a manner similar to lithostones.

**Photolithography** permits the creation of an ink-receptive image on stone or plates by photographic means. Plates are available commercially presensitized or may be coated with a light-sensitive emulsion to receive photographic imagery. Photosensitive emulsions for stone can be created from a mixture of powdered albumin, ammonium dichromate, ammonia, and water. Metal plates are usually photosensitized with diazo compounds, as are commercially sensitized plates.

Positive/negative presensitized offset plates are a safer alternative to the traditional photoprocess. They eliminate the need for vinyl lacquer and hand diazo coating, and in the washout procedure require less solvent (such as lithotine). Positive/negative presensitized offset plates use a water-based processing chemistry. For further information contact:

**Cookson Graphics Ultrapos**
**383 Dwight Street, P.O. Box 348**
**Holyoke, MA 01041, 413-538-9624**
    **or**
**340 Shore Drive**
**Hinsdale Industrial Park**
**Hinsdale, IL 60521, 312-920-9205**

**Precautions** in lithography include:

♦ Effective ventilation, both general dilution and local exhaust, for etching, lacquering, photoprocessing, and cleaning.

♦ Protective attire: gloves; impermeable apron; flexible, hooded goggles for work with acid and alkali concentrates; and a toxic dust mask for work with ammonium dichromate.

♦ Avoidance of solutions containing phenol, dichromates, and chrome alum.

♦ Avoidance of nitric acid, which is highly toxic by inhalation. Unfortunately, acid gas cartridge masks and gloves may not provide total protection. Symptoms of nitric acid toxicity include fever and chills, cough, shortness of breath, headache, vomiting, and rapid heart beat. Nitric acid–nitrogen oxide combinations may cause pulmonary dysfunction and symptoms of emphysema with chronic exposure. Use these agents as sparingly as possible and only with effective exhaust ventilation. Wear nonvented goggles, an impermeable apron, and neoprene, Buna-N, or NBR rubber gloves, which, while not impervious, will decrease skin contact. Keep containers tightly covered or capped when not in use.

Figure 7.28.—Timothy Sheesley, Corridor Press, artist and Tamarind master printer. I minimize my exposure to vapors and chemicals with an exhaust ventilation system, purchased from the Grainger Industrial catalogue #380. The system consists of a Power Cat 400 XLH, which has a two-speed fan (300 and 450 cubic feet per minute). This portable blower has an exhaust duct attachment and a fume exhaust nozzle. The cost of this system was $377 ($227 for the Power Cat and $150 for the ducts). When necessary, I wear an organic cartridge respirator and gloves for added protection.

◆ Vinyl plate lacquers often have xylene as their primary ingredient, along with diacetone alcohol, di-isobutyl ketone, isophorone, and resins. Use only with effective ventilation and nitrile gloves. If you need a solvent, use acetone instead of lacquer thinner.

Litho-Sketch plate, a less expensive and simplified alternative, is a paper that is responsive to drawing with litho crayons and tusche, which are moistened between each inking with a solution called Litho-Sketch plate solution. Print quality does not equal that of plates or stones, but affordability makes Litho-Sketch plate attractive to introduce lithography to young people. However, because the Litho-Sketch plate solution and solvent-based inks present the same potential hazards as lithography in general, we think it unwise to introduce lithography processes to children. Use of acids, solvent-based inks, and other potentially hazardous chemicals in lithography requires caution and safety measures.

**Offset lithography** dominates commercial printing. An image is reproduced photographically onto a thin metal plate, which is prepared by procedures similar to those described previously. Mounted on a roller, the inked image is transferred ("offset") onto a rubber-blanketed roller, from which it is printed onto paper at high speed. While outside the scope of this book, we note that the offset printing industry is not without hazards. For example, lithographic film developer/replenisher may contain up to 30 percent formaldehyde (GAIU, p. 3). In addition, two chronic health problems have been associated with offset lithographic printing: bone marrow damage attributed to the combination of glycol ethers and organic solvents, and respiratory allergies and illnesses attributed to inhalation of powdered gum arabic. As commercial printers discover the savings in ventilation requirements, worker safety, and waste disposal when they switch to water-soluble inks, we speculate that computerized ink-jet printing processes (or yet-to-be-developed methods) may overtake offset lithography as the process of choice.

## UNIQUE PRINTS

A **monoprint** is a one-of-a-kind impression from a preexisting plate or matrix such as wood or lino cut, etched plate, collagraph, silkscreen, stencil, or prepared lithostone. Each impression is inked uniquely or modified after printing by drawing, painting, embossing, or collaging. As with most printmaking today, numerous procedural options provide a virtual infinitude of multimedia expressions (color plates 7.20–7.22).

No matrix is used to produce **monotypes,** transfer prints, or chine collé, which are methods associated with unique prints. Any flat, relatively nonporous surface (e.g., glass, enameled table tops, Plexiglas, plastic sheeting or laminate (Formica), metal, wood,

**Figure 7.29.—S**helly Lependorf: *Diptych*, 1990, monotype with water-based ink and collage, 30″ × 40″. I roll water-based printing ink on a Plexiglas plate, then add and delete color with brushes, rags, and rollers. When the image is complete, the ink is dried. I print with dampened Arches Cover on a Brand etching press, and complete the print with collaged elements and graphite.

Figure 7.30.—Merle Spandorfer: *Dance of the Taro*, 1989, monotype with water-based ink on Arches Cover, accented with water-soluble crayons, 20″ × 26″. After rolling out black Speedball water-based printing ink on a Plexiglas plate, I removed ink with water-dampened Q-tips, brushes, and paper towels to achieve the light areas. The plate was dried with a hair-dryer and printed on dampened Arches Cover paper with an etching press. Water-soluble crayons were used for accents.

lithostone, or cardboard sealed with gesso) may serve as the nonmatrix plate. Some artists choose to "grain" the surface of the plate by sanding, filing, abrading, or etching it lightly to create "teeth" to improve ink retention. Whatever its nature or manner of preparation, the plate material remains physically unaltered by the printing process and can be cleaned and used repeatedly for new prints.

Monoprints and monotypes made with water-soluble inks predate modern concern with making art safely; William Blake, Gauguin, Picasso, and Mark Tobey made monotypes with water-based colors. The directness of the process, its correctability, and its simple transfer to paper make monotypes especially attractive to artists of all disciplines.

For a monotype, positive method, an image is created directly on the plate with water-soluble printing inks such as Createx monoprint colors or Speedball water-based block printing inks; egg tempera, watercolor, and/or gouache paints; or dyes, water-soluble pencils, or crayons. Various implements may be used, including pens, brayers, rags, and brushes, with as much color and imagery as one desires.

One may also work negatively by rolling or spreading ink on the plate and selectively removing it with a drawing tool (with or without water). Modifications and corrections to the drawing or painting on the monotype plate—by wiping an area clear

Figure 7.31.—Esther Rose Fisher: *Toxicity*, 1988, monotype, 33″ × 44″. I work on a Plexiglas plate with Caran d'Ache Neocolor II Aquarelle water-soluble crayons, which can be very opaque. I also use a few drops of Kodak Photo-Flow in water and Windsor Newton transparent watercolors. I work until I achieve the desired effect; the plate is allowed to dry thoroughly and then printed on dampened paper.

Figure 7.32.—Marion B. Alexandri: *Sanibel*, 1988, monotype and mixed media, 18″ × 24″. French windows in my studio make it difficult to install exhaust ventilation, and I dislike face masks. So I switched from oil to acrylic paints, and from oil-based to water-based etching inks. I often heighten my monotypes with oil sticks that I rub and blend without turpentine. I also draw with litho pencils and crayons into a beeswax base on paper, but I rub the wax into the paper to avoid vapors from heated wax.

with brushes, Q-tips, sponge, or rag dipped in water —provide a flexible, positive/negative approach.

Once a desired image is achieved, a **print** is usually made with an etching press or by placing paper over the inked surface and rubbing the back of the paper to lift the ink or paint from the plate. Usually a single print is made, but a second, fainter pull can be produced by repeating the process. A wide variety of papers can be used, including Japanese papers, Arches Cover, Rives BFK, and Stonehenge.

There are many advantages to printing with water-soluble inks. First, water is the only solvent required. Moreover, the plate can be inked for several days to weeks before actual printing. Ink should be dry prior to printing and dampened paper placed on the plate. The dried ink is quickly moistened and prints can be obtained by rubbing or using an etching press. Some artists do not clean the plate after printing one monotype. Instead, they build the next image on whatever ink is left over. In summary, for this most direct of printing methods, experimentation with inks, materials, implements, and procedures provides a good way to develop appropriate artistic strategy.

To make **transfer prints,** also known as trace drawings and transfer drawings, one or more colors are rolled evenly over the surface of the plate with a rubber roller. Paper is placed over the inked surface, and a drawing on lightweight mylar (impervious to water if you are working with paper dampened for printing) or tracing paper is placed over it. With a pencil or other drawing implement, the drawing is traced, pressing lines and areas onto the inked surface to lift ink only from those pressed areas. Some smudging throughout other areas does occur. Depending upon the amount of ink lifted by the first print, several transfer prints can be made from one inking by placing subsequent sheets of paper on the plate at slightly different locations. Dampened and well-blotted paper will help release water-soluble inks that have begun to dry. To prepare the plate for additional transfers, either wipe the plate clean or lightly spray the inked surface and roll on new ink.

All monoprinting methods can be expanded with paper registration and additional colors and imagery. Thin objects such as papers, fabrics, yarns, leaves, or flower petals may be positioned on the plate either before or after inking for a monotype collage print.

**Chine collé** is a paper collage technique used to obtain added color or texture in the print. Torn or cut lightweight paper placed on the inked matrix becomes laminated to the final printing paper (e.g., Arches Cover, Stonehenge, Rives BFK) when run through a press. The inked image prints on both the chine collé paper and the printing paper.

To achieve lamination, use methyl cellulose, an ideal, nontoxic archival adhesive that can be purchased from Light Impressions, Rochester (800-828-6216); Talas, New York City, (212-736-7744); and a few well-stocked art supply stores. Library, potato, arrowroot, wheat, and rice pastes are less expensive adhesives that also work well for chine collé techniques.

These unique printing methods respond beautifully to water-soluble inks and paints, which provide safer, as well as exciting, means for creative printmaking. Some plates yield more than one print from each inking or preparation, but exact replication is rare. Combinations of these safe techniques also provide many possibilities for creative expression.

# PHOTOGRAPHIC AND ELECTRONIC PRINTMAKING PROCESSES

Intaglio, lithography, and screen printmaking techniques permit **photographic image transfer.** The many procedures for incorporating photographic im-

Figure 7.33.—Michael Wommack, pulling a chine collé print using Japanese mulberry paper with methyl cellulose on an inked etching plate.

agery combine the excitement and magic of photography with the extensive, colorful, and graphic options of printmaking.

Printmakers who use photoprocesses of any kind are referred to chapter 9 to familiarize themselves with safety and handling procedures for the many complex chemicals used. This is especially important if you use a darkroom or other enclosed environment, where hazards can be intensified unless specific precautions are observed. Some methods, such as cyanotype and dichromate printing (described in chapter 9, "Nonsilver Processes") can be accomplished without a darkroom and incorporated into printmaking.

Lights, such as carbon arc lamps, may also present significant health hazards. If you use strong studio lamps to transfer images, be sure to read the chapter 9 section "Studio and Darkroom Lighting" to ensure your safety.

In addition to traditional photographic procedures, printmakers may encounter chemicals unique to printmaking processes. For example, developing images on metal etching and lithography plates may require xylene, an undesirable aromatic solvent that

may be inhaled or absorbed through unbroken skin. It should be used only with nitrile-gloved hands and local exhaust ventilation (see, e.g., Appendix I, concerning KPR).

The proliferation of **photocopiers** in the workplace and home has led creative artists to explore "xerography" as a method of making esthetically substantive and expressive prints. Colored inks, known as toners, are available, and through the color separation process, full-color registration and reproduction, as well as various color effects, are possible. The procedures of this photoimaging technique, including use of digitized laser color, are explored in Chapter 9, "Xerographic Processes."

**Computer programs** for "drawing" and "painting" have been designed to help artists achieve a wide range of imagery and patterns. In addition, a scanner can read original drawings, paintings, and photographs and feed a digitized translation into the computer, where images can be altered or manipulated. The chapter 8 section "Computers" provides information on safe use of computers in art.

Once computerized imagery, suitable for your printmaking intentions, has been achieved, it may be printed with daisy-wheel, dot-matrix, ink-jet, laser, pen-plotter, thermal, or videographic printers or printing transfer methods. The printout may be the end product, or it may be incorporated into traditional printmaking techniques in several ways:

• Transfer via Kodalith to a screen for printing with water-soluble inks

• Negative image projected onto a photosensitized intaglio or lithographic plate

• The base for overprinting with any printmaking techniques

The possibilities are so great that the imagination, visual knowledge, and sensitivity of the artist are the only limitations.

## Printmaking Papers for
## Water-Soluble Inks

In general, best results for printing with water-based inks are obtained with papers that are heavyweight (235 to 320 grams), possess 100 percent rag content, are hot press (smooth surface), and are sized. Lighter-weight and wood pulp papers may warp, textured surfaces may compromise clarity of the print, and unsized papers may absorb ink unpredictably. We have experienced success with lightweight Japanese papers, and we encourage experimentation. The following papers are appropriate for most water-soluble ink printing (the alphabetical list is not meant to be exhaustive):

- Aquaprint Silk Screen: 100 percent rag; 235, 290 grams

- Arches Cover: 100 percent rag; 250, 270, 300 grams

- Dansk: 100 percent rag, 320 grams

- Fabriano Tiepolo: 100 percent rag, 240–290 grams

- Gallery 100: 100 percent rag, 250 grams

- Giovanni (formerly Italia): 50 percent rag, 310 grams

- Inomachi: 100 percent kozo, 180 grams

- Mirage: 100 percent rag, 310 grams

- Okawara: 100 percent kozo, 60 grams

- Rives BFK: 100 percent rag; 240, 250, 270, 280, 300 grams

- Stonehenge: 100 percent rag; 250, 320 grams

- Utrecht: 100 percent rag, 250 grams

For proofing prints:

- Domestic Etching: 50 percent rag, 175 grams

- Smooth newsprint

Printing, especially relief and silkscreen, on **textiles** with water-soluble inks is appropriate for young people and professionals alike. Natural fibers, such as cotton, linen, and silk, react differently. Preshrunk cotton retains its shape well, linen tends to draw up and buckle from moisture but will press flat, and silk stretches when wet and dries taut. Synthetic fabrics tend to resist dyes but receive pigments fairly well. Most fabrics are sized prior to sale, which will also affect the way they take ink. We recommend experimentation with prewashing and testing to achieve the qualities you desire.

**Papermaking** A number of artists make their own **handmade papers** for printing. Papermaking can be an art in itself, a satisfying alternative to some of the more hazardous printmaking procedures. Its clean, refreshing appearance, however, can be deceptive; there are several hazards.

Linters, the bundles of fibers used to make paper, are a source of lint and dust. Excessive handling without respiratory protection can cause breathing problems; wear a dust mask.

■■

Stored papers and fibers not treated with preservatives may harbor invisible molds, which, if inhaled, may cause symptoms similar to hay fever or flulike illness. Those containing preservatives may provide a source of allergen for sensitized individuals, inducing similar symptoms.

■■

For dyes to color papers, observe precautions for working with dyes outlined in chapter 6, "Coloring Agents."

■■

Sizings and alkalis, especially lye, are caustic and may cause dermatitis or stinging sensations upon contact with skin. Wear rubber gloves and protective clothing; use exhaust ventilation.

■■

Retention aids used to adhere coloring agents, fillers, and other additives to fibers are usually ca-

**Figure 7.34A.—Carol Moore:** *Magic Carpet #5 ASHANTI*, 1985, handmade papers of linen, cotton, abaca, 24″ × 28″. I turned to papermaking as a medium in 1981 when, as a printmaker, I experienced coughing and other adverse reactions in the print studio. Papermaking has solved most of my needs on this issue, but it is a physically demanding medium, and challenges arise as each new form of experimentation progresses.

Figure 7.34B.—Carol Moore, artist in residence at the Haverford School, teaching papermaking.

tionic polyacrylamides that are eye, skin, and respiratory irritants. Mix powders into solution in a glove box, and handle solutions carefully while wearing eye and hand protection. It is especially important to avoid exposure to acrylamide monomer, a neurotoxin that may contaminate polyacrylamide powders.

There are also several physical hazards associated with papermaking:

Electric shock from using electric equipment in the presence of water. Use a ground-fault circuit interrupter in the electric outlet with a circuit breaker that will pop open if short-circuit shock conditions develop. Wear rubber-soled shoes.

■■

Back, arm, and hand strains and sprains may occur from lifting heavy buckets and tubs of water and wet pulp. Use smaller containers, and observe orthopedic lifting procedures: both hands, bent knees, and straight back.

■■

Arm and hand tenosynovitis may develop from repetitive movements and long-term immersion of hands in cool water. Wear lined rubber gloves, and take frequent breaks to flex and stretch muscles and joints.

■■

Ear damage and hearing loss can result from chronic exposure to noisy beaters. If possible, separate noisy equipment in a closed, sound-absorbing room, and provide and wear ear protection.

Papermaking is a highly flexible medium that continues to grow and expand, so remain alert to possible chemical and physical hazards.

We are confident that, as an increasing number of printmakers choose to adopt the safer use of water-soluble inks and the many processes they have already spawned, options for safer printmaking will continue to expand. As a result of employer liability and environmental concerns, we would not be surprised to see a substantial reduction in oil-based printing in both schools and the workplace.

Increased thoughtfulness and selectivity regarding the prints we produce and the procedures and materials we employ will enrich us all. With so many creative developments in printmaking in the past ten years, there is no reason to avoid new methods and alternatives or to continue to jeopardize our health with old practices.

**G**raphic designers and illustrators employ many and diverse materials when advancing from the initial idea phase to the finished piece. When progressing from rough sketch to layout, from comprehensive to camera ready art and mechanicals, a wide array of markers, pens,

# Graphic Design and Illustration

pencils, inks, paints, dyes, papers, adhesives, fixatives, photochemicals, and solvents can be utilized. New products that come on the market for the purpose of making jobs easier or more streamlined are often quickly put to use. Ideally, all such tools, materials, and processes should be functional as well as safe; many, however, have their own set of potential hazards, safeguards, and safer alternatives.

◆ ◆ ◆

Figure 8.1.–**B**etz Green: *Medieval Wagon*, illustration, 1990, pencil on paper, 10½" × 16½". I use B pencils, water-soluble color pencils, and watercolor. These nontoxic materials are easily transported and cleaned, leaving no dust or vapors to be inhaled. This is important to me since I have young children whose health must be protected.

Figure 8.2.–**D**onna R. Neuman: *Rabbit Jaw Specimen*, illustration, 1989, inkwash on paper, 5" × 8". For medical/scientific illustrations I use light-fast, nontoxic, water-based India ink, distilled water, colored pencils, water colors, several 0–3 soft bristle brushes, and a hard pencil (3–6H).

Designers and illustrators generally have the option to choose their own media. These media are addressed separately in chapters 5, 6, 7, and 9. For example see chapter 5 concerning drawing tools and materials, chapter 6 concerning pigments, paints, and solvents, chapter 7 concerning printing inks and procedures, and chapter 9 concerning photographic processes.

We presume that readers have read chapters 1 through 4, so that concepts of acute and chronic illness, sensitizer, and general and specific irritant are understood; that principles of studio design, including proper ventilation and safe storage and disposal of art materials, are known; and that the need for personal protective attire, emergency procedures, and medical attention are recognized and observed. This is not a how-to chapter on graphic design, but how to do what you already know in the safest way possible (color plates 8.1–8.17).

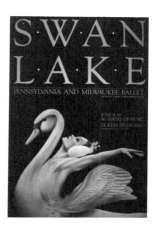

Figure 8.3.–**J**oseph Scorsone: *Swan Lake*, poster, 1988, 36" × 24". The original is charcoal drawing on Canson paper. I provide the printer with the drawing plus color separations cut from Rubilith or Amberlith film. I indicate colors on the films with PMS (Pantone Matching System) numbers. I work in a well-ventilated studio and use a wax coater instead of rubber cement.

## Tools of the Trade: Potential Hazards and Safer Alternatives

**D**esigners and illustrators should evaluate all tools and materials for comfort as well as efficacy. Properly designed equipment can prevent undue fatigue, tenosynovitis, related inflammatory disorders, and mishaps such as errors, cuts, or pinches. In addition, materials with the least toxic properties should be used. To achieve one's visions and intentions, potentially hazardous substances should be used only with appropriate precautions and safeguards.

## DRAWING AND PAINTING MATERIALS

For many graphic designers, colored "magic" **markers** and felt-tipped pens are the principal "tools of the trade." Markers containing water-based or alcohol-based inks are increasingly popular. Therefore, markers containing xylene are no longer necessary. If you must use xylene-containing products, work only in an exhaust-ventilated area. All **paints** and painting processes should be used with appropriate precautions (see chapter 6). Avoid products containing toxic metal pigments (identified in chapter 6).

An **airbrush** creates special effects through an adjustable, pen-shaped spray device (color plate 8.15). Using compressed air, airbrush techniques permit a smooth hairline as well as a finely divided broad shading spray in retouching, illustration, and other painting and graphic design tasks. Unfortunately, airbrushing also creates potential hazards when particulates, mists, or solvents enter personal breathing zones.

Airbrushing with water-soluble inks and paints that contain nontoxic pigments is safer than spraying with oil-based aerosols. Therefore, avoid both oil-based paints for air-brushing and spray paints that contain aliphatic hydrocarbon solvents such as benzine, mineral spirits, and petroleum ether. Petroleum distillates are irritating to skin and mucous membranes; if inhaled, swallowed, or absorbed in large quantities, they may cause pulmonary edema, pneumonia, and central nervous system problems.

Also, avoid water-based spray paints that contain mercury (e.g., phenylmercuric acetate as preservative), cadmium, cobalt, or lead. Instead, try synthetic pigments that have outstanding coloring power and are not associated with acute health problems. These synthetic colors are found in many art products, including paints for airbrushing.

All spray procedures require a spray booth or effective exhaust ventilation (see chapter 3, "Appropriate and Adequate Ventilation"). For environmental protection, the ventilation system should have a frequently cleaned or disposable filter that traps particulates. Spray booths vented directly to the outside are preferred in multiemployee studios. The cost of installing a spray booth or local exhaust system is quickly balanced by improved employee health, creativity, and productivity (see color plates

Figure 8.4.—**I**ris Brown (Bernhardt Fudyma Design Group): *Merrill Lynch Perspectives*, 1987, photostats of typographic elements, inking of design elements with Rapidograph pens, photocopy and computer-generated gradation, 11½″ × 9″. (Photo: Dennis Gottlieb.)

Figure 8.5.—**K**ristine Herrick: *Temple University*, corporate signature, 1983, pencil, water-based felt markers, and ink, 13″ × 10″. Roughs were executed in 4H pencil on 100 percent cotton rag felt marker paper; comprehensives were made with water-based black felt markers. The camera-ready art was executed with technical pens and water-based F&W nonclogging ink on Denril multimedia paper. My studio is large and well ventilated with three windows on two different walls. A 24-inch heavy-duty, three-speed exhaust fan vented directly to the outside draws fresh air in through the other windows.

flammable. Use fixative sparingly; most artwork can be protected safely and effectively by covering it with tissue paper or acetate.

**Aerosol sprays** containing fixative, paint, or adhesive must be used with caution to decrease the amount of potentially hazardous airborne materials. Read labels before purchase to avoid spray products with undesirable ingredients. Obtain a Material Safety Data Sheet for the product; observe all safety precautions recommended by the manufacturer. Aerosol cans often contain large amounts of propellant gas (isobutane) and other solvents that are flammable and potentially explosive upon puncture or exposure to heat. Keep all spray cans away from flames and heat, including direct sunlight, radiators, and heat vents. Whenever possible, use these products in a spray booth or outdoors.

Self-adhering **masking** films, papers, and tapes are used by airbrush (and bristle brush) painters to prevent contact of paint with selected areas of the artwork. These art materials are not associated with adverse health effects. Thus, masking film, clear or colored (for use in photography), can also serve as a protective overlay; some films are even receptive to pencil. To prevent seepage of paints and inks under the mask, apply a thin band of acrylic medium along the edge of the tape or mask. After painting over it, carefully cut through the paint and acrylic film at the edge of the mask, and lift the mask from the base. This method, and designated nontoxic masking fluids (Grumbacher, Hunt/Speedball Red Ruby), are not associated with health hazards. Avoid, however, liquid rubber masks if they contain hexane, heptane, or other solvents.

Figure 8.6.—**Jon Ellis: *Get a Monkey off Your Back***, poster created for the Heart Association of America, 1989, acrylic on Fabriano illustration board, 17" × 14". When my seemingly inoffensive and odorless acrylic paint tubes stated, 'Caution, contains cad-mium sulfide, a suspected cancer agent when inhaled,' I really woke up. I installed an exhaust fan and purchased a respirator for use during airbrushing. Eventually I perfected a drybrush blending method and stopped all airbrushing.

8.2–8.3). When effective ventilation cannot be achieved, wear a respirator with an organic vapor cartridge and spray prefilter, especially when using oil paints.

Some works of art, especially those created with dry drawing mediums, require setting with a **fixative** spray. These sprays often contain toluene and xylene, which dissolve and suspend plastic or varnish particles that provide the protective coating. Acute inhalation, absorption, or ingestion of these solvents may cause dizziness, headache, euphoria, nausea; weakness, irregular heartbeat, altered lung, liver, and kidney function, and irritation of the eyes, skin, and respiratory tract. Chronic exposure may alter kidney, liver, muscle, and nervous system function. Furthermore, xylene is combustible, while toluene is

## ADHESIVES

Graphic designers and illustrators frequently use adhesives and glues, including potentially hazardous spray adhesives and rubber cement. There are, however, safe alternatives, including waxers, adhesive transfer products, tapes, and glues.

In general, artists should avoid **spray mounting adhesives.** If alternatives (as discussed further in this section) are unsatisfactory, use only spray products labeled "Does not contain chlorinated hydrocarbons." Discard older cans that may contain these potentially harmful substances. (The presence of 1,1,1-trichloroethane and methylene chloride in spray mounting adhesives has been curtailed by the Consumer Product Safety Commission.)

**Rubber cement** is the designer's most common glue. Unfortunately, this strong, natural rubber ad-

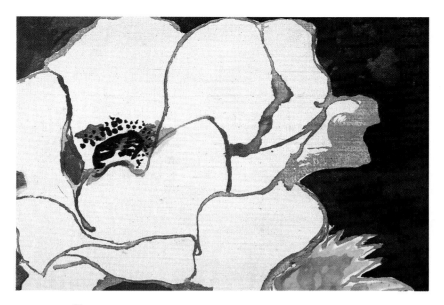

Plate 7.2.—Robert Kushner: *White Anemone*, 1989, color woodblock on silk, 20½" × 23". Some colors are printed and allowed to dry, while others are kept wet during printing in this water-based version of traditional Chinese woodblock printing. (Courtesy Crown Point Press, San Francisco and New York.)

Plate 7.1.—Janis Provisor: *Long Fall*, 1989, color woodblock printed on silk, 32½" × 21⅞". This work was completed in China, where I collaborated with printers in Suzhou and Shanghai with ink comprised of hide glue, tree resin, and earth pigments mixed with water. Blocks are carved from pear or boxwood. (Courtesy Crown Point Press, San Francisco and New York.)

Plate 7.3.—William T. Wiley: *Eerie Grotto Okini*, 1982, color woodblock, 22" × 30". This print was made from 26 blocks using 85 colors. Collaborators included Todashi Toda, a third-generation Ukiyo-e-style printer; Shunzo Matsuda, a woodcarver; and project co-ordinator Hidekatsu Takada from Kyoto, Japan. Only water-based inks are employed, and no printing press is used in this highly refined method of Japanese woodblock printing. Because it demands precision and technical expertise, this method is always executed by professional artisans who learn their craft through apprenticeship to a master printer. (Courtesy Crown Point Press, San Francisco and New York.)

Plate 7.4.—**A**nne Skoogfors: *Temple to Blue*, 1990, woodcut, 61″ × 39″. As an oil-based printmaker, I found I could no longer tolerate noxious vapors from inks and solvents. I thought I would have to stop printing altogether, but instead began to experiment with water-soluble printing inks and was thrilled with the results. I have enhanced my technique and no longer deal with toxic vapors.

Plate 7.5.—**E**laine Soto: *The Coming*, 1989, linocut, 15″ × 10″. I stopped etching with acids when I developed a cough and respiratory irritation with each usage. I decided to use wood and linoleum cuts with Speedball water-based inks. With my studio in my home, I no longer worry about adverse effects of my art materials on me, my animals, or my environment.

Plate 7.6.—**R**udolf Bikkers: *Icon XXXVI*, etching, aquatint with oil-based inks. Safe working conditions are a top priority in our printmaking studio at Ontario College of Art in Toronto, where I teach and do my printing. Students are instructed on all the dangers they would be exposed to if they do not adhere to our strict safety rules. This piece involves techniques that are hazardous, requiring safety precautions such as local exhaust ventilation over etching trays, printing area, and drying racks; eyewash station within reach; and use of gloves, mask, and goggles.

Plate 7.7.—**N**aomi Limont: *North Cascades National Park*, oil-based collagraph, 26″ × 34″, detail. I used a constant rotating motion with pressure when inking my large collagraph landscapes. As a result I developed carpel tunnel syndrome in my right hand. I now create plates that are printed by someone else.

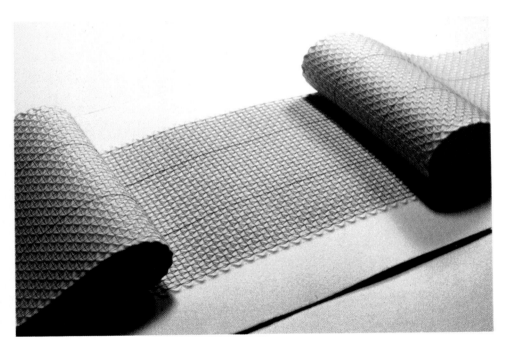

Plate 7.8.—**H**erb Appelson: *Big Ribbon* (detail), hand-threaded embossment, 14″ × 40″, detail. To avoid the hazards of nitric acid, mineral spirits, and various inks, I began to experiment with paper embossments and colored threads. The paper is embossed to produce a matrix; sections are folded to achieve added sculptural relief. With an overlay of various threads, I can work with a broad range of color to get effects that are both tactile and painterly.

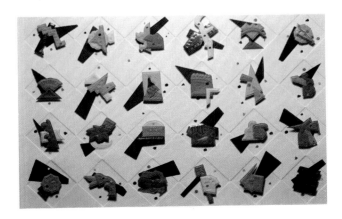

Plate 7.9.—**A**rt Salazar: *Rock and Roll*, 1986, inkless collagraph with colored papers, 24″ × 32″. Forms are cut from mat board, embossed onto cut colored papers, and each affixed with Elmer's glue to an embossed backing paper.

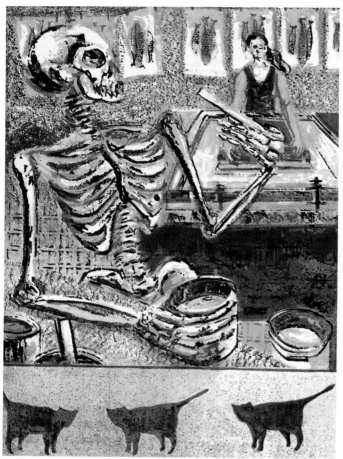

Plate 7.10.—**B**arry Wilson: *Untitled*, 1989, water-based screen print, 27″ × 20″. While no title was provided, this might be called "Silkscreen in the Old Days."

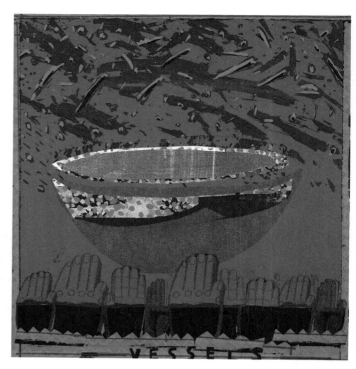

Plate 7.11.—**J**udy Stone Nunneley: *Vessels*, 1987, water-based screen print and collage printed on black Arches, 21″ × 22″. I teach at Minneapolis College of Art and Design, where we switched to water-based inks in 1986. After much experimentation, we developed an excellent system of printing with water-based inks that meets the standards and demands of a college print studio.

Plate 7.12.—**K**aren Weil: *October Party*, 1988, water-based screen print and oil pastels, 8″ × 8″. I printed for eight years with oil screen inks when anticipation of pregnancy induced a change. Eventually I found a suitable water-based ink and use packaging tape as my stencil.

Plate 7.13.—**N**ancy Callahan: *Elephant Cabbage Too*, 1986, water-based screen print, 22″ × 16″ × 7″. My last oil-based screen print required 50 stencils and two years to complete, during which I developed back pain, numbness in my fingers, and excessive fatigue. It caused me to change to water-based printing inks, which I've found to be faster, cheaper, and healthier. They have allowed me to increase production and have given me the flexibility for more experimentation due to their easy cleanup. Fortunately, all my symptoms experienced with oil inks and solvents have disappeared with no permanent damage. I can't imagine returning to my old methods.

Plate 7.14.—**J**ane Gregorius: *Striped Vase with Balloons*, 1988, water-based screen print monotype, 60″ × 44″ × 2″. Oil-based screen printing was my primary medium and one that I had taught for 15 years before becoming ill. In 1987, I was granted a sabbatical to research water-based methods to teach to my students at Cabrillo College in Aptos, Calif. It was difficult to change from the comfort of a known method, where I could focus on my imagery, to a new method where technique demanded primary attention. It was a remarkable year in which I learned about resources, equipment, supplies, and, finally, how to translate my own work into the new, nontoxic method. I am extremely happy that I changed and feel that my students are benefiting immensely. My health is back to normal, and I wish I'd done it years earlier.

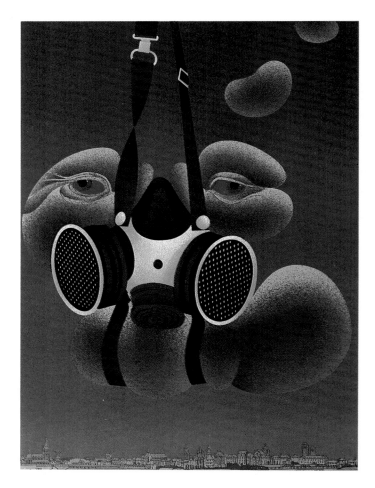

Plate 7.15.—**R**udolf Bikkers: *Mask*, 1989, oil-based screenprint, 34″ × 26″. This screen print was produced in an edition of 115. The combination of many colors and large edition number make the production process potentially very hazardous. In the setup used, vapors were extracted quickly in a well-ventilated room with exhaust ducts directly over the printing area and drying racks. For cleanup I also wore gloves, mask, and goggles near exhaust ventilation.

Plate 7.16.—**V**ivian Bergenfeld: *Celestial Passages*, 1989, pochoir and marbling on heavyweight watercolor paper, 36″ × 52″. One day while I was marbling with carragheen moss, water, and oil paints mixed with a few drops of mineral spirits, my pure white, long-haired Persian cat leaped into the tray and became covered with green paint. I tried washing her with soap and water, but it seemed to spread further into her fur. I called Emergency at the University of Pennsylvania Veterinary Hospital and was advised to take her immediately to my neighborhood animal hospital because absorption of organic solvent through her skin could prove fatal. The attending veterinarian agreed, clipped away all painted fur, and washed her. I have since switched to water-based marbling paints and keep my cat out of the studio! The other areas of color in this print are achieved by cutting a waxed-paper stencil, registering, and applying acrylic paints with a stencil brush.

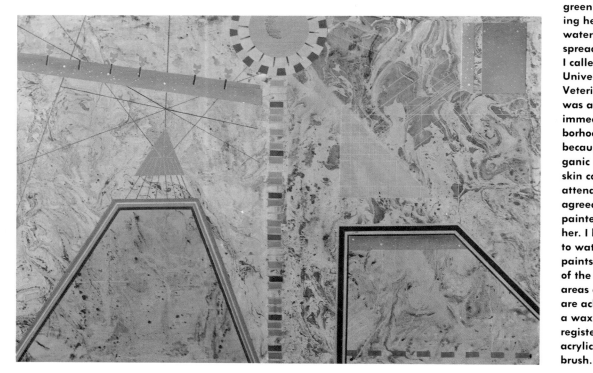

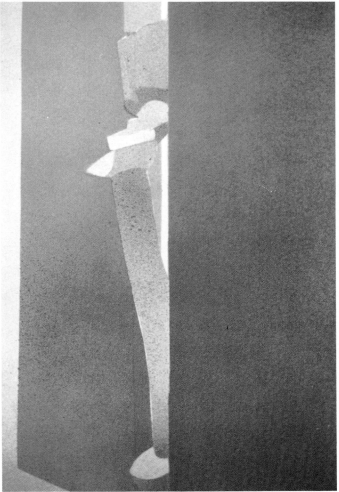

Plate 7.17.—Margot deWit: *Form VI*, 1987, pochoir on Rives paper, 26″ × 19″. The pochoir technique allowed me to decrease the use of hazardous materials and yet be able to produce multicolored print surfaces in different layers. With acetate template stencils secured with metal pallets, I spray a quick-drying water-based paint over the art work. I work outside dressed in a long-sleeved apron, rubber gloves, and a mask with filter for airborne paint particles. The stencils are discarded after the edition is completed, requiring no cleanup.

Plate 7.18.—Timothy P. Sheesley: *Ocean Fire*, 1988, lithograph, 30″ × 22¼″. In my lithographs, I like to take advantage of different litho processes. For example, *Ocean Fire* began with a drawing on transfer paper, was transferred to stone, then counteretched with citric acid solution, and reworked with addition of tusche diluted with water. Subsequent colors were printed from directly drawn, ball-grained litho plates. When a strong solvent is required in processing, I prefer acetone over lacquer thinners or other strong solvents. I use an exhaust system with ducts that pull the vapors directly away from the working surface. Also, because the ventilation is not 100 percent effective, I wear an organic vapor mask as well as gloves for additional protection. A critical step to reduce fire hazard and the quantity of low-level vapors in the press room is to make sure all soiled rags are placed immediately in an air-tight metal container.

Plate 7.19.—Enid Mark: *Pages from a Summer Album: Antibes*, 1987, lithograph, collage, 22″ × 30″. The image was prepared from hand-drawn and photographic lithography plates. The photo work is done in a darkroom equipped with local exhaust (Nutone) and lit with a Thomas Duplex Super Safelight that helps to prevent eye-strain. The lithographic printing is done at Corridor Press. Collaged elements, depending upon materials, are adhered with a 1:12 solution of methyl cellulose, polyvinyl acetate, rice starch (of which I mix only a two-day supply to avoid thymol as preservative), UHU stick, and/or Gudy-O Tape transfer.

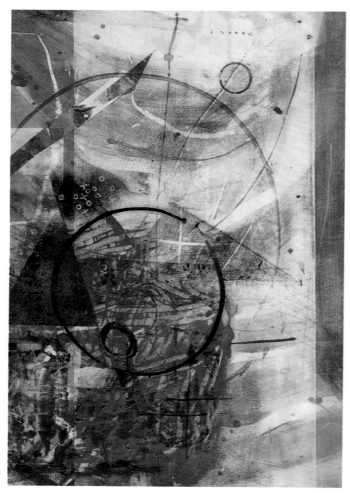

Plate 7.21.—**D**an Welden: *Canyon Series*, monotype with water-based media, 15″ × 11″. I use Createx monotype medium and colors, Caran d'Ache water-soluble crayons, ordinary brushes, a plate of thin plastic, and Arches or BFK printing paper. I let a plate dry overnight or even for several days and still get an excellent print because damp paper, under the pressure of the press, picks up most water-based pigments even when they are dry.

Plate 7.20.—**M**itch Lyons: *Count on Me*, clay monoprint, 43″ × 39″. Clay printing is a process I developed using a wet clay slab as my matrix. I mix kaolin clay, permanent pigments, and distilled water to a slip consistency and 'paint' it on the slab. Layer upon layer is built up until the desired colors and images are achieved. The entire surface is rolled with a rolling pin between applications to keep the surface flat at all times. Finally, I take a dampened paper or canvas and place it over the wet slab, and apply pressure with the rolling pin to transfer the image. Cleanup is simply with soap an water." (Photo: Jon McDowell.)

Plate 7.22.—**M**arcie Feldman: *Untitled*, 1989, monotype with water-based printing inks on BFK Rives paper, 29″ × 42″. I approach monoprinting in a painterly way, using Speedball water-based printing inks directly on Plexiglas. With the advantage of nontoxic, water-soluble inks, I no longer fear the skin rashes that used to develop from my use of turpentine, Lithotine, and other solvents. Also, setup and cleanup are easier. I've found minimal sacrifice in the medium's result and maximum gain for my health.

Figure 8.7.—**L**anny Sommese: *Arts Festival*, program, 1984, cut paper, pen, and ink, waxer used as an adhesive, printed offset, 6″ × 9″.

Figure 8.8.—**T**homas White, using a waxer.

hesive often contains n-hexane, a well-established neurotoxin via skin absorption or inhalation. Both central and peripheral nervous system damage have been well documented among artists using one-coat and two-coat cements and cement thinners for many years. Early symptoms include fatigue, cramping, loss of appetite, numbness and/or weakness in the extremities, and weight loss. Unsteady gait, decline of initiative and intellectual ability, recurring recent memory loss, difficulty in performing fine or rapid movements, and spasticity in the legs have been observed. The adverse effects of n-hexane may occasionally be misinterpreted as evidence of multiple sclerosis (McCann, *AHN* 2:6).

In a few products, n-hexane in rubber cement and thinner has been replaced by heptane, a safer solvent that takes longer to dry. Heptane, however, is not completely innocuous. Recurrent, low-dose exposures lead to skin and respiratory irritation. Massive exposures can cause dermatitis, nausea, chemical pneumonia, and narcosis. General dilution ventilation may be insufficient to properly disperse more than very small amounts of n-hexane or heptane vapors. Rubber cement, rubber cement thinner, and spray adhesives containing these ingredients should be used only with a spray booth or effective local exhaust ventilation (chapter 3) that draws vapors away from the source and user. In addition, keep containers closed when not in use and wear protective gloves.

Dried rubber cement discolors, deteriorates, and can be difficult to remove. Due to these disadvantages and with potential health hazards so significant, we recommend avoidance of rubber cement. Suitable alternatives for adhesion are available and described in a following section.

Other potentially hazardous glues include contact cement, cyanoacrylate (Super Glue), epoxy, model cement, and plastic welds. All should be used with skin protection and effective local exhaust ventilation. Contact cements are extremely flammable and may contain acetone and n-hexane.

Cyanoacrylates, which contain fast-drying plastic solvents, will bond skin; epoxy resin glues contain uncured epoxy resin (suspected carcinogen) and various hardeners. Model and plastic cements typically contain toluene or xylene, while plastic weld glues use solvents to dissolve a thin layer of each surface to be joined.

The hot-melt glue gun delivers melted glue adhesive. There are two types of glue guns: trigger-operated and thumb-fed. The trigger-operated device is the safer of the two, for it does not require a finger to push the solid glue stick into the inlet tube. The glue stick melts in three to five minutes; accidental

Figure 8.9.—**F**rank Baseman: *Tyler Glass*, poster, 1985, photogram and water-soluble ink, 22″ × 17″. The original was a photogram of actual broken glass on photographic paper. To enhance contrast between light and dark areas, a photostat was made from a black-and-white print and water-soluble ink was used to fill in some areas. It was positioned with wax on mechanical board; printed offset lithography. In the darkroom I use gloves or tongs, and wash my hands after exposure to chemicals.

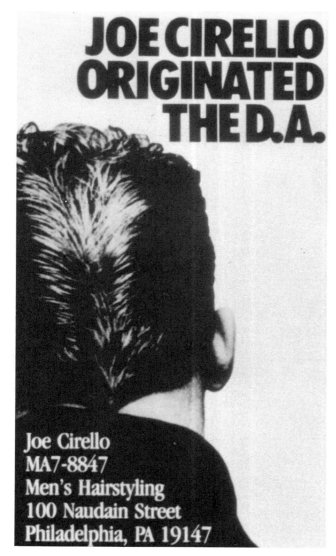

JOE CIRELLO
ORIGINATED
THE D.A.

Joe Cirello
MA7-8847
Men's Hairstyling
100 Naudain Street
Philadelphia, PA 19147

Figure 8.10.—Frank Base-man: *Joe Cirello*, business card, 1986, 2" × 3½". The original was a black-and-white photograph. Photocopies were used for size and scale on mechanical boards, then waxed for po-sitioning. It was printed by offset lithography. Wax is an excellent substi-tute for rubber cement, spray mount, or other potentially hazardous ad-hesives.

Figure 8.11.—Rona Ginn Klein removes backing from ATG (3M) tape to adhere decorative papers. Adhe-sive transfer tape is an easy-to-apply, nontoxic al-ternative to spray adhe-sives and rubber cement. Unlike spray adhesives, the tape can be repositioned as well as applied in small areas.

burns may occur if hot glue or heating elements of the gun contact skin.

**Safe Adhesive Alternatives** Hand-held and automatic electric **waxers** have many advantages over rubber cement and spray adhesives. These de-vices roll a thin, uniform coat of nontoxic, pressure-sensitive, chemically inert wax onto the back of materials to be mounted (e.g., paper, board, art-work, photos, plastic and film). Melted wax within the appliance is always ready for efficient applica-tion and immediate, freely movable pasteup. The coating is not sticky, so coated items can be placed aside or stacked. Waxed items require no drying time and do not adhere until burnished into posi-tion. These timesaving devices eliminate mistakes and maximize flexibility in placement and reposi-tioning.

When a design is complete, the properly posi-tioned pasteup items are burnished down for a firm bond. A temporary overlay sheet prevents abrasion or tearing while burnishing. The wax provides se-cure adhesion, yet waxed items can be lifted and repositioned as often as necessary. If lifting becomes difficult, soften waxed items by warming briefly with an iron or blow-dryer.

We do not recommend rubber cement thinner to remove excess wax. Use instead a kneaded rubber eraser, tissue, or cheesecloth to remove traces of wax. Large amounts of wax should be scraped away first (color plate 8.8). Unlike cement thinner, these alternatives will not smear or stain artwork or paper. To clean mechanicals and artwork, use an eraser appropriate to the surface. Electric erasers are good for stubborn spots.

Select a waxer with a shockproof plastic body and handle, keep it away from water, unplug it when not in use, and follow manufacturer's di-rections. Most waxers require little maintenance. Waxers and wax bars can be purchased at art supply stores.

**Adhesive transfer products** also provide safe substitutes for rubber cement and spray adhesives. These mounting aids accomplish adhesion tasks quickly and efficiently, with no mess or odor. Available in varying lengths and widths, they require no moisture, glue, heat, or activator. Large rolls may be cut to any size and shape for small pasteup jobs or for mounting large photos and artwork. Small rolls are most efficiently used in conjunction with a hand-held dispenser. The adhesive permits accurate positioning and repositioning of materials before pressure is applied for a firm, long-lasting bond that does not stain, dry, or discolor with age.

**Double-coated tapes** stick on both sides for permanent attachment. They are available in various sizes; in paper, film, or foam; and with or without a liner. These tapes are most efficiently used with a dispenser mounted on a worktable or weighted with a nonslip base.

**Glue sticks** are handy for last-minute adhesion on paper, board, photos, cloth, or other porous surfaces. Glue sticks are packaged in convenient roll-on or rub-on dispensers that provide quick and easy application. Many glue sticks are safe, nontoxic, and odorless; check labels before purchasing or using.

**Water-soluble glues,** with sponge-tip dispensers that may be remoistened if left uncapped, provide a good bond on paper, cloth, and other porous surfaces but may cause wrinkling due to water content. **Synthetic resin** water-soluble glues (Elmer's, Sobo) irreversibly bond and present no significant health hazards. Other water-based glues include nontoxic methyl cellulose and grain pastes such as wheat, rice, arrowroot, or starch. These are safe adhesives when mixed in small quantities and stored in a refrigerator to retard spoilage. Commercial mixtures may contain pungent thymol as a preservative.

## CUTTING TOOLS

**Knives, razor blades,** and other sharp-edged tools must be handled and maintained properly to assure accurate and accident-free cutting.

- Never take your eyes off the knife (and your hand) while cutting.
- Keep the noncutting hand away from the direction of the cut.
- Always cut *away* from your body.
- Do not use knives or blades to pick, scrape, or pry. The point of the blade may snap off and cause injury.
- Blade manufacturers suggest wearing safety goggles while cutting.
- Use tweezers instead of a blade tip to position elements during pasteup.
- Replace blades frequently to cut effectively.

Always cut against a steel (not aluminum, wood, or plastic) ruler or straightedge with a nonskid cork or rubber backing to prevent slippage. (The backing also raises the straightedge from the surface and helps to prevent the knife from skidding.) If you are right-handed, cut along the right side of the straightedge; if left-handed, cut along the left side.

**Figure 8.12.—R**ona Ginn Klein: *Abie's Lion's Land—Denizens II*, 1990, collage, 22″ × 30″. ATG adhesive transfer tape is applied to the back of selected decorative papers leaving its backing intact. After cutting the desired shapes, the backing is removed and the pieces pressed into place on a second sheet of paper. This procedure is a variation of the reverse appliqué fabric Mola technique used by the Cuna Indians of Panama.

Never cross hands or put excessive pressure on the knife in an attempt to cut through with one stroke. Several light strokes with even pressure produce a better (and safer) cut.

Some consider razor blades to be the most effective cutting tools. Those who prefer razor blades should use only single-edged blades; double-edged blades are difficult to use, easily broken and hazardous.

By contrast, knives with handles (X-Acto, Mat/Utility) promote safety, accuracy, cutting ease, and control. They are available in a wide assortment of model designs and styles. Current safety features include:

- Safety grip handles
- Rubberized handles for better control and comfort
- Roll-resistant design
- Retractable blades
- Safe and easy blade-changing mechanisms
- Protective safety caps

The round handles of some knives and cutters tend to roll off desks and drawing tables. Prevent this with inexpensive plastic or rubber guards (found in office supply stores) that fit snugly around handles.

Knives should be stored safely when not in use by covering with plastic safety cap. If stored vertically, the blade should point down. (For convenient storage, try the X-Acto Knife Station. This balsa wood block, which mounts on desk or drawing board, protects and cleans blades while providing easy access to knives.)

Designate a covered container to store used blades or wrap securely with tape prior to disposal so no one risks cuts when emptying the trash. Replacement blades can be purchased in safety dispenser packs where blades can be stored before and after use.

**Paper cutters** should be equipped with the following safety features:

- Protective shields, guards, and rails to keep fingers out of the path of the blade
- Tension springs that prevent the blade handle from dropping or flying up
- Locks or safety catches that keep the blade handle in a "down" or lowered position

Figure 8.13.—**S**tephen F. Schwartz: *Elf on a Mouse,* illustration, © 1988, *Highlights for Children,* reprinted with permission; brush and India ink on paper, 9" × 7½".

**Mat cutters** are precision instruments to be used only according to manufacturer's directions. Change blades often since dull blades cut unevenly and may cause accidents. Keep the noncutting hand away from the path of the blade.

## CORRECTIONS AND CLEANERS

Perfection is the ultimate goal on the road from rough layouts and comprehensives to mechanicals and print production. To get clean, precise, and accurate work, designers and illustrators use various products to correct, touch up, clean, and erase.

Rubber cement **thinner,** commonly used by graphic designers, poses a significant health hazard. It can be used to:

- Thin and reduce rubber cement
- Remove wax residues (use kneaded eraser, tissue or cheesecloth instead of thinner)

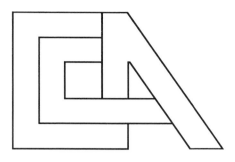

Figure 8.14.–**C**athy Spandorfer: *Cheltenham Center for the Arts*, logo, 1991, computer-generated design using a Macintosh II with the software program Quark XPress. Designing on the computer has allowed me to eliminate many traditional and labor-intensive methods. From the development of preliminary designs to the refinement of finished art, I can do my work more easily, faster, and at lower cost.

Figure 8.15.–**T**homas Porett: *Anaheim Matrix 7*, 1990, digital imaging, 10″ × 13″. This image was created from a video sequence digitized in a frame buffer. The composite image was then sent to a thermal printer. Regarding hazards in computer-related art, I remain current by reviewing manufacturers' announced plans and product specifications for their attempts to reduce electromagnetic radiation.

◆ Clean dirt marks and smudges from mechanicals and artwork (use erasers instead)

◆ Loosen and separate paper that has been glued or mounted with rubber cement

◆ Remove self-adhesive tapes and films

If you must use rubber cement thinner, use it sparingly with local exhaust ventilation and nitrile or polyvinyl alcohol (PVA) gloves. Store thinner in leak-proof and evaporation-proof **dispensers,** which, when used properly, reduce the likelihood of vapor inhalation. Small dispenser openings allow precise, controlled use of thinner without clumsy splashing. Dispensers come in various sizes, either flat or cone-shaped. Flat dispensers topple over; cone-shaped types are more stable and preferred. Keep closed when not in use.

For corrections and touch-ups, small jars of water-based, **opaque white paint** are diluted with water to desired consistency. Many brands are available that can be used with a brush, pen, or airbrush on various surfaces, including photos and acetate.

**Correction fluids,** supplied in small plastic bottles with a serviceable brush, are handy for quick corrections and touch-ups. Correction fluids containing trichloroethane or trichloroethylene should be avoided. A safer alternative is water-based correction fluids, correction films or tapes (see Appendix I under trichloroethane).

## COLOR IMAGING PROCESSES

There are several ways that graphic designers make color proofs of work before the project is actually printed. One of the most common is called Color Key, a light-sensitive acetate material that is available in the process colors as well as several other colors. An ortho negative of the desired image is sandwiched with a piece of Color Key and exposed to high-intensity light. The positive image is developed out with a chemical rubbed onto the surface of the Color Key material with a soft cotton wipe. The finished image is washed in water and allowed to dry, giving a transparent colored image on clear acetate.

Another common proofing material is I.N.T., or image and transfer material. This process creates a customized rubdown image, using an ortho negative exposed onto the I.N.T. with a strong light. After exposure, the protective acetate is peeled off the material, and the image is developed out using a

Figure 8.16.—Anne Seidman: *Ghoti*, 1989, 3-D animation, 16″ × 20″. The image was generated at Ohio State University on a VAX 11-780 using special software written by William Koloymjec and John Donkin. The animation depicts a voyage through an infinite tubespace.

developing fluid and soft cotton wipes. After the image is completely developed, the I.N.T. is washed in water and allowed to dry. When dry, it can be rubbed onto any surface, giving an opaque image that is available in a variety of colors. Gloves and exhaust ventilation should be used during this process. Check Material Safety Data Sheets for information on the chemicals used.

Both of these processes are widely used by students in graphic design programs, since students are required to produce pieces that look as much as possible like printed pieces for their portfolios. Many design studios also use these processes for showing clients the appearance of a finished piece, particularly for high-budget projects (color plate 8.6).

## COMPUTERS

Many designers use computers to generate, store, and print images, symbols, artwork, and graphics as well as text and numbers. Modern computer programs provide an astounding array of design possibilities, including drawing and painting software developed specifically for graphic designers, illustrators, and other visual artists. Images can be created electronically on the screen, or external artwork can be introduced into the computer using scanning hardware. Once an image is in computer memory, drawing and coloring programs provide endless ma-

nipulative possibilities. Any image can be elongated, expanded, or reduced; repeated; fragmented and reassembled; curved and squiggled; colored and recolored; or transformed into a sequence of "stills" or animated video. Once a satisfactory image or design is achieved, it may be printed or stored for later access and modification (color plates 8.16–8.17).

Use of computers may reduce exposure of graphic designers to the hazardous materials and procedures of traditional design work. Moreover, since the design process with a computer can be dramatically accelerated, computer printouts need not be the end product but rather a starting point for subsequent development using traditional media options.

**Computer Hardware**   The video display terminal (VDT) is a key element of personal and office computers. The terms *personal computer* and *VDT* are often used interchangeably; *VDT* will be used here for simplicity. The display portion of a VDT is often referred to as a *monitor*; the actual image is viewed on the screen. VDTs are based on the familiar display of the television-like cathode ray tube (CRT). The CRT-based unit remains the dominant form of VDT in use today; over 40 million units have reportedly been sold worldwide.

The CRT display provides brilliant colors and high image contrast. The purchase price of the personal computer version ranges from less than $1000 to more than $10,000, depending on performance and options. Pricing factors include screen size, image clarity, color or monochrome, storage capacity, computation speed or speed of generating images, and quality of printer and scanner. Images can be printed with single or multicolor impact or non-impact printers. Depending on the type of printer used, the medium can, for example, be paper, mylar, acetate, magnetic video tape, or photographic film.

Potential health effects of the VDT[1] have been studied since the 1960s in response to repeated complaints of headaches, eyestrain, backaches, and stiff necks; dizziness, nausea, irritability, and stress; skin rashes and dermatitis; and swollen ankles and sore hands and wrists. Many of the potential problems stem from lack of good *ergonomic* design, i.e., poor physical-anatomical relationships of individuals to their work environment. Ergonomic problems may be prevented with corrective measures described later in this chapter.

Eye fatigue; irritation with itching, redness, soreness, stinging, or watering; blurred vision with double images, temporary nearsightedness, or loss of accommodation; halos, colored fringes, and tinted vision; and eyeglass and contact lens problems have all been associated with VDT use. VDTs have also been linked to cataract formation. Thus, some clinicians suggest an eye exam prior to use of a computer and annually thereafter.

"Eyestrain" is often related to long periods of viewing VDT images whose resolution and contrast are inferior to those of more conventional images. By contrast, high-quality images are clear, sharp, and adjustable for contrast and brightness. Whenever possible, select a screen (color or monochrome) that is visually comfortable. Extended viewing of green or amber monochrome screen characters may produce (upon looking away) tinted vision in pink or pale blue-violet, respectively. Known as *afterimage,* this phenomenon is due to fatigue of the color-receiving cones in the retina. It is temporary, but annoying to some individuals. Designers who work with color screens may experience ongoing afterimage flashes of complementary colors.

The screen image should be visible without distortion. Bifocals may cause problems with VDT use; vision through the lower part of prescription glasses may require the user to tilt the head back or to place the VDT low on the support surface, leading to fatigue, strain, and distortion.

Recently marketed computer monitors have screens with reduced glare, but a filter can also be mounted over the VDT screen. In addition, a personal desk lamp helps provide balanced illumination between the screen and printer. Orient the screen for comfortable viewing. (Some are adjustable with a swivel movement.) Obtain a desired height with adjustable furniture or, more simply, with a box or other prop.

Use a chair that can be adjusted to your build. The chair should support the back in alignment with the head and neck. Adjust desk height, keyboard, and VDT screen height (and angle) so that these components are arranged comfortably relative to your personal seat, hand, and eye angle and level. Use a document holder to keep original material positioned for comfortable viewing alongside the computer. Repetitive movements and uncomfortable shifts between printed copy and the screen can increase tension.

Prolonged sitting in a poorly designed or adjusted chair may reduce circulation, leading to swollen feet, ankles, or legs. Adjust the chair height, or put a block under your feet so that when your feet are flat, the backs of your knees rest comfortably on the seat.

Many computer monitors emit high-pitched tones near 16 kilohertz (a kilohertz is 1000 cycles per second). Because many women can hear beyond 20 kilohertz, some observers suggest that females more easily detect these tones. Lack of awareness of tone, however, may still not prevent irritation. Two solutions are available. The expensive option is to switch to a monitor that does not emit the tone. The other approach involves use of foam earplugs that might provide some relief from high-pitched sounds without excessive interference to ordinary conversation.

Some observers have associated skin rashes and dermatitis with VDT use. The rash appears on the cheeks, forearms, and wrists; it disappears when VDT use is terminated, then reappears upon return to work at the computer. There is, as yet, no proof that computers cause skin rashes. There is speculation, however, that relates the condition to buildup of static electricity around the computer and in the work environment. (Use of a humidifier, antistatic plastic mats under the chair and keyboard, or a grounded antiglare screen will decrease the level of static charge.) Short-term rashes are not linked to long-term health effects, but chronic rashes and irritation can result in permanent scarring. If the rash or irritation persists, see an industrial hygienist, toxicologist, or dermatologist. Other chemicals in the workplace may be the culprit.

Hand and wrist fatigue and pain are associated with poor alignment between arms, hands, and keyboard. Observe your keyboard use for a few days to determine, if possible, whether you have tension in your hands, arms, and shoulders extraneous to the actual pressure required to operate the keyboard. Tension increases and aggravates fatigue and pain, so make a point to reduce it. Low-grade hand discomfort is associated with repeated pressure on certain areas. If you habitually rest the heels of your hands and wrists on the desk during pauses, circulation may be impaired and muscle tenderness may develop. Place the keyboard at the edge of your desk or on a foam pad to free your hands from this pressure.

Laser printers are increasingly used with VDTs. The laser beam itself is fully contained within the printer cabinet; it should not cause problems during operation. Since the laser printer relies on a photoconductive drum and the xerographic process, it contains a toner-fusing assembly, which can get quite hot. The user must not touch the fusing assembly when removing occasional paper jams.

Despite strong rumblings in the press concerning important ongoing studies,[2] there is, as yet, no proven link between long-term use of VDTs and long-term, adverse health effects. Current knowledge does not support the conclusion that modern VDTs emit harmful amounts of x-ray or microwave radiation. VDTs do emit a comparatively low-level radiation in the so-called radio frequency range from about 15 kilohertz up through the microwave region. Some studies suggest that sufficient exposure to low-level radiation at these frequencies, especially in the upper range, could adversely affect humans. This issue, however, remains controversial. (The effects of radiation above several hundred kilohertz can be greatly reduced in front of the VDT by means of commercially available CRT screen shields.)

An important new element of controversy surfaced in the 1980s with reports that extended exposure to 60-hertz magnetic fields might be a factor in unusual occurrences of tumors or reproductive problems.[3] (Magnetic fields and electric fields are the two primary components of electromagnetic radiation.) The reports grew out of an examination of the effects of 60-hertz magnetic fields generated by high-voltage, overhead power transmission lines. Although it had been known that CRTs emanate low-level 60-hertz magnetic fields, it had not been suspected that 60-hertz fields related to household wiring and various appliances could also be associated with adverse human health effects. (Although rather small, the magnitude of VDT fields may be comparable to those of nearby transmission lines.)

The 60-hertz, or "extremely low frequency" (ELF) magnetic field appears at varying low levels around many VDTs, shows some variation between models, and falls off with distance from the VDT. Unlike the higher frequencies mentioned previously, there is no practical way to inhibit ELF magnetic fields with an external shield.

To minimize potential risk in the event that ongoing studies verify an adverse health effect, some authorities recommend that users work at a minimum distance of 28 inches from the front of the VDT. Furthermore, users who work in an environment surrounded by VDTs should note that magnetic fields emanate more from the sides and rear of most VDTs than from the front. Thus, users should keep a distance of at least four feet from the side or rear of neighboring VDTs.

In December 1990, a draft report by the U.S. Environmental Protection Agency concluded that there is enough evidence of a possible link between cancer and low-level magnetic fields from power lines and appliances to warrant new research.[4] Because of current studies and expected new studies on the long-term effects of VDT magnetic radiation, we are likely to see new findings and design standards in the future.

**Exercises To Reduce Strain** Exercises can reduce strain, one of the primary aggravations of VDT use. The San Francisco Board of Supervisors approved an ordinance (also in December 1990) that would *require* employers to give workers 15-minute breaks or alternative tasks after every two hours of work on a VDT.[5] We suggest that you:

Take periodic visual breaks, remove glasses if appropriate, and focus on objects in the distance. Exercise your eyes by rolling them from side to side, up and down, and all around.

■■

Periodically stretch and flex your shoulders, neck, head, arms, hands, and fingers. Choreograph a creative arm dance when reviewing images.

■■

Press chin to chest, lift, turn head slowly from side to side, back to center, look up at the ceiling, and then repeat.

■■

Bring shoulders up to ear level, then press downward. Roll shoulders in a circular movement forward several times and backward several times and then alternately, somewhat like swimming front and back crawl with your shoulders alone.

■■

Wiggle and stretch toes and feet frequently, stand up and stretch at least once each hour, and take a break for walking, shakeout, and stretching every two hours. Without losing much work time, you may be able to periodically kneel on a pillow or pad at the computer to stretch the hip and back area.

If discomfort, swelling, or inflammation persists, medical evaluation should be sought.

## Hazard-Free Workplace for
## Designers and Illustrators

We would all like to work in a hazard-free environment. If, however, potentially hazardous products and methods must be employed, they should be used as safely as possible. There are several ways to set up a design studio for safety and creativity. Some may require changing materials and procedures; others may require new equipment and changes in the work space, especially those that improve ventilation. Although the goal is increased health and safety, the results may often expand creativity and productivity as well.

Organization of the workstation supports a desire to design safely. Rotary organizers, shelves, trays, troughs, and taborets (compartmentalized tables) will keep frequently used supplies and tools close by. Knowing how to locate essential equipment facilitates effective work, especially when deadlines approach.

Graphic designers often work long hours. Under such conditions, one's exposure to regularly used materials may exceed permissible exposure limits. Learn to take a break, especially when working with known chemical hazards. In addition, remember that vulnerability to accident, injury, and illness is highest when fatigue is greatest.

Graphic designers employed by a business or agency, as differentiated from self-employed workers, should be especially familiar with employer and employee responsibilities designated by the Occupational Safety and Health Administration (OSHA) of the U.S. Department of Labor. Some responsibilities, appropriate for readers of *Making Art Safely,* have been summarized in chapter 2, "Health and Safety Programs," and chapter 3, "Evaluating Classroom and Professional Studios for Safety and Ventilation."

# ERGONOMICS

In addition to appropriate physical layout and up-keep of the studio, proper application of principles of **ergonomics** (the relationship between individuals and their work equipment and environment) is essential to improve working conditions. Recognize the importance of physical comfort in the studio, for it lessens strain while enhancing worker health and productivity. As noted previously, chairs, desks, and lighting should be adjusted to accommodate body needs.

Some of the basic features of a good chair include:

* An adjustable seat—adjustable in height and angle relative to the floor. When sitting, the hips and legs should be roughly at right angles; the feet should be flat on the floor. Chair height and seat angle, as well as location and tilt of the backrest, help achieve this posture.

* An adjustable backrest—adjustable in depth, tilt, and vertical position. The backrest should make contact with your back about four to six inches above the seat and should allow you to sit up straight.

* A swivel seat allows continued use of the backrest when turning. Although casters can be useful, they should be avoided since they can cause calf muscle strain as one tries to keep the chair in a fixed position. Wool and rayon covers, rather than vinyl or synthetics, are best for the seat since they are porous and allow body heat to escape.

Drafting tables should have height and angle adjustments. When the tabletop is in a horizontal position, its height should be adjusted such that your forearms are at right angles to the body when sitting upright. This position is the least fatiguing since it places the least strain on muscles of the back and arms. A stable base and safe tilt lock should be provided to avoid accidental table collapse.

Vision may suffer from **lighting** that is too intense, too dim, poorly located, or otherwise inappropriate. Poor lighting can contribute to eyestrain, headaches, and tension. Work areas that are over-illuminated with ceiling-mounted fluorescent lighting can induce eyestrain associated with extended time in bright sunlight: the irises of our eyes remain contracted to reduce light entering through the pupil. It is better for general illumination to be dim rather than too bright if most designers have individual lighting at their desks. A floating-arm, adjustable lamp attached to the drafting table or workstation should be used to provide maximum flexibility and accommodation to personal needs and varying tasks. Whether one uses standard incandescent, fluorescent, high-intensity lighting, or a combination of all is a matter of personal choice. Each has its visual, temperature, and economic advantages.

Fluorescent tubes produce a good strong light that gives off comparatively little heat. They are more expensive initially, in contrast with incandescent bulbs, but cost less to operate and last longer. Fluorescent lighting is not as "warm" as incandescent lighting and tends to create a harsh feeling, although "warmer" yellow tubes are available. Lamp assemblies containing both types of lighting are also available. Fluorescent tubes tend to flicker when wearing out; this may be due to a faulty ballast component. Melted tar, or more importantly a strong noxious odor, can emanate from a faulty ballast; under these conditions, the ballast must be replaced.

Halogen bulb lamps can provide an intense source of light. Although more efficient to operate than incandescent lighting, the purchase price of halogen bulbs is considerably higher.

Proofing areas and light-tables sometimes produce too much light that can be tiring and harmful. If possible, obtain a system that permits light adjustment to suit individual needs. As we age and are affected by presbyopia (loss of flexibility in the lens of the eye), our needs for light, as well as magnification, increase. Illuminated magnifiers are available to aid in precision work.

## EQUIPMENT

A number of machines, such as copiers, cutting machines, photostat cameras, and projectors, may be employed in graphic design studios. Mechanized or hand-operated equipment that provide actions such as rotating, reciprocating, and transverse motions, bending, cutting, punching, and shearing should be adequately guarded to prevent injury. This equipment should be used with care to keep fingers, hands, loose clothing, and long hair away from moving parts.

Photostat machines require a variety of photographic and printing chemicals (see chapter 9). Single-bath, self-contained photostat diazo processing systems are relatively safe because they require little or no contact with photochemicals. Wear rubber gloves to prevent skin contact with wet stats. Exhaust ventilation must be used in studios containing photostat machines. Obtain Material Safety Data Sheets on the chemicals used.

Copiers are discussed under "Xerographic Processes" in chapter 9. Many manufacturers now provide systems serviced by prepackaged cartridges that require no contact with copier solutions, toners, printing inks, or solvents. The increased safety is worth the expense. Exhaust ventilation remains important wherever copiers and photostat machines are used.

Electric equipment should be safely wired and grounded, with protection from hot and ultraviolet bulbs and heating elements. Where lasers or high-intensity photo lamps are used to expose images, they should be contained so that potentially harmful beams will not cause eye damage. Equipment should be placed in the studio so that it may be used in a convenient and efficient manner close to related tasks.

The potential hazards associated with graphic design can be dealt with effectively through appropriate ventilation, ergonomic improvements, care during equipment operation, and substitution of safer alternative methods and materials.

As the complexion of design changes with the burgeoning use of computers, significant front-end expense is often required just to stay abreast of the latest advances. To keep costs low, management may overlook the necessity of worker safety and well-being, especially with regard to ventilation. Through advocacy groups within the workplace, in state and local committees for safety and health (COSH), and in the Graphic Artists International Union (GAIU), efforts are made to provide and disseminate information for employers, employees, and members. Becoming and remaining informed is an asset on many levels: for oneself, one's colleagues and family, and one's employer and community.

While you work safely and creatively as a graphic designer or illustrator, let management hear one message loud and clear: What's good for the worker may prove even better for the boss.

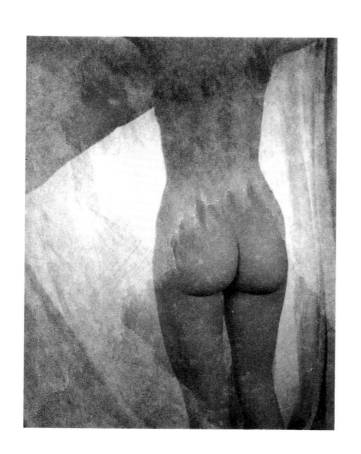

*9*

P hotography, which has captured the imagination of amateurs and professionals for more than 150 years, can be irresistible. The seduction lies in the myriad possibilities inspired by aiming a lens at a subject, the simplicity of pressing the shutter release, and the thrill of

*Photography*

watching the image develop in the darkroom. As both creators and viewers, our appetite for representation can be insatiable.

Photography, however, is not a risk-free activity; inappropriate exposure to photochemicals confers significant **health** risks. Thus, photography offers substantial challenges to working safely. Some photochemicals, in-

◆ ◆ ◆

cluding solvents, acids, and alkalis, are **individually** (and predictably) hazardous. However, photographers using many different kinds of agents also face the specter of chemical **interactions** with unpredictable exposures and adverse health effects. While many photographers (perhaps unwittingly) do not experience health problems, a significant minority do encounter adverse effects that profoundly alter both work and life-style.

Fortunately, awareness of potential health problems among photographers is increasing. For example, a survey commissioned by the National Press Photographers Association suggested that 74 percent of professional photographers were already attempting to prevent occupational illness by using protective equipment (gloves, tongs, etc.) and darkroom ventilation (Shaw and Rossol, 1989).

All who work in a photographic studio (including those who do not handle photochemicals) should participate in a training program. When everyone is equally informed about hazards in photography, safe work practices are more likely to be observed. The importance of handling photochemicals appropriately should be emphasized, and specifics of electric hazard, fire, and general accident prevention should be included in the training curriculum. Periodic refresher sessions are also recommended (see chapter 2, "Health and Safety Programs").

We stress that this is not a chapter on "how to do" photography. You will not find the basics of camera use, film processing, or darkroom setup here. Instead, we address how to do photography as safely as possible in an overview that we hope will enhance health and safety awareness among all photographers (color plates 9.1–9.18).

Essential preconditions for safe photography include:

◆ Awareness and prevention of hazards

◆ Use of substitutes for hazardous photochemicals and equipment

◆ Effective ventilation in photographic darkrooms and studios

◆ Safe handling of photographic materials and processes with protective attire and safeguards

Prior reading of chapters 1 through 4 will aid understanding of basic issues of toxicity and safety of concern to photographers today.

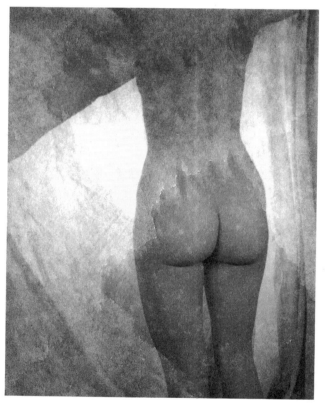

Figure 9.1.—**R**obert Asman: *Untitled*, 1990, toned gelatin silver print enlarged from paper negative, 19½" × 15½". For the past ten years I have been a professional darkroom technician, working exclusively with black-and-white processes. I work in a well-ventilated darkroom with fresh air pumped in and stale air vented by exhaust fans. Handling prints in solution is best done with tongs. With larger prints, where tongs aren't feasible, I use gloves. The standards and practices of yesterday, when darkroom technicians and photographers had chemically stained fingers, is unacceptable today. One's hands should remain clean and dry to prevent print staining as well as skin absorption.

## Resources

**I**n addition to emergency numbers listed in chapter 2, Kodak's 24-hour hotline for health, safety, and environmental emergencies, 716-722-5151, and ILFORD Photo's 24-hour hotline for medical emergency, 800-842-9660, should be posted in all studios where photographic procedures are used. Kodak Material Safety Data Sheet (MSDS) information is available by calling a toll-free number 8 A.M. to 6:30 P.M., Eastern time, Monday through Friday: 800-242-2424.

For technical information on color or black-and-white photography call ILFORD Photo at 201-265-

6000, or write to W. 70 Century Road, Paramus, NJ 07653.

Polaroid Resource Center's customer service number is 800-225-1384.

The Chemical Referral Center of the Chemical Manufacturer's Association (CMA) also maintains a hotline for information regarding common and uncommon chemicals in the visual arts: 800-262-8200. CMA, however, may direct inquiries on photochemicals to Kodak, ILFORD, or to the Center for Safety in the Arts.

## Potential Hazards of
## Photography

Ingestion, inhalation, and skin absorption of photochemicals can adversely affect the skin, eyes, lungs, nerves, intestines, and reproductive organs. Some symptoms, such as headache and depression, can be subtle, unrecognized, or considered a "normal" result of photographic work.

### PHOTOCHEMICALS IN THE DARKROOM

Many photochemicals are potential irritants, sensitizers (incite allergic reactions), or both. In Appendix I, "Art Material Chemicals," we list photographic processing ingredients alphabetically, identify their potential hazards and uses, and recommend precautions and substitutions. As you **read** labels for ingredients, consult this list. Follow precautionary information on labels and observe all recommendations, especially indicators such as "Caution," "Danger," and "Warning." (Photochemicals sold in the United States must conform to the Art Hazards Labeling Act of 1988.) We recommend safe disposal (as described later) of any chemicals in your studio that predate this improved labeling.

In poorly ventilated darkrooms, processing trays containing mixtures of developers, toners, bleaches, and fixes may release gases and vapors in concentrations sufficient to cause adverse health effects if inhaled for long periods. The following gases and vapors may be particularly hazardous:

**Ammonia** gas arises from all ammonia solutions, including washing aids, hypo eliminator, photo rinses, bleach fixes, and replenishers. Ammonia irritates the eyes and respiratory tract and may cause chronic lung problems or even death.

■■

**Chlorine** gas arises from addition of heat or acid to hypochlorite (bleach) or potassium chlorochromate (an intensifier). Chlorine gas is intensely irritating and potentially lethal.

■■

**Formaldehyde** is a preservative and hardener found in stabilizers, wetting agents, final baths, Kodalith developer, prehardeners, replenishers, and retouching dyes. Its vapors are highly irritating and may sensitize respiratory tissues. Formaldehyde is also a suspected carcinogen. Frequently overlooked symptoms include shortness of breath, eye and skin irritation, headaches, dizziness, nausea, and nosebleeds. Photochemical manufacturers have begun to reduce or replace formaldehyde in their products, but as of this writing, an effective and safe alternative has not been found.

■■

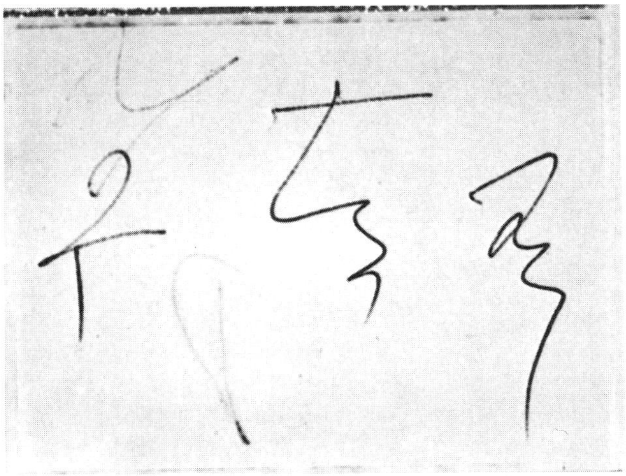

Figure 9.2.—Nancy Hellebrand: *Untitled*, 1989, gelatin silver print, 22″ × 29″. I see a strong connection between personal safety and earth safety (ecology). I use resin-coated paper, which requires less processing time and significantly less water for the final wash. This means less time over trays in the darkroom and fewer chemicals down the drain. When I leave the darkroom temporarily during a session, I cover the trays with waxed paper directly on the chemicals to control vapors. I use tongs instead of my hands for holding prints. When I use packets of dry chemicals, I cut open the short end of the packet and place it against the bottom of an inverted mixing bucket, turning both over together. I leave the packet over the powder in the bottom of the bucket until the dust subsides. Whenever possible, however, I use premixed liquid solutions instead of powdered chemicals.

**Hydrogen selenide,** a sepia toner, irritates the eyes, nose, and throat. Low-level exposures can cause nausea, vomiting, diarrhea, metallic taste, odor of garlic on the breath, dizziness, and fatigue.

■ ■

**Nitrogen dioxide** forms from nitric acid etching and carbon arc lamps. It is highly toxic by inhalation, and symptoms include fever and chills, cough, shortness of breath, headache, vomiting, and rapid heartbeat. Prolonged high exposure can be fatal. Chronic exposure may cause pulmonary dysfunction and symptoms of emphysema.

■ ■

**Ozone,** which has a sweet odor, is also released from carbon arc lamps and from photocopy machines. Ozone irritates the eyes as well as the upper and lower airways.

■ ■

**Phosgene** (carbonyl chloride) arises from chlorinated hydrocarbons (solvents), flames, lighted cigarettes, and ultraviolet light. The odor and irritant properties are not strong enough to give warnings of hazardous concentrations. Therefore, we must learn the sources and avoid them, since phosgene severely irritates the eyes and respiratory tract. When inhaled, phosgene decomposes in the lungs to form carbon monoxide and hydrochloric acid, which may lead to pulmonary edema or "water on the lungs." Symptoms of pulmonary edema include cough, foamy sputum, and increasing shortness of breath. Moderate exposures can cause symptoms of dryness or burning sensations in the eyes and throat, chest pain, and shortness of breath.

■ ■

**Sulfur oxides,** especially **sulfur dioxide,** arise by heating or by combining acids with any compound containing sulfur, such as ammonium thiosulfate, persulfates (reducers), sodium sulfite, sodium thiosulfate, sulfuric acid, and sulfamic acid. Sulfur dioxide gas is a powerful lung irritant that can cause chronic lung problems.

Three especially hazardous gases and their sources deserve separate recognition:

**Carbon monoxide** is odorless and therefore difficult to detect. It may arise from incomplete combustion in carbon arc lamps or may form in the body following methylene chloride exposure. In sufficient quantities, it can kill by asphyxiation (oxygen deprivation).

■ ■

**Hydrogen cyanide** is formed from heating or combining acid with cyanides (reducers), potassium ferricyanide, or thiocyanates. Cyanide poisoning has received wide press attention; its lethal potential is well known.

■ ■

**Hydrogen sulfide** ($H_2S$) poisoning can also be rapidly lethal. Its characteristic rotten-egg odor may not be detected by some. For others, the smell causes olfactory fatigue, which may decrease awareness as exposure increases. Unfortunately, there is no reliable indicator of the presence of hydrogen sulfide. Photographers should know that $H_2S$ arises from sulfide toning solutions and by combining acids with potassium sulfide or sodium sulfide. Low-level exposure may cause digestive and nervous system damage, and symptoms include headache, dizziness, excitation, digestive upset, and unstable gait.

The **skin** is also vulnerable to photochemicals, especially if one processes film and develops photographs with unprotected hands. Wear appropriate gloves when handling and mixing chemicals; use tongs to transfer photographs through the developing process. This greatly reduces skin exposure, absorption, and irritation. Protection is critical in the presence of cuts or breaks in the skin, for even minor scrapes increase the risk of chemical dermatitis and absorption into the body.

Severe cases of dermatitis have ended careers in photography (Tell 1989). Dermatitis (chemically related) and dermatosis (occupationally related), however, can be treated with corticosteroids, but effective treatment may also require avoidance of *all* routes of exposure to photochemicals during the healing process.

**Ingestion** of toxic substances in photography should be rare, but stories persist about photographers who identify or check their solutions by taste. Because they often suffer no evident adverse effects, they may assume tasting is relatively safe. This may be true if the dose and duration of exposure to a substance is minimal. However, even small amounts of acid, alkalis, organic solvents, and other agents used in photography can cause airway or gastrointestinal irritation. To avoid pain or discomfort of the tongue, lips, or mouth, **do not identify or check the strength of photographic solutions by tasting!**

Adverse health effects may also arise from exposure to the products of chemical decomposition. For example, photographic fixers have no hazardous ingredients listed in Section II of the Material Safety Data Sheet, but under decomposition in Section VI, sulfur dioxide, which can cause airway irritation or bronchitis, is listed as a potential hazard. Similarly, decomposition of sulfide toners may produce hazardous amounts of hydrogen sulfide gas.

Finally, some color developing agents are suspected carcinogens. These include benzotriazole, hydroquinone, p-methylaminophenol, p-phenylenediamine, pyrogallol, 6-nitrobenzimidazole, potassium dichromate and ammonium dichromate powders, thiourea, and formaldehyde (Shaw and Rossol 1989). We recommend minimizing or avoiding exposure to these agents by substituting agents not associated with the development of tumors in animals or humans.

In summary, photographers must evaluate all photochemicals and **the total context of the photographic process** for potential health and safety hazards.

## STUDIO AND DARKROOM LIGHTING HAZARDS

Sources of light, both natural and artificial, are essential to photography. Some are safer than others. **Carbon arc lamps,** once commonly used by photographers, create light through combustion (burning), producing several potentially hazardous by-products, such as:

- Carbon monoxide, a potent chemical asphyxiant released during incomplete combustion

- Nitrogen oxides and ozone, which can damage cells of the airway and may increase permeability of the lungs

- Ultraviolet light, which can cause sunburn and eye problems

- Potassium ferricyanide, which, upon decomposition, may release highly toxic hydrogen cyanide gas

Thus, avoid carbon arc lamps and substitute photofloods, quartz/halogen lamps, sunlamps, or sunlight. If you must use carbon arc lamps, use an overhead canopy hood exhausted to the outside. Protect your eyes (and your subject's) from intense ultraviolet light by wearing NIOSH-approved welding goggles with shade #14.

**Quartz and halide lamps** are relatively safe, but one must avoid skin contact with quartz bulbs. Skin oils and acids may cause a heated bulb to explode (Shaw 1983). Handle quartz and all high-powered light bulbs with clean gloves or a dry, clean cloth.

**High-intensity photoflood,** flash, and stroboscopic lights can cause chronic eyestrain known as aesthenopia, associated with eye discomfort and headaches. Photographers and models should protect eyes from direct exposure to strong studio lighting by wearing polarized sunglasses or visors or by using shields that redirect or scatter light rays.

Minimize **electrical and fire hazards** associated with studio lighting. Be certain that wiring is in good condition and that electrical equipment is properly grounded by a three-pronged plug and outlet. If an adapter is used, make sure the ground wire is attached to a dependable ground. Do not overload circuits, and do not work in the presence of excessive moisture or water on the floor.

Darkrooms present a different lighting challenge, for darkness reduces visibility and increases the possibility of accidents, eyestrain, and chronic eye discomfort. Most safelights have either coated bulbs or filters or sleeves placed over regular incandescent or fluorescent bulbs. For optimum visibility, use the brightest safelights that procedures allow. Place lights six feet away from work areas; enhance light dispersal by painting the ceiling and walls white.

The presence in the darkroom of both electricity and water increases **electric shock hazard.** Keep exposure areas and electrical enlargers dry; separate them from processing, developing, and other wet areas. Establish a corridor between enlargers, paper cutters, and contact printers along one wall, and developing trays along the opposite wall. Install handy foot switches to operate enlargers safely and efficiently. Wear rubber-soled shoes and use nonskid, shock-resistant floor runners for additional safety.

All electric devices and outlets used in the darkroom should have a ground-fault circuit interceptor (GFCI) in the electric outlet. GFCIs are wall outlets with a circuit breaker that pops open when short-circuiting develops.

> I was told that a lab tech[nician] "would have been fried" if it were not for the interceptor. Apparently a tank heater shorted out and the lab tech got a mild shock before the breaker popped. But without the interceptor he could have been killed.
>
> —Jim Mone, Associated Press Photographer (Tell 1988)

When used as directed, insulated processing drums, "phototherm" water jacket trays, and trickle-flow water jackets are also designed to prevent electric shock in the darkroom.

Figure 9.3.—Jobo Autolab ATL-3. This automatic photoprocessor can handle most standard film and print processes, including E-6, C-41, black-and-white, EP-2, Cibachrome, microfiche, lith, and x-ray film. Virtually any film format from disk up to 16″ × 20″ sheet film, roll film lengths up to 25 feet (8 meters), and prints as large as 20″ × 24″ can be processed in an ATL-3. With 15-liter chemical containers, up to 240 rolls of 35mm film can be processed without refilling. One loads the tank or drum, puts it on the Autolab, presses the start button, and film or prints are processed as directed. A local exhaust device should be used with the processor. (Courtesy Jobo Fototechnic, Inc.)

Figure 9.4.—Jobo color processor CPP-2 is the economical alternative to the ATL-3 automatic processor, for it does not have a microprocessor for automatic processing. The user pours in and drains chemicals for each step in the process, while the unit provides electronically controlled temperature, bidirectional rotation with variable rotation spreads, and fast, even wetting of the film or paper with the Jobo-Lift. A local exhaust device should also be used. (Courtesy Jobo Fototechnic, Inc.)

In photography, various alternatives can be used to reduce exposure to potential hazards. For example, **instant processes** are safer than the standard color processes discussed later. In 1947, Edwin H. Land produced Polaroid, the first instant camera and film, where processing of negative and positive images "is accomplished in a single step by incorporating a photosensitive emulsion, an image-receiving system, and a minute amount of totally enclosed viscous processing reagent into a film assembly" (Sturge 1977, p. 259) (color plates 9.1–9.2). Early black-and-white and later color versions required a two-step process in which coating was applied to preserve the print. Most films today are in color and precoated, which renders them hazard-free if the mylar envelope (containing dyes and caustic chemicals) remains intact. If the envelope is broken or stripped away, exposure to developer/reagent (highly caustic potassium hydroxide) may irritate the skin and eyes. Repeated exposure may induce allergic responses.

**Video still-cameras** do not require a darkroom or photochemicals, and are an excellent, safe alternative.

## AUTOMATED PHOTO PROCESSING

Increased use of automatic **photographic processing units,** which accommodate black-and-white as well as color film and paper processes, has greatly reduced exposure to potentially hazardous agents (color plates 9.3–9.4). The artist must, however, premix the photochemicals. Buy concentrates instead of powders. Use exhaust ventilation and wear protective goggles, gloves, and respirator with acid gas cartridge to mix chemicals according to manufacturer's instructions, taking precautions recommended in chapter 4 ("Safe Handling of Toxic Substances").

Use containers with both "tight-fitting" plastic lids and "floating" lids to decrease evaporation and oxidation, respectively. Transfer photochemicals to the processing unit according to the manufacturer's directions. Place rolls of film (or an exposed photograph) into the special drum attached to the processing unit; process according to directions.

An automated, computer-controlled system maintains appropriate temperatures, agitation, and sequence of developers, stop baths, and rinses. With fewer variables than the typical darkroom, automated systems can produce fully professional films, transparencies, or prints. Moreover, the fully enclosed process reduces exposure to chemicals and vapors. Obviously, these contained processing units do not permit one to visually participate in the developing process or to make creative decisions between exposure and finished print. Thus, many photographers prefer traditional darkroom creative engagement. Manual systems are cheaper (several hundred dollars) than automation (thousands of dollars), but they require timing of each step, individual setting of water temperatures, and greater likelihood of undesirable chemical exposure.

## DARKROOM SAFETY AND VENTILATION

Many photographers maintain darkrooms at home, most often in bathrooms or kitchens. In general, hazards in home photography can be avoided with adequate space and effective ventilation.

General dilution ventilation is usually adequate for hobbyists who undertake black-and-white processing. Greater hazards associated with toning, intensifying, and color processing, however, require installation of local exhaust systems.

If small children are present, or if living and working space cannot be separated, the home is *not* an appropriate site for a darkroom. We suggest photographers share the costs of a safe workplace located elsewhere.

The safe studio:

**1.** Maintains effective ventilation

**2.** Maintains a source of accessible, running water

**3.** Observes local regulations governing air pollution and disposal of potentially hazardous waste

**4.** Maintains workspace sufficient to separate wet and dry areas

**5.** Has walls covered with acid-resistant, water-based paint (available from industrial suppliers)

Because the primary route of entry of photochemicals into the body is through inhalation of vapors and gases, appropriate and adequate **ventilation** is the key factor to safety in photography. Its fundamental importance cannot be overstated. In addition to learning the health risks associated with photochemical exposure, photographers must also learn to identify poor ventilation so that it may be corrected.

Figure 9.5.—**D**avid Lebe: *Scribble # 15*, 1987, gelatin silver print, 46″ × 30″. I have installed a light-tight exhaust fan in a window at the end of the sink where I place my chemicals. At the opposite end there is a light-tight vent that provides fresh replacement air over the chemicals.

Appropriate darkroom ventilation incorporates both general dilution and local exhaust to exchange air at an acceptable rate per hour. For general dilution ventilation, which circulates fresh air through the darkroom, Kodak recommends ten changes of air per hour (or 170 cfm). Effective local ventilation should trap contaminants close to their source and exhaust them where they will not contaminate other people. For example, a window exhaust fan should not vent near windows that may draw contaminants back into a building. Rooftop exhaust blowers are preferred. They should prevent entry of rain and snow and have a stack high enough (eight to ten feet) to release vapors and fumes above the heads of anyone working on that roof.

**Figure 9.6.—Joseph Nettis:** *Spanish Woman*, 1957, gelatin silver print, 8″ × 10″. Being concerned about ventilation, I installed an exhaust fan and two fresh-air intake vents in my darkroom. I don't use toners or other exotica. I use tongs to transfer prints while developing and have replaced acetic acid with water.

One can calculate the size of an exhaust fan needed for a darkroom. Multiply room height, length, and width to obtain volume in cubic feet. Then multiply cubic feet by 15 (changes per hour) and divide by 60 (minutes). The resulting number, cubic feet of air per minute, is the standard by which fans are rated in their ability to move air.

$$\frac{H \times W \times L \times 15}{60} = \text{cubic feet of air per minute}$$

For example, a room that is 10 by 12 and 8 feet high requires a fan that draws 240 cubic feet of air per minute:

$$\frac{8 \times 10 \times 12 \times 15}{60} = 240 \text{ cfa/m}$$

The intake vent, which permits "clean" air to enter a space, should be mounted opposite the location of the exhaust fan or vents. It should be large enough to permit air flow that corresponds to the draw of the fan.

**Hooded ventilation systems** for darkrooms are available, but photographers often lean over trays with their heads under the hood. In this position, the fan in the hood draws concentrated vapors directly into one's personal intake area. Plexiglas shields attached to the hood provide better protection. Slots at the bottom allow one to reach into the processing trays, while permitting air inflow at a level well below one's personal intake. Ventilation hoods made for kitchens are not adequate for darkroom use. In general, ceiling-mounted hoods without base shields or lateral slots provide safe, adequate ventilation in the darkroom only if photographers and artists keep their heads out of the path of the vapors drawn into the hood.

A **lateral slot ventilation** hood reduces personal intake by placing the hood in front of, rather than over, the user. Horizontal slots draw contaminated air simultaneously from work and personal intake areas. Mounted behind and slightly above developing trays, with appropriately spaced vents that draw vapors from each tray, a lateral slot system draws clean air across the processing trays (away from one's face) while removing unwanted vapors, gases, and dusts from their sources. It also reduces the concentration of airborne contaminants that would otherwise be recirculated by the general dilution system.

Every commercial or institutional darkroom should have a lateral slot exhaust system in operation.

Amateur photographers, who occasionally use a home darkroom for standard black-and-white processes only, can probably work safely with general dilution exhaust or informal cross-ventilation from an intake vent to an exhaust fan. For greater safety, however, it is possible to fashion a duct ventilation system inexpensively using a flexible exhaust duct of the type used to vent laundry dryers. The flexible duct, long enough to extend from a window- or wall-mounted exhaust fan past the developing trays to the end of the developing process, is attached with duct tape to the fan and sealed with tape at the other end. It is then mounted slightly behind and above the developing trays. Holes or slits are cut into the flexible duct according to the width of each tray. If the duct jerks around when the fan is turned on, more holes are needed to allow adequate flow of air demanded by the fan, which should be large enough to provide a minimum of 10, preferably 15, air changes per hour. This can happen only if there is also an intake vent, opposite the fan and duct, that is large enough to allow clean air to flow into the darkroom commensurate with the pull of the fan.

To prevent gas or vapor accumulation, periodically check and clean both intake and outtake vents of all systems to eliminate dust and debris. To assure the absence of accumulation of unwanted substances, periodically measure concentrations of airborne chemicals in darkrooms. (Call a reputable industrial hygienist.) With appropriate ventilation in place, one can really get serious about doing photography safely.

## PERSONAL SAFETY

During photoprocessing, the primary goals are to avoid skin contact with chemicals and to avoid inhalation of vapors and dusts.

- Wear a protective apron, smock, or lab coat in the darkroom, and leave it there to avoid contamination of other work and living spaces. Launder work clothes and towels separately and frequently.

- If ventilation is not adequate, use a NIOSH-approved respirator in the presence of acids, gases, ammonia, formaldehyde, dust, paint mist, or organic chemicals.

  **a.** Use a NIOSH-approved toxic dust respirator when mixing or handling powdered photochemicals.

  **b.** Use goggles and respirator with appropriate cartridge if working with materials (e.g., ammonia) that emit irritating gases and/or vapors (see also chapter 4, "Eye, Face, and Respiratory Protection").

- Wear protective gloves when mixing or handling photochemicals:

  **a.** For acids: neoprene, Buna-N

  **b.** For alkalis: rubber, neoprene, Buna-N, butyl rubber

  **c.** For organic solvents:

    - Alcohols: rubber, neoprene, butyl rubber, polyvinyl chloride (PVC), nitrile

    - Aromatic hydrocarbons: nitrile, polyethylene

    - Chlorinated hydrocarbons: polyvinyl alcohol (PVA), nitrile, polyethylene

    - Ketones: butyl rubber, natural rubber, neoprene, polyethylene

    - Lacquer thinner (aromatic hydrocarbons, alcohols, petroleum distillates): polyvinyl alcohol (PVA)

    - Paint thinner (petroleum distillates, aromatic hydrocarbons, alcohols): nitrile, polyethylene

    - Petroleum distillates (kerosene, mineral spirits, naphtha); nitrile, natural rubber, neoprene, Buna-N, and polyethylene

    - Turpentine: Buna-N, PVA, and nitrile

  **d.** Photochemicals: Thymol: neoprene, latex/neoprene, butyl rubber

  **e.** Formaldehyde: natural rubber, neoprene, Buna-N, butyl, PVC, polyethylene

In general, wear the heaviest, longest, and best-lined gloves that enable you to work effectively. Keep gloves clean, especially on the inside, by washing whenever necessary. Keep a second pair of clean gloves to use when others become damaged or contaminated. The following regimen for daily cleaning of gloves worn during photoprocessing is recommended:

**1.** Soak for 30 minutes in a strong solution of laundry detergent in hot (120°F) water.

**2.** Rinse thoroughly with hot water and test for small punctures by filling the glove, holding it closed at the wrist, and squeezing. Water will squirt from any holes that may be invisible but present and permeable. Discard gloves if warranted.

**3.** If intact, hang to dry and reuse.

Wear gloves and use separate tongs in each bath to transfer prints along the developing process. Barrier hand creams may be used in addition to gloves, but creams alone are not appropriate because none are impervious to all photochemicals. Used alone, creams may also contaminate solutions and leave deposits on film and photographs.

Observe good personal hygiene (see chapter 4). Do not smoke, eat, or drink in the darkroom. After working with chemicals, even with gloves, wash hands thoroughly during breaks and before eating. Use soap, preferably pH 5.10 to 5.5; then apply a soothing hand lotion or cream.

To avoid exposure to potentially hazardous powders and concentrated solutions, purchase and use the least toxic premixed photochemical solutions available. Read labels completely; follow directions precisely. Many photochemical solutions are diluted from more concentrated liquid or dry forms of the chemical. Whether liquid or dry, concentrates pose significantly greater hazards than dilute solutions. If you must handle concentrates, take meticulous precautions. Liquid concentrates can splash onto skin or into the face and eyes. Always pour liquid concentrates, especially acids and alkalis, into water. **Never** pour water into the concentrates, for doing so may cause sudden exothermic (heat-generating, potentially explosive) reactions. Work slowly to prevent splashing.

Dry, powdered chemicals pose a hazard because dust can be inhaled. Some powders can strongly irritate the skin. If it is not possible to use premixed solutions, ladle powders gently in a ventilated but nonbreezy environment to avoid breathing dust clouds. Work under a fume hood or with a glove box, or wear a toxic dust respirator.

When labels do not explain safe handling procedures, obtain expert advice and review Material Safety Data Sheets.

OSHA requires eye wash stations and showers wherever chemicals are mixed in commercial darkrooms. This precaution is appropriate because acid-containing solutions, for example, may cause temporary blindness. Immediate flushing of the eyes for 20 minutes is essential emergency treatment for any chemical splash. In the home darkroom, have a flexible spray attachment on the sink so that eyes, if contaminated, can be flushed with continuously running water. After rinsing, always seek medical attention. If powders or solutions splash on skin, rinse immediately and continuously, also for 20 minutes.

Keep a first aid kit in the darkroom and studio.

## SAFE DISPOSAL AND STORAGE OF PHOTOCHEMICALS

Take stock of the photochemicals you have on hand. **Avoid** all suspected carcinogens, including:

- Ammonium dichromate powder
- Benzotriazole
- Chlorinated hydrocarbons
- Chromic acid
- Formaldehyde, formalin
- Hydroquinone
- 6-Nitrobenzimidazole
- p-Methylaminophenol
- p-Phenylenediamine
- Potassium dichromate powder
- Pyrogallol
- Selenium
- Sodium chloropalladite
- Thiourea
- Trichloroethylene

If you find any of these ingredients in your photochemicals, dispose of them safely (see later in this chapter and chapter 3, "Safe Disposal of Dangerous Materials").

Second, dispose of any chemicals that contain heavy metals such as arsenic, cadmium, lead, or mercury.

Always be prepared to clean up **spills** promptly and appropriately. *Prior* to using chemicals, read Material Safety Data Sheet, section VII, which provides information on equipment and procedures for cleanup, so you can have appropriate equipment available. Keep MSDSs in a specific place for dependable, accessible reference.

Poor-quality negatives or prints should not be immediately tossed in the trash. They should be fixed and rinsed thoroughly before disposal. This will prevent spread of undesirable chemicals beyond the immediate environment.

**Safe Disposal** Following printing sessions, **discard** solutions one at a time to avoid chemical reactions that generate chlorine or sulfur dioxide gas. If frequent use requires solutions to be maintained in processing trays, cover the trays when not in use to prevent gaseous by-products (e.g., sulfur dioxide) from diffusing throughout the darkroom. **Dispose** of photochemicals with a frequency relevant to the volume of work. Small-scale processing that requires one-half to one gallon of each chemical may require disposal of chemicals only once or twice a week. If local regulations permit, flush each solution separately down a drain with generous amounts of cool running water.[1] Run the water for a few minutes after each chemical has been discarded to assure thorough dilution and drainage. Larger quantities of photochemicals may cause problems for the local waste treatment system, so check with the local department for appropriate disposal procedures. If you have a septic tank, do not fill it with bleaches, cyanide compounds, or color photochemicals, which can disturb the breakdown process in the tank. Never dump photochemicals into the environment. For further information, call Kodak at 800-242-2424 for publication J-4, "Safe Handling of Photographic Chemicals". Local laws may require you to specifically package wastes for pickup by the local waste management authority or a private toxic waste removal firm.

**Safe Storage** Store unused chemicals in a cool, dry, well-ventilated place in original, labeled containers. Keep them tightly closed and near the mixing area. (Do not store chemicals in glass bottles, which may explode under pressure.) Corrosives should be kept on low shelves to prevent breakage and splashes at eye level. Do not keep photochemicals beyond expiration dates on labels or Material Safety Data Sheets. If a label is defaced, illegible, or absent, and you are not sure what is in the container, do not use it, but discard it safely. When adding more than one agent into another receptacle, label it legibly with name, date, and concentration. Unopened chemicals usually have a longer shelf life than those that have been exposed to air, and concentrates usually last longer than diluted solutions. When children are present, store photo materials in safety cabinets (see chapter 3).

If you must **transport** photochemicals, use the original containers, with care to avoid spillage and breakage.

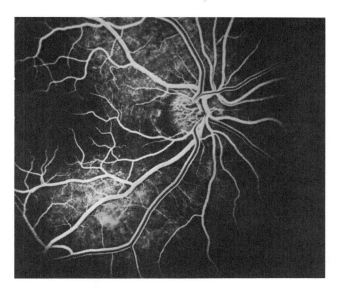

**Figure 9.7.—M**erle Spandorfer: *Tapetum*, 1973, gelatin silver print on canvas, 61" × 46". The image was derived from a fluorescein angiogram of my own eye, taken at Wills Eye Hospital, Philadelphia. A professional photo lab printed the image on a commercially photosensitized canvas.

The precautions and safe procedures recommended here may be difficult or impossible for some individuals to undertake. Yet another way to avoid potential hazards in photography is to rely on a **professional photographic processing laboratory.** People working in custom developing and printing labs are usually photographers themselves, and are responsive to working with artists and their specifications to achieve the best possible print.

Associated Press photographer Judy Tell, editor of *Making Darkrooms Saferooms*, shares the conviction that safety in photography need not be expensive or involve compromise of quality. Use of safe procedures and avoidance of hazardous chemicals can maintain health and perhaps lengthen or improve one's creative years.

## Photographic Processes

**T**he following sections, "Color," "Black-and-White," "Historic" and "Non-Silver Processes", identify those chemicals and procedures (unique to each process) that clearly require safeguards. When possible, we suggest and describe safer alternatives. Recall that this is not a how-to book; we presume our readers know the basic procedures for making photographs. (For information on the toxicity and hazards of specific photochemicals, refer to Appendix I.)

These sections are followed by a passage on **xerography,** which, along with the computer graphics discussed in Chapter 8, offers safe methods for creating and reproducing images.

## COLOR

As visual communication has radically expanded and proliferated since 1950 in both electronic and print media, there has been a corresponding increased use of color photography and processes (color plates 9.5–9.8). In this period of growth, significant adverse health effects have been associated with some of the chemicals used in color photography. Associated Press photographer Charles Krupa provides an example. When his bureau began shooting all photographs in color, Krupa was exposed daily to color-processing chemicals. He wore surgeon's latex gloves to protect his hands, but the gloves were permeable and actually aggravated exposure of his skin. He developed an incapacitating dermatitis with extreme swelling and painful cracking of the skin, requiring topical and oral steroids for relief. Other photographers using color chemistry have also reported skin rashes (Tell/Krupa 1988).

Color processing requires a developer that couples dyes in layers of silver particles suspended in gelatin. Images are registered in both color and silver. The silver halides are reduced or bleached out, and the image fixed. The complex organic chemicals required for dye coupling in the development process have been recognized as causes of allergic contact dermatitis in up to 25 percent of unprotected photography workers. The incidence, moreover, increases with duration and frequency of exposure (Tell 1988). Lichen planus, characterized by itchy raised purple spots on the skin, has also been associated with exposure to several color developers. Thus, especially in color photography, unprotected hands should never be placed in the baths, and splashes on skin should be immediately washed with water. Splashes in the eyes should be rinsed continuously for 20 minutes and medical attention sought. To emphasize precautions:

Wear goggles and rubber or neoprene safety gloves while mixing solutions.

■■

Use a glove box, fume hood, and/or an approved respirator when mixing powders into solutions.

■■

Figure 9.8.–**A**rthur Salazar: *Land of Plenty*, 1990, photomontage, 24″ × 36″. My black-and-white photographs are commercially processed in multiple prints. I then cut these prints and reassemble them to create a new photograph which is drymounted with heat onto a larger format backing.

Add concentrates to water, never water to concentrates.

■■

Use tongs (a separate pair for each tray) to transfer photographs through the developing process.

■■

Work in a cool darkroom operated with lateral slot local ventilation and general dilution exhaust ventilation.

■■

Clean up spills promptly to prevent electric shock and fire hazards. Color processes require precise water temperatures controlled by electric heating devices. For electrical devices used in or near water, maintain ground-fault interceptors in all relevant wall outlets.

■■

Immediately flush eyes for 20 minutes if chemicals are spilled or splashed into them.

■■

Know when and how to reach emergency medical attention.

Sulfur dioxide gas is an agent of concern in color processing. Sulfur dioxide is produced when sulfites from developer mix with acid from the bleach, or when acid from the bleach contacts sulfites in the fixer. Inadvertent mixing can occur when prints are transferred from one step to another, or when developer, bleach, and fixer are mixed during disposal. Sulfur dioxide gas is a lung irritant that may contribute to chronic lung problems (McCann, *AHN* 4:1). To prevent significant inhalation of sulfur dioxide, insert a water wash step between developer and bleach and between bleach and fixer. In addition, dilute developer, bleach, and fixer with water before disposal in the sink (if allowed by local, state, and federal regulations).

**Developers** for both negative and color reversal processing are complex solutions that contain more agents than black-and-white developers. Chemicals that color, couple, accelerate, restrain, facilitate penetration, prevent fogging, control contrasts, and preserve the solution are likely to be present. Color developers may contain: phenidone, hydroquinone, phenylenediamine, benzyl alcohol, citrazinic acid, ethylene diamine, ethylene glycol, hydroxylamine hydrochloride, hydroxylamine sulfate, potassium bromide, potassium carbonate, potassium iodide, potassium thiocyanate, sodium thiocyanate, sodium carbonate, sodium hydroxide, and TBAB (tertiary-butylamine borane). (Check Appendix I for details on toxicity, health effects, and precautions with these substances.)

Skin and respiratory irritation and allergy are commonly associated with exposure to color developers. Both inhalation of and skin contact with powders or concentrated solutions pose the greatest hazard. In addition, most color developers are moderately toxic by ingestion. **Tasting, therefore, is not an appropriate way to identify these solutions.** Whenever possible, avoid catechin, pyrogallol, and paraphenylenediamine compounds. Less toxic phenidone is preferred.

Cibachrome, one of the first color processes available outside photo laboratories, achieved widespread use before its health hazards were acknowledged. The Cibachrome developer 1A is alkaline and potentially corrosive to skin, eyes, and mucous membranes. Cibachrome developer 1B contains ethoxydiglycol and hydroquinone, which are toxic by ingestion and skin contact. Developers 2A, 2B, and 2C also pose hazards. Avoid use of Cibachrome developers without effective local and dilution ventilation; handle and mix these products with care. Many photographers avoid exposure by using developing tanks.

Proper use of developers requires strict compliance with label instructions. When handling chemicals and mixing solutions, use a glove box or wear an impermeable apron, goggles, elbow-length rubber or neoprene gloves, and an approved respirator with organic vapor/acid gas cartridge. During development, wear gloves, use tongs, and introduce a **water rinse** between developer and subsequent acid baths. Pregnant women, as well as men and women planning a family, should avoid all exposure to color developers.

**Bleaches** (to remove silver deposits) typically contain potassium ferricyanide, ammonium or potassium bromide, and potassium or sodium thiocyanate. Color bleaches may contain potassium ferricyanide, potassium alum, or other salts used as buffers and organic solvents as accelerators. Ethylene diamine, a contrast agent, may also be included. Cibachrome bleaches may also contain acids and 2-ethoxyethanol, an agent associated with adverse reproductive effects. Local exhaust ventilation, preferably a lateral slot exhaust system, is recommended. Use elbow length safety gloves and goggles to protect against splashes.

**Toning** is a process that gives slightly different colors to photographs. Toners for color photography are the same as for black-and-white photoprocessing, where silver deposits are replaced by other metals or metal compounds that may present significant hazards.

Available, effective toners include cobalt chloride, gold chloride, iron compounds, lead salts (acetate, nitrate, and oxalate), mercuric chloride, palladium chloride, platinum chloride, selenium sulfides, uranium nitrate, and vanadium chloride.

Toning can be accomplished by immersion in a single solution toner or by multistep bleach and redevelopment. Toners that require dilute acid baths between bleach and developer may release irritating sulfur dioxide gas from sulfur compounds such as hypo-sulfide and sodium thiosulfate. Gold toners that require heating may also emit sulfur dioxide. Other toner ingredients, in both liquid and gaseous forms, can also irritate and occasionally sensitize skin and mucous membranes. These include ammonium persulfate, copper sulfate, gold chloride, palladium chloride, platinum chloride, selenium oxide, and vanadium chloride.

Do not use selenium toners, which are prone to decomposition. They may emit toxic gases such as hydrogen selenide, ammonia, and sulfur dioxide. Do not use toners that contain lead, mercury, or uranium. Replace them with less hazardous compounds: copper sulfate, gold chloride, iron salts, platinum chloride, and vanadium chloride. Use sulfide toners only with effective local exhaust ventilation. Avoid toners that require heating.

Wash prints thoroughly after the fixing step (prior to toning) to prevent contamination of bleach and toning baths. Similarly, if intermediate acid baths are employed, rinse prints thoroughly (to remove acid) before transferring them to toner solutions.

To protect skin, eyes, and respiratory system from contact with toner powders and solutions, use pre-mixed solutions whenever possible. Otherwise, observe personal and darkroom precautions for mixing solutions as outlined above.

**Stabilizing baths** usually contain a weak solution of formalin (a suspected carcinogen), ethylene glycol (a wetting agent), and fluorescent dyes as brighteners.

**Hardener** stop bath, employed after the first or second developer, contains potassium chrome alum, 6-nitrobenzimidazole nitrate (restrainer), 2,5-dimethoxytetrahydrofuran (solvent), and formalin or succinaldehyde (preservative). All these agents have been associated with adverse skin or respiratory effects.

**Neutralizers** may contain hydroxylamine sulfate, potassium bromide, glacial acetic acid, preservatives, and buffers.

The hazards of bleaches, stabilizers, hardeners, and neutralizers are significant and similar to those of developers. In specific processes, Ektachrome requires an acid stop bath that, when contaminated with developer solutions, may produce sulfur dioxide gas. Cibachrome processing likewise produces sulfur dioxide gas along the developing path, even when bleaches are heated. Potassium ferricyanide (Farmer's reducer) and thiocyanates, which are low hazard if used correctly, may release highly toxic hydrogen cyanide gas if heated, contaminated with acid, or exposed to ultraviolet light.

When using bleaches, stabilizers, hardeners, and neutralizers, employ all safeguards relevant to developers. Use water rinses between steps to prevent the formation of toxic gases. Cover bath trays between printing sessions.

Photographers employ **retouching, coloring,** and **dye transfer** to correct or alter negatives and prints and to make prints from color-separated negatives.

These procedures involve controlled application of dyes and pigments, bound with lacquer varnishes and thinned with solvents, to lighten or darken areas of the print. Retouching medium, or dope, is a special varnish used to prepare the gelatin surface of the negative. Dope is brushed, wiped, or sprayed on with an air brush. It contains natural gums (arabic, damar, or tragacanth) combined with solvents (juniper oil, turpentine, toluene, or xylene).

Layers of binding or reticulating lacquers may be applied to the photo surface. Lacquers contain natural or synthetic resins suspended in solvents such as turpentine, mineral spirits, toluene, or xylene. When applied with an airbrush or aerosol spray, lacquers can be inhaled. Effective ventilation and proper use of a respirator protective gear can minimize potential health hazards (see "Drawing and Painting Materials" in chapter 8).

Retouching by brush or airbrush often employs water-soluble azo, quinone, and/or phthalocyanine dyes available in liquid or powder. One-step premixed retouching fluids combine these dyes with pigments in lacquer and solvent mixtures as noted previously. Water-based retouching fluids are preferred over lacquer- or oil-based products. Retouching products may also contain ethylene glycol, iodine, methyl alcohol (methanol), polyether alcohol, sodium thiosulfate, thiourea, trichloroethylene, and dilute acetic acid, which may accelerate dye penetration.

Removal of excess dyes may be accomplished with polyether alcohol, ethylene glycol, 14 percent ammonia solution, or a dye bleach containing potassium permanganate and sulfuric acid. Significant health hazards are associated with all of these products (See Appendix I).

In dye transfer, the photographer makes color-separated negatives or matrices using strongly alkaline hydroquinone/amine developers. Prepared separation positives are soaked in solutions of cyan, magenta, and yellow dyes, and each matrix absorbs dye in proportion to the thickness of the gelatin deposits. Dye absorption is followed by two dilute acetic acid rinses. Each dye is then applied to (printed on) the receiving paper, sometimes with application of heat as well as pressure. The matrices are then rinsed in a clearing bath of dilute ammonia before the procedure is repeated.

Both the substances and procedures of color retouching and dye transfer pose significant health hazards. Avoid products containing chlorinated hydrocarbons, thiourea, or trichloroethylene (suspected carcinogens). To avoid inhalation of dye, pigment, and resin powders, use a fume hood, glove box, or dust cartridge respirator. Whenever possible, substitute premixed solutions. Avoid highly toxic lead, manganese, and cadmium pigments; do not use paints containing lead and manganese driers.

To decrease exposure to solvents during retouching or dye transfer, assure effective ventilation (or work in a spray booth), eliminate excessive heat or flame, wear appropriate gloves, and wear a mist/paint prefiltered, organic vapor cartridge respirator. To avoid inhalation or ingestion of dye powders, do not breathe onto dye cakes to moisten them for application. Do not point brushes with lips. Keep dye transfer baths covered whenever possible, and dispose of them carefully and properly.

**Negative and print cleaners** should be avoided because they offer significant health hazards and reduce the life span of the film. Cleaners contain solvents such as methyl chloroform, dichlorodifluoromethane (Freon 12) and trichlorofluoromethane (Freon 11). The fluorocarbons (Freons) should be used only with effective dilution ventilation. Methyl chloroform, a flammable chlorinated hydrocarbon, should be used only with appropriate gloves and either local exhaust ventilation or an organic vapor cartridge respirator.

**Tray cleaners** usually contain concentrated alkalis or acids, which are corrosive to skin, eyes, mucous membranes, and the digestive system. Repeated inhalation of vapors can aggravate chronic lung conditions. Ingredients include hydrochloric acid, nitric acid, sulfuric acid, potassium dichromate (a suspected carcinogen), potassium permanganate, sodium hydroxide, and sodium sulfite.

Use tray cleaners slowly and carefully, with good ventilation. Wear goggles, elbow-length rubber or neoprene gloves, and an acid-proof apron.

# BLACK-AND-WHITE

Black-and-white photography has become highly refined and accessible to amateurs as well as professionals (color plates 9.9–9.10). It is often presumed to be a safe image-making procedure and is even taught in some elementary schools. There are, however, potentially significant health hazards during handling of concentrated powders and solutions and throughout the developing process.

To reiterate precautions:

Wear goggles and rubber or neoprene safety gloves while mixing solutions.

Use a glove box, fume hood, and/or an approved respirator when mixing powders into solutions.

Add concentrates to water, never water to concentrates.

Figure 9.9.—**S**tephen Perloff: *Welcoming to Freedom*, 1989, gelatin silver print, 12½" × 16". I work with basic black-and-white chemistry, transferring prints with tongs and letting them drain thoroughly before insertion into the next bath.

Use tongs (a separate pair for each tray) to transfer photographs through the developing process.

∎∎

Work in a cool darkroom operated with lateral slot local ventilation and/or general dilution exhaust ventilation.

∎∎

Clean up spills promptly to prevent electric shock and fire hazards.

∎∎

Immediately flush eyes for 20 minutes if chemicals are spilled or splashed into them.

∎∎

Know when and how to reach emergency medical attention.

**Basic Procedures** Print and film **developers** contain a formidable array of complex chemicals. Common ingredients include hydroquinone, mono-methyl-para-aminophenol sulfate (metol, elon), phenidone, and other agents used for special effects. Developing solutions also contain sodium sulfite as preservative, potassium bromide as a defogging or restraining agent, and toxic alkali (borax or sodium carbonate) as accelerators. Potentially hazardous agents in developers, however, are not limited to these. For all developers, we recommend that you identify ingredients from the product label or Material Safety Data Sheet. Assess toxicity and health hazards in Appendix I, "Art Material Chemicals," and observe all appropriate precautions and substitutions.

Skin and respiratory irritation and allergy are commonly associated with exposure to black-and-white developers. Both inhalation and skin contact with powders or concentrated solutions pose significant hazards. In addition, most color developers are moderately toxic by ingestion. Tasting is not an appropriate way to identify these solutions. Whenever possible, avoid catechin, pyrogallol, and para-phenylenediamine compounds. Less toxic phenidone is preferred.

**Stop baths,** which halt the developing process, most often contain potassium chrome alum (a hardening agent) and sodium sulfate (an antiswelling agent) in dilute solutions of acetic acid, which may irritate skin, eyes, and respiratory passages. Continued inhalation of acid vapors may cause chronic bronchitis. Therefore, insert a water rinse between the developer and stop bath. Clean up spills immediately, always cover acid baths between sessions, assure adequate ventilation (local exhaust and general dilution), wear gloves and goggles when handling concentrated solutions (e.g., acetic acid, alum), always add acids to water (never the reverse),

and store all chemicals on low shelves to prevent accidental spillage. Remember that **water** is a safe, useful alternative to traditional stop baths.

**Fixing baths** or **hypo** contain sodium thiosulfate or ammonium thiosulfate as silver halide solvents, acetic acid as neutralizer, and sodium sulfite as preservative. Acid-hardening hypos may also contain potassium aluminum sulfate (alum) and boric acid as buffer and antisludging agent, respectively.

Again, sulfur dioxide release can be avoided by inserting a water rinse between stop bath and hypo. After use, discard hypo solutions since they will decompose or contaminate other baths. By contrast, **hypo clear** and **washing solutions,** which contain sodium chloride (ordinary table salt), quickly and safely remove hypo chemicals from negatives and prints. Always prepare fresh hypo clear and washing solutions for each new developing session.

**Additional Procedures** Hardening preserves film negatives before (prehardening) or after development, but prior to intensification or reduction. Do not use formalin (a solution of formaldehyde, methanol, and sodium carbonate). Substitute safer succinaldehyde hardeners, avoiding inhalation of powder and skin contact with solutions.

To save water and time, some photographers use **hypo eliminators** to oxidize thiosulfates in fixer and to remove undesired residual chemicals. Hypo eliminators are most often dilute solutions of potassium permanganate, persulfates, iodine, potassium perborate, ammonia, hydrogen peroxide, and sodium hypochlorite (household bleach), which can irritate and occasionally sensitize skin and mucous membranes. When heated or mixed with acids, sodium hypochlorite may form chlorine gas, a potent respiratory irritant. Furthermore, hypo eliminators with potassium compounds can **explode** on contact with some combustible and flammable substances. Use hypo eliminators only with caution and appropriate safeguards.

To confirm that hypo chemicals have been thoroughly washed from photographs, **hypo testing** can be done with silver nitrate in 28 percent acetic acid, a solution that is corrosive to eyes, skin, and mucous membranes. Use silver nitrate or acetic acid only with personal protection. Do not use highly toxic mercuric chloride for hypo testing.

Because silver ions can poison plant and animal life, **silver recovery** should be strongly considered. Although it may only be profitable for some large-scale photography studios, recovery should nevertheless be undertaken to protect the environment. Indeed, silver recovery is now required by law in some locales. Electrolytic recovery devices that are attached to a hypo filter system or are immersed in the hypo tank are available. In some systems, the cartridge that collects the silver is returned to a manufacturer who buys the silver and supplies a new cartridge. Chemical silver recovery methods are not recommended since they produce irritant sulfur oxides. Metal displacement methods using steel wool or zinc are relatively safe and less expensive but can be time-consuming. Kodak Publication J-9, "Silver Recovery," presents the options Kodak offers.

**Silver tests** employ dilute solutions of sodium sulfide to detect silver residues in frequently used fixing baths. If performed properly, these tests pose no significant health hazards. However, hypo and silver testing chemicals may also include potassium bromide, potassium permanganate, selenium powder, and sodium hydroxide. These agents require careful handling and personal protection.

**Retouching,** altering, and correcting of negatives and prints involves controlled application of dyes and pigments, bound with lacquer varnishes and thinned with solvents, to lighten or darken areas of the print. Retouching medium or dope is a special varnish used to prepare the gelatin surface of the negative. Dope is brushed, wiped, or sprayed on with an airbrush. It contains natural gums (arabic, damar, or tragacanth) combined with solvents (juniper oil, turpentine, toluene, or xylene).

Layers of binding or reticulating lacquers may be applied to the photo surface. Lacquers contain natural or synthetic resins suspended in solvents such as turpentine, mineral spirits, toluene, or xylene. When applied with an airbrush or aerosol spray, lacquers can be inhaled. Effective ventilation and proper use of a respirator can minimize potential health hazards (see "Drawing and Painting Materials" in chapter 8).

Retouching is accomplished in several ways. In spot touching, a neutral gray dye is applied to correct pinhole and dust spots in the negative. Retouching by brush or airbrush employs Crocein Scarlet dye or red and black opaque, water-soluble paints to cover broader areas. Other water-soluble azo, quinone, and/or phthalocyanine dyes (for brush or airbrush) are also available in both liquid and powder forms. For chemical retouching, Farmer's reducer (potassium ferricyanide) is used to remove dark spots. One-step premixed retouching fluids combine these dyes with pigments in lacquer and solvent mixtures as noted previously. Water-based retouching fluids are preferred over lacquer- or oil-based products. Retouching products may also contain ethylene glycol, iodine, methyl alcohol (methanol), polyether alcohol, sodium thiosulfate, thiourea, trichloroethylene, and dilute acetic acid, which may accelerate dye penetration.

Removal of excess dyes may be accomplished with polyether alcohol, ethylene glycol, 14 percent ammonia solution, or a dye bleach containing potassium permanganate and sulfuric acid. Significant health hazards are associated with all of these products (see Appendix I).

Both the substances and procedures of retouching pose significant health hazards. Avoid products containing chlorinated hydrocarbons, thiourea, or trichloroethylene (suspected carcinogens). To avoid inhalation of dye, pigment, and resin powders, use a fume hood, glove box, or dust cartridge respiratory mask. Wherever posssible, substitute premixed solutions. To decrease exposure to solvents during retouching or dye transfer, assure effective ventilation (or work in a spray booth), eliminate excessive heat or flame, wear appropriate gloves, and wear a mist/paint prefiltered, organic vapor cartridge respirator.

**Intensifiers** increase the density of a negative by adding silver to the original silver deposited on the negative (or print). Potentially hazardous metals that also may be found in intensifiers include chromium, copper, mercury, and uranium. Intensification may involve bleaching as a first step; bleaches may contain highly toxic mercuric chloride, potassium cyanide, or sodium cyanide. **Do not use** intensifiers that contain mercury, uranium, potassium cyanide, or sodium cyanide. Keep all solutions cool; observe all personal and darkroom safeguards.

**Reduction** is the opposite of intensification. Selective removal of silver deposits from a photograph relies on various combinations of potassium ferricyanide, sodium ammonium thiosulfate (Farmer's reducer), iodine, potassium cyanide, potassium permanganate, sulfuric acid, ammonium persulfate, ammonium bromide, and thiourea (a suspected carcinogen).

Some agents in reducers are associated with adverse health effects. Avoid reducers that contain highly toxic potassium cyanide; substitute potassium ferricyanide, but do not expose it to heat, acid, or ultraviolet light. Keep acids and reduction solutions separate because mixing them may produce highly toxic, potentially lethal gases. Observe all personal and darkroom safeguards when using reducers, maintain effective local exhaust and dilution ventilation, and wear protective gloves, aprons, and goggles. Control bath temperatures (cool) and keep baths covered when not in use.

**Toning** for black and white is a redeveloping process that gives slightly different shades to photographs. Many toners replace silver deposits with other metals or metal compounds. Available, effective toners include extremely toxic chemicals and some that are relatively safe if used appropriately.

**Do not use** lead acetate, lead nitrate, lead oxalate, mercuric chloride, platinum chloride, selenium sulfide, or uranium nitrate. Selenium toners may emit highly toxic gases such as hydrogen selenide, ammonia, and/or sulfur dioxide during decomposition. Instead, **substitute** cobalt chloride, gold chloride, iron compounds, or vanadium chloride.

Toning can be accomplished by immersion in a single solution toner or by multistep bleach and redevelopment. Toners that require dilute acid baths between bleach and developer may release irritating sulfur dioxide gas from sulfur compounds such as hypo-sulfide and sodium thiosulfate. Gold toners that require heating may also emit sulfur dioxide. Other toner ingredients, in both liquid and gaseous forms, can also irritate and occasionally sensitize skin and mucous membranes. These include ammonium persulfate, copper sulfate, gold chloride, palladium chloride, platinum chloride, selenium oxide, and vanadium chloride.

To avoid the significant hazards associated with emission of hydrogen sulfide gas, use sulfide compounds only with effective local exhaust ventilation. Avoid toners that require heating. Use premixed solutions if possible, and observe all personal and darkroom precautions for mixing solutions safely.

Wash prints thoroughly after the fixing step (prior to toning) to prevent contamination of bleach and toning baths. Similarly, if intermediate acid baths are employed, rinse prints thoroughly (to remove acid) before transferring them to toner solutions.

Protect skin, eyes, and respiratory system from contact with all toner powders and solutions.

**Negative, film,** and **tray cleaners** are the same for black and white as described previously for color processes.

## HISTORIC PROCESSES

In 1826, Joseph Nicéphore Niépce observed that an image cast through a pinhole *camera obscura* exposed a pewter plate coated with light-sensitive chemicals. Louis Jacques Mandé Daguerre refined this technique using silver coatings on copper plates, while William Henry Fox Talbot, working independently, invented the negative from which multiple prints could be made. Announcing their discoveries in 1839, these men galvanized research and development in photography that continues even today.

The essence of black-and-white photography is silver halide crystals suspended in gelatin applied to plastic film or paper. However, other metals, such as iron and chromium, possess light-sensitive quali-

ties that, in the presence of additional agents, make effective images.

In this section we explore some of the historic silver processes, including daguerreotype, wet-plate collodion, ambrotype and tintype, salted paper, casein paper printing, albumen, and printing-out-paper processes.

The earliest methods of **daguerreotype** used toxic vapors of iodine and heated mercury to develop the image. With photographic experimentation, safer methods were developed. The Becquerel method simplified procedures and eliminated use of mercury vapors. Nevertheless, the vapors of iodine, bromine, and chlorine, which are used as plate sensitizers and accelerators, intensely irritate eyes, skin, and the respiratory tract. Therefore, daguerreotype printing should be undertaken with awareness of the hazards, effective ventilation (especially important during heating steps), and personal safeguards including gloves, barrier apron, nonvented goggles, and respirator with acid gas cartridge.

**Wet-plate collodion** negatives result from an historic process rarely used today. A glass plate is coated with collodion, a viscous form of potassium iodide (or potassium bromide) plus cellulose nitrate (potentially explosive in air) in a solution of alcohol and ether. Before this coating dries, the plate is sensitized in a solution of silver nitrate (producing silver iodide), exposed in the camera, developed in pyrogallic acid or (safer) ferrous sulfate, and fixed with hypo or potassium cyanide. The collodion emulsion is flammable and explosive, while pyrogallic acid and potassium cyanide have undesirable health effects. Avoid the process if you can, and observe appropriate precautions if you indulge.

**Ambrotype** and **tintype,** also known as ferrotypes, are direct positive corollaries to the collodion negative. In ambrotypes, highly toxic mercuric chloride or nitric acid is added to the developer. The back of the glass is blacked out with paint, varnish, or fabric to highlight the positive image. By contrast, tintypes highlight a positive image on black lacquer-coated metal plates. Unfortunately, the toxicity of the developer and lacquer solvents make both procedures unsafe. If you must engage in them, do so with extreme caution.

**Salted paper** printing was introduced by William Henry Fox Talbot's calotype process. The paper is "salted" with a solution consisting of citric acid, sodium chloride or ammonium chloride, or both, and a sizing agent: gelatin, arrowroot starch, or resin. Bleached shellac with gelatin may also be used as sizing. Do not use bleached shellac if plain gelatin or arrowroot is available. If shellac is used, the vehicle should be denatured alcohol instead of methyl alcohol. The prepared papers are sensitized with silver solution: silver nitrate alone or in combination with citric acid, tartaric acid, ammonia, lead nitrate, or uranium nitrate. Do not use lead or uranium nitrate; use silver nitrate and avoid inhalation of dust. The sensitized paper is exposed to the image with sunlight or an artificial source of ultraviolet light, developed and fixed with hypo or potassium bromide, and then toned with gold chloride or potassium chloroplatinite.

**Casein paper printing** is a related process where salted paper is immersed in a solution of casein paint and ammonia prior to treatment with silver nitrate. For added stability, the paper can also be immersed in dilute citric acid. The prepared paper is exposed to light and developed and fixed as in salted paper printing.

Photographic papers prepared with egg white in salt solution are known as **albumen** prints. They preceded Kodak's developing-out papers of the 1880s and **printing-out papers** (P.O.P) of the 1890s, which are still available for historical procedures of contact printing and toning. The glossy albumen-prepared papers are sensitized with a solution of silver nitrate, exposed with a contact negative, and rinsed in water. They are usually toned with gold chloride, fixed in hypo plus sodium carbonate, and finally rinsed in water. Even with effective ventilation and avoidance of skin contact, albumen printing must be undertaken with caution.

The term **P.O.P., printing-out paper** arose in the 1890s to distinguish gelatin chloride P.O.P. from previously introduced developing-out papers (D.O.P.). The only gelatin chloride printing-out paper now available in the U.S. can be obtained from Chicago Albumen Works (P.O. Box 379, Front Street, Housatonic, MA 02136, 413-274-6901). This P.O.P. is sold presensitized and can be printed out, washed, toned, and fixed as in salted paper printing.

# NONSILVER PROCESSES

Iron, or ferric processes include cyanotype, palladium, platinum, kallitype, and Van Dyke printing; chromium processes include gum dichromate (bichromate), carbon/carbro, and oil/bromoil printing. In nonsilver processes, chemicals containing lead, mercury, and uranium should be avoided. Dichromate is a suspected carcinogen and should be used only with great caution and respect for its potential long-term effects. Some chemicals and procedures outlined here pose significant fire and/or explosive

Figure 9.10.—**M**artha Madigan: *Peaceable Kingdom*, 1984, gold-toned printing-out paper, 60″ × 72″. Following ten years of working with large-scale cyanotype, I experienced fatigue, sensitive nasal passages, puffy eyes, and sore throat. Nitrile gloves, a respirator, and improved ventilation were to no avail. Despite working outdoors, vapors released during coating, drying, and developing my mural-sized works may have contributed to my symptoms. Even though I was in the middle of a large installation project, I made the difficult decision to stop using cyanotype altogether. Within a few months I felt significantly better and, committed to working in nature without a camera, explored the use of printing-out paper. The single-weight paper is fragile and must be toned with gold chloride, which gives an exquisite neutral brown-purple tonal range. I wear nitrile gloves and have proper ventilation. For a mother who has nursed three babies, these health considerations are very real.

hazards, while others present health hazards by all routes of entry. One must, therefore, have ardent esthetic or historical reasons to engage in most of the following processes. If you choose one of these methods, please be a wise wizard (alchemy and sorcery reign here), and observe all personal and environmental precautions. Wear appropriate gloves, barrier aprons, and goggles; assure effective ventilation; and prevent fire and explosion.

In iron, or **ferric processes,** such as cyanotype, palladium, platinum, kallitype, and Van Dyke printing, specific light-sensitive chemicals are brushed or sprayed onto various papers in a dark environment. A contact negative is applied, and the image is exposed to strong light (sun; carbon arc, quartz, halide, mercury, or sunlamps). The print may then

Figure 9.11.—Martha Madigan: *Vinespot*, 1981—88, solar photogram with gold-toned printing-out paper, 20″ × 24″. *Vinespot* consists of images that are direct solar photograms made by exposing light-sensitive paper to nature without use of a camera. This process utilizes precoated paper, which eliminates the use of light-sensitive chemicals.

undergo a series of developing, fixing, and toning procedures. Wear appropriate gloves, barrier aprons, and goggles; assure effective ventilation; and prevent fire and explosion. Avoid use of carbon arc lamps and avoid spraying.

**Cyanotype** (cyan blue prints) utilizes a blue iron compound known as Turbell's blue (color plate 9.11). Paper is coated with ferric ammonium citrate and potassium ferricyanide. A contact image or negative is applied and exposed to light to form a pale image of ferrous ferrocyanide. The print is washed in water to remove soluble salts unreduced by exposure. Upon drying, the ferrocyanide oxidizes to deeper blue tones in a process that can be accelerated by placing the print in an oxidation bath of dilute hydrogen peroxide or potassium dichromate. Because small amounts of cyanide may be released during drying and exposure to intense ultraviolet sources, use of an acid/gas cartridge respirator is recommended, even when working outdoors. Toners include lead acetate and 5 percent ammonia, followed by a saturated solution of tannic or gallic acid.

Avoid lead acetate and carbon arc lamps. If possible, avoid potassium dichromate, but if you use it, mix powders wearing an approved respirator with toxic dust cartridge or, preferably, use a glove box.

Since their introduction in 1879, **palladium and platinum** prints have been recognized for their permanence, tonal richness, delicacy, noble presence, and, unfortunately, for their potential adverse health effects. These metals, when inhaled as powders or absorbed through the skin as liquids, have been associated with long-term health effects. Some palladium salts are suspected carcinogens; platinum salts may cause a severe form of asthma known as platinosis. Palladium and platinum procedures are similar. For palladium printing, the sensitizing solution is ferric oxalate plus sodium chloropalladite or palladium chloride. For platinum prints (platinotype), potassium chloroplatinite or platinum chloride are used. Combination sensitizing solutions contain ferric oxalate, oxalic acid, potassium chlorate, and potassium chloroplatinate (or sodium chloropalladite). These solutions are brushed on paper, a contact negative overlaid, and the image exposed to light (sun; carbon arc, quartz, or sunlamp). As a result, ferric salts are reduced to ferrous salts. The print is immersed in potassium oxalate developer, which dissolves ferrous salts and reduces palladium or platinum salts to a metallic state. Several hydrochloric acid baths remove any remaining ferric salts to produce a clarified image in the palladium or platinum deposits. During the developing process, images may be toned by heating the potassium oxalate

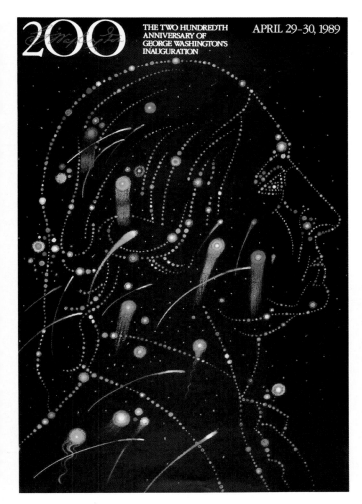

Plate 8.1.—**M**ilton Glaser: *The 200th Anniversary of George Washington's Inauguration*, poster, 1989, colored pencil on black charcoal paper, 17" × 11".

Plate 8.2.—**L**anny Sommese: *Penn State Art Graduate Program*, poster, 1990, cut paper, spray painted, 26" × 24". The image was created with cut paper, the halo spray painted on acetate using a spray booth and a mask. Type was adhered with a waxer, and the posterprinted by offset lithography.

**Plate 8.3.—L**anny Sommese: *Central Pennsylvania Festival of the Arts*, poster, 1983, 36" × 24". The original drawing was pen and ink, spray painted on acetate using a spray booth and a mask. Camera-ready art overlays on boards were printed in three-color offset lithography.

Design in Texas ★ A Retrospective

AIGA ★ Philadelphia

**Plate 8.4.—F**rank Baseman: *Design in Texas*, poster, 1987, water-soluble ink, color photography, 36" × 24". The Indians and arrows were drawn with water-soluble Koh-I-Noor rapidograph ink. Photostats were made and positioned on a mechanical board with wax. Photocopies were used to position photographs on the mechanical boards; final poster was printed with offset lithography. Whenever possible I use colored pencils, water-based markers, pens, and ink.

Plate 8.5.—Frank Baseman: *Chouette, Ltd.*, identity and stationery, 1987. The original art was created by photocopying a piece of lingerie. Additional photocopies were used for size and scale on mechanical boards, waxed for positioning, and printed with offset lithography. Advances in photocopy technology have eliminated much of the chemical exposure associated with photostat preparation.

Plate 8.6.—Iris Brown (Bernhardt Fudyma Design Group): *Bernhardt Fudyma 15th Anniversary Self Promotion*, 1989, page makeup with a Macintosh SE with Quark XPress, $5\frac{7}{8}'' \times 5\frac{7}{8}''$. The days of darkrooms and odorous toxic chemicals are just about over for us. Only three years ago, the only way to apply color type to a comp/presentation was by using chemicals, either to develop a Color Key in-house or order rubdowns (Identicolor) out-of-house. When we purchased an Omnichrom machine, which adheres stock colors to photocopy or laser toner through a heat process, I learned to forget Color Key. With the purchase of a photocopier that reduces or enlarges, we make fewer stats and fewer trips to the darkroom and chemical processor. With a Macintosh and scanner, we no longer need X-Acto blades and rubber cement for cutting and pasting. A large automatic waxer generally replaces spray glue, and the purchase of a spray booth minimized further hazards. Most of these tools have become integrated into the creative process only in the last three years. What was once laborious and hazardous is now less time-consuming and more effective. (Photo: Dennis Gottlieb.)

Plate 8.7.—Iris Brown (Bernhardt Fudyma Design Group): *Maxwell House Messenger*, magazine cover, 1988, original made from black-and-white prints, photostats of typographic elements, photocopy, 11¾" × 16".

Plate 8.8.—Stephanie Knopp: *Stoudt Brewery*, logo, 1989, pen and ink on mylar. As an art student I frequently ignored safety rules, cleaning silkscreens with mineral spirits and bare hands, or eating lunch while working with solvents. Now I try to work responsibly to promote good work habits in my students. For collage illustrations and type mechanicals, I have replaced rubber cement with hot wax adhesive. I find it safer and easier to reposition with wax. Adding petroleum jelly to the wax creates a stronger, more permanent bond. Excess wax is easily removed with a clean cloth rubbed gently across the edges of the paper. Most of my photostats are made at a printer's shop, to save time and to decrease exposure. Lately I have been using black-and-white photocopies; the quality is so good that they can frequently be used in place of photostats.

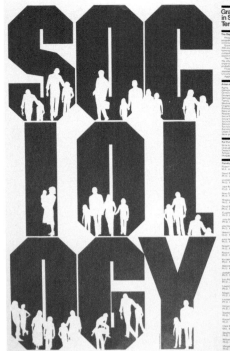

Plate 8.10.—**K**ristine Herrick: *Graduate Studies in Sociology*, poster, 1981, pencil, water-based markers, ink, 21″ × 16″. The roughs were executed with 4-H pencils, and comprehensives executed with water-based felt markers, on 100 percent rag felt marker paper. The camera-ready art was prepared with technical pens, sable brush, and water-based (F & W nonclogging) ink on heavyweight natural vellum paper. Mechanicals were applied on hot press illustration board with Daige waxer used for adhesive.

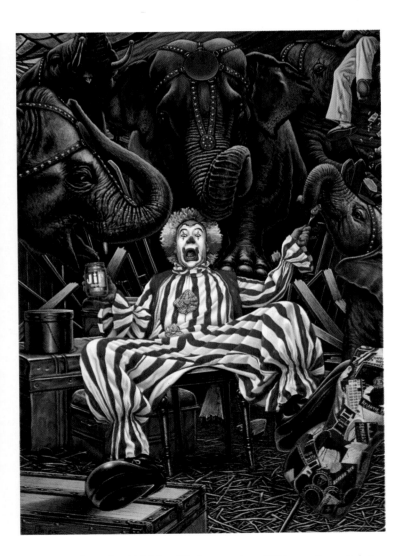

Plate 8.9.—**J**on Ellis: *Wild for Jif*, circus poster, 1991, acrylic on Fabriano illustration board, 17″ × 14″. I no longer airbrush colors containing cadmium, but have perfected a technique of dry brush blending and glazing.

Plate 8.11.—**K**ristine Herrick: *Graduate Programs in Art*, poster, 1984, Prismacolor pencils, water-based markers, pen and ink, watercolor, 23″ × 17″. Roughs were executed with Prismacolor pencils on layout paper, comprehensives prepared with water-based felt markers on felt marker paper, and camera-ready art executed with technical pens, brush, and F&W nonclogging water-based ink on Denril multimedia paper. Full-color illustration was executed in watercolor pigments, and the mechanical executed on hot press illustration board with technical pens, water-based inks, nonrepro blue pencil, and Daige waxer used for an adhesive. Typesetting and photostats were purchased through commercial vendors.

Plate 8.12.—**J**on Ellis: *Grandpa's Garden*, magazine cover, 1986, acrylic on Fabriano illustration board, 17″ × 14″. I no longer accidentally dip my paintbrush into my coffee, nor do I drop into bed after a 26-hour marathon work session without first scrubbing encrusted acrylic paint residues from under my fingernails. I love my art passionately, and I equally love my affair with those smooth, vibrant, sensuous acrylic paints.

Plate 8.13.—Joseph Scorsone: *Zero Moving Dance Company*, poster, 1986, 24″ × 18″. The original is charcoal on Canson paper. I provide the printer with the original drawing plus color separations cut from Rubilith or Amberlith film. I indicate colors on the films with PMS (Pantone Matching System) numbers. I work in a well-ventilated studio and use a wax coater instead of rubber cement.

Plate 8.14.—Joseph Scorsone: *Summer Art in Rome*, poster, 1986, pencil on Canson paper, 22″ × 30″. To replace rubber cement and spray adhesives, I encourage my students to use Twin-Tac, a double-sided, pressure-sensitive permanent adhesive made by Color Vu. It comes in 18″ × 24″ sheets that may be cut to size.

Plate 8.15.—Stephanie Knopp: *Philadelphia Flower Show*, program cover, 1987, photocopies, photostat, water-based markers and dyes, 11" × 8½". The flowers are color copies that I hand-colored with water-based markers and dyes. Watercolors and pastels were applied to a textured blue mat board. When I work with pastels, I wear a respirator to avoid inhalation of dust. I always wash my hands after working and avoid eating or drinking in the studio.

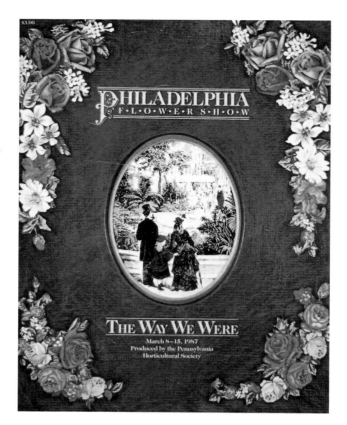

Plate 8.17.—Steven Berkowitz: *EC(s) TASIS*, compact disk cover, 1990, a black-and-white photograph, computer processed, 4½" × 4½". The subject is a ripple of water photographed with 35mm film. A black-and-white print was computer-scanned; the palette was manipulated to create a subtle beige-mauve-black tonality. The image file was color-separated in the computer and output to an image setter. The final design was produced by offset lithography. The original black-and-white image also provided a basis for development of synthetic music recorded on the disc. The main problem I have encountered with computers is eyestrain. One solution is to make the background dark and the text lighter if the software permits. I also have neck strain from long sessions, which can be alleviated by adjusting chair height and monitor angle.

Plate 8.16.—Thomas Porett: *Anaheim Matrix 2*, 1990, digital imaging, 10" × 13". This image was created with an array of high-performance video recording and computer-related equipment, which provides outstanding color and fine picture resolution. The original image was recorded on a Sony tape recorder and then fed into a Sony "frame grabber," which permits specific scenes to be frozen digitally. After positioning the individual frames to form the desired overall image, the result is printed using a high-performance Mitsubishi color printer.

Plate 9.1.—**M**arty Fumo: *Fairmount Park, Sunday Afternoon*, 1983, modified Polaroid SX-70 Integral film, 3¼" × 3¼". I wait at least 24 hours after exposure to allow the emulsion to harden thoroughly. Wearing latex gloves to avoid contact with caustics, I cut across the black backing (negative) and carefully lift it up and slowly pull it away from the positive. Then I wash away the white pigment (titanium dioxide) on the back of the positive with a cotton swab and slow-running lukewarm water. This leaves an extremely fragile positive image formed with metalized dyes on a very thin film of mylar on acetate as a pseudo-transparency. After drying with a blow-dryer, I may further alter the image by applying dyes to the emulsion with a fine brush, felt-tip pens, or other small implements. I avoid skin contact with chemicals and run the exhaust fan in my darkroom while I'm working.

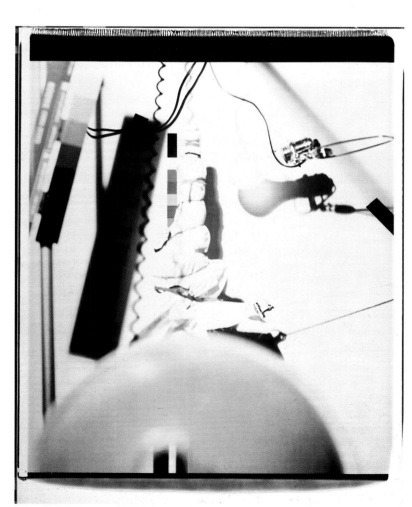

Plate 9.2.—**W**illiam Larson: *Untitled: Still Life*, 1980, Polacolor print made in the Cambridge Studios of Polaroid Corp. under a grant from Polaroid, 24" × 20".

Plate 9.3.—Marty Fumo: *Antiope Is Still Sleeping*, 1990, Ecta-chrome on Fuji-chrome SG Paper, 4" × 5". I use a Jobo-ATL-2 color-processing unit that is completely enclosed. Wearing a protective mask and gloves, I mix the chemicals from concentrates (no powders) into two-gallon storage tanks with floating lids that keep them closed at all times. I transfer the chemicals into the processing unit as needed. Exposed photopaper or film is placed into a drum at-tached to the processing unit, and processing begins automatically with no further exposure to chemicals or their vapors. All I have to do is dry the film (or print) to complete the process.

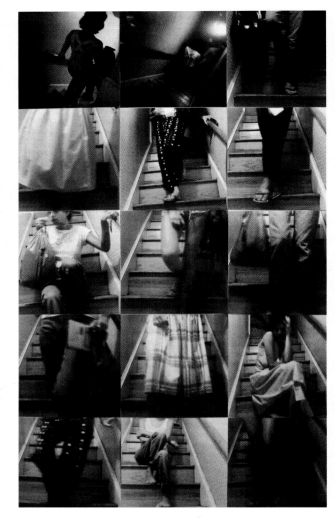

Plate 9.4.—Jill Saull: *The Stairway, One Year: A Time Piece*, 1986, Type C print, made with a Durst Color Transport Processor, 105" × 45". I have experienced dizziness, loss of memory, and chemical stains on my fingers from working in small, un-ventilated darkrooms with open trays of color chemistry. With a Durst Color Transport Processor that I use at the Maryland Institute College of Art, contact with chemicals is avoided. This device also provides constant temperature and agitation control, which are vital in color printing. The exposed print is in-serted in one end, goes through the developer, stop bath, and bleach fix and exits the other end as a finished print. When I have to mix chemicals, I wear rubber gloves and a particle mask.

Plate 9.5.—David Graham: *Cocoa Beach, Fla.*, dye-coupler photograph, 20" × 24".

**Plate 9.6.—William Larson:** *Tucson Garden Series,* 1981, Type C print, 16" × 20". Early in my career I was cavalier about the conditions under which I worked and about the chemistry used in making color photographs. I developed respiratory and allergic problems. Although I improved conditions in my lab, my health didn't follow, and I finally sent most of the work to a professional lab to process and print. Only then did I see significant improvement in my health.

**Plate 9.7.—David Graham:** *Mrs. Jackson & Friends,* 1987, dye-coupler print, 20" × 24". In my early years of photography I had virtually no safety concerns. Ventilation was never considered or discussed. When I think back on closets and cramped basements that served as darkrooms, I get queasy. Only a few years ago I thought my darkroom was made safer by placing a tiny exhaust fan, vented to the attic, in the far corner of the room. With the birth of my daughter, I finally became concerned about ventilation and vapors. I then decided to move my entire color printing operation to a friend's well-ventilated darkroom.

Plate 9.8.—**M**erle Spandorfer: *Hotel in Suzhou*, Type C color print, 10″ × 8″. After working many years in a poorly ventilated darkroom, I learned the dangers of continual exposure to color chemistry. I now use a professional custom color lab for processing and printing to my specifications.

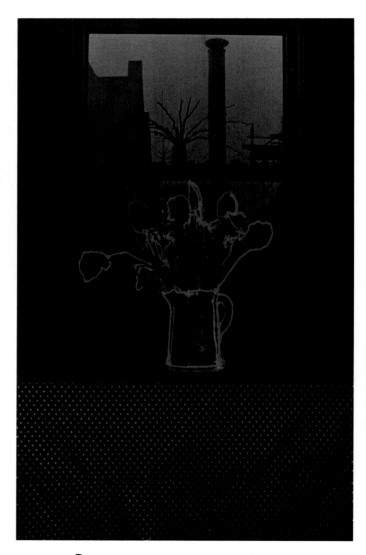

Plate 9.9.—**D**avid Lebe: *Tulips Outlined in Red*, 1982, hand-painted gelatin silver print, 40″ × 30″. I haven't spent much time considering safety procedures, though I've always read labels and have improved ventilation in my darkroom. Perhaps I haven't been too concerned because I work in the darkroom sporadically, and primarily in black-and-white. Colors are added later with water-based paints.

Plate 9.10.—David Lebe: *Parrot Tulips II*, 1987, hand-painted gelatin silver print, 20" × 30". I use latex gloves (from the supermarket) for my work with bleach and toners. I always add chemicals to water instead of the other way around to prevent splashing the concentrates.

Plate 9.11.—Sarah Van Keuren: *By the Sea*, 1989, cyanotype and gum bichromate print, 9½" × 7½". Since 1980, I have used and taught cyanotype, brownprint, gum dichromate, and palladium processes at the University of the Arts. I stress the hazards of chromium salts, brownprint solutions, ferric oxalate, etc. All students of nonsilver processes observe safety standards. We use latex gloves to eliminate contact between skin and chemicals. My students mix only as much of a solution as they will actually use. We prepare solutions in nonspill, unbreakable, recycled plastic dispensing bottles that permit drop-by-drop measurement into 45-ml graduated cylinders. Exhaust fans are located above the coating area and above the long sink. We use a fume hood in a separate room for potentially hazardous procedures. Goggles are always available for eye protection, and I wear a dust mask when mixing dry chemicals for the classes. In a farewell handout, I warn students against using these processes in domestic settings where children could be exposed, eating areas contaminated, or sewage disposal systems damaged.

Plate 9.12.—Merle Spandorfer: *Blue Mist Echo*, 1978, gum bichromate on canvas, 67" × 64". I worked with gum bichromate printing for ten years, starting with small works on paper and progressing to canvases up to nine by eight feet. I was well aware that the process, which uses ammonium bichromate, was hazardous, although I did not know I was dealing with a suspected lung carcinogen. I wore gloves, used a respirator, and frequently worked outdoors, since my studio was not well ventilated. After developing serious health problems, I made a total change in my approach to doing art. I'm fully recovered now and my good health is precious to me. My art has taken new directions, and working with nontoxic materials has become a top priority.

Plate 9.13.—Catherine Jansen: *Michael*, detail from *Michael's Room*, 1990, Sharp color copier transferred to cloth and stitched, 8' × 4'. The computer-digitized Sharp color copier offers a wide range of vibrant, soft, translucent colors as well as variation of image size. This color copier uses fluorescent lights and precoated paper.

Plate 9.14.—Catherine Jansen: *Landscape with Pins*, 1988, xerography on cloth, 25″ × 20″. After eight years of cyanotype, brownprint, and gum print processes, I am now devoted to color copier procedures, which are clean, safe, and capable of full color.

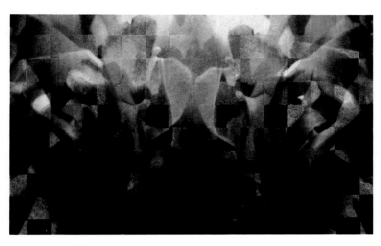

Plate 9.16.—Mary Stieglitz Witte: *Floral Weave*, 1989, photo translation via color laser copier, interwoven mirror images, and transferred with heat to fabric, 16″ × 20″.

Plate 9.15.—Mili Dunn Weiss: *Life Model*, color xerography, 14″ × 20″. Color photocopying presents no known significant health hazards. A number of companies now produce color copiers: I have used the Xerox 6500, Canon Laser Copier, and Sharp cx 5000. I do my creative work primarily in my studio, where the basic preparation is a collage of my drawings, found images, and real objects. Manipulation with the color copier makes possible various color changes, increasing the dominance of a particular color as well as the range of values from light to dark.

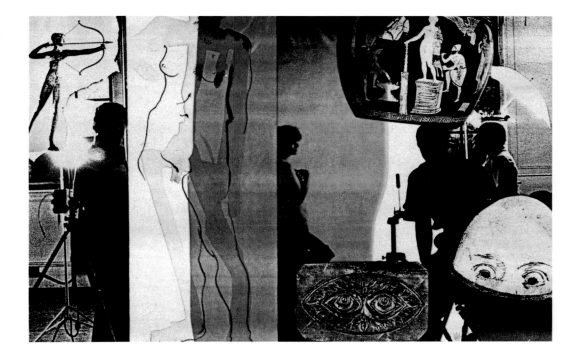

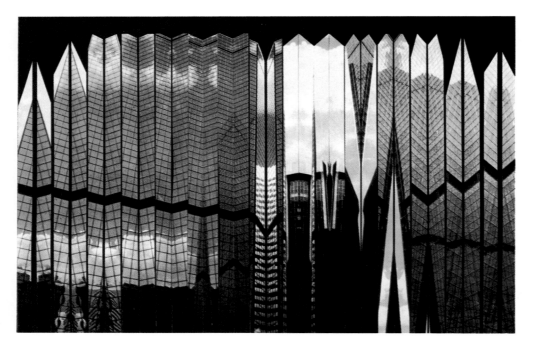

Plate 9.17.–**Mary Stieglitz Witte:** *Denver Reflections,* **1989, photo, color laser copier collage, 16″ × 20″. I turned to this technology after abandoning my darkroom because a color laser copier eliminates direct exposure to chemicals. This print was created by translating an original color transparency to both positive and negative prints with a Canon Color Laser Copier. The two prints were cut into strips, alternating negative and positive parts to achieve the image. I have also made heat transfers of images from a special transfer paper to other materials including fabric. It is a dry, enclosed process with no residual substances produced or emitted.**

Plate 9.18.–**J. L. Steg:** *Self-Portrait,* **1988, photocopy and encaustic, 30″ × 48″. I used a Panasonic copier machine in a well-ventilated room. Copies are worked with mineral spirits and gloss spray in an exhaust-ventilated area.**

developer, or by adding mercuric chloride for brown, lead oxalate for black, or potassium phosphate for blue-black tones. Glycerin or glycerol also provides toning effects but are not recommended due to hazard of explosion when mixed with potassium chlorate from the sensitizing solution.

Palladium and platinum printing should be undertaken with great care. Powders should be mixed in a fume hood, in a glove box, or with adequate skin and respiratory protection (i.e., gloves and an approved respirator with toxic dust cartridges). Premixed solutions are preferred. Gloves and respirator should also be used for blow-drying prints.

Preferred substitutions[2] include:

• Citric acid in place of potassium oxalate (as developer)

• Ethylenediamine tetra-acetic acid tetrasodium (EDTA) in place of hydrochloric acid (in the clearing bath)

• Potassium phosphate in place of lead, mercury, or uranium compounds (as toners) (To avoid the possibility of explosion, do not use glycerin plus potassium chlorate for local toning.)

**Kallitype** and Van Dyke printing are similar to platinum but use silver. The kallitype coating contains ferric oxalate, oxalic acid, and silver nitrate. After exposure of the image, silver nitrate is reduced by the ferrous salts to metallic silver, creating a brown image. The developer may also include sodium potassium tartrate (Rochelle salts) and borax; potassium dichromate may be added to control contrast. The clearing bath contains potassium oxalate; the fixing bath consists of hypo and ammonia. A hypoclearing bath may be used prior to toning with gold chloride. If you handle potassium dichromate, gold chloride, and/or silver nitrate, wear gloves, goggles, and respirator. Use a spray booth when spraying silver nitrate solutions.

The **Van Dyke** process uses a premixed solution that contains ferric ammonium citrate, tartaric acid, and silver nitrate. It is developed in water and fixed in hypo. Gold chloride toner may be used after developing. In terms of safety, Van Dyke is preferred over kallitype, but similar precautions must be taken.

**Carbon** printing is a chromium process achieved with special paper coated on one side with carbon-pigmented gelatin. This surface is sensitized with a solution of ammonium dichromate or potassium dichromate (suspected carcinogens). Ammonia may be included as a preservative, and acetone or isopropanol may be added to accelerate drying time. After drying, the pigmented surface is exposed to strong light with a contact negative. The carbon paper is then placed gelatin side down onto transfer paper coated with plain, insoluble gelatin that may contain formalin or thymol dissolved in methanol as preservative. While still in contact, the two sheets are immersed in warm water to increase adherence. The carbon paper backing is peeled away, and the print immersed in alum or 5 percent sodium sulfite solution to remove dichromate stains prior to hardening in dilute formalin. Do not heat or add

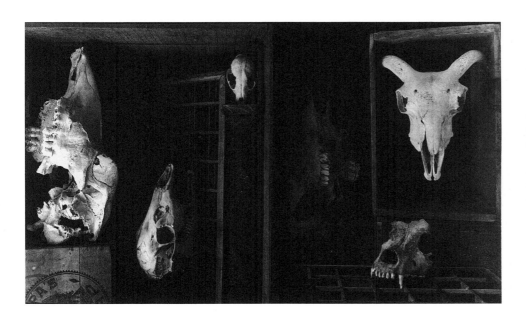

Figure 9.12.—**B**ruce Katsiff: *The Golden Section*, 1989, platinum/palladium print, 12″ × 20″. Platinotype procedures require (1) consistent use of gloves and a dust mask, (2) purchase of premixed solutions, (3) substitution of citric acid for potassium oxalate developer, (4) substitution of EDTA for hydrochloric acid in the clearing bath, and (5) use of the sun or xenon as sources of light rather than carbon arc lamps (eye protection still required).

acid to sodium sulfite solutions. Wear appropriate gloves, barrier aprons, and goggles, assure effective general and slot exhaust ventilation, and prevent fire and explosion. Avoid use of carbon arc lamps. Use respirator if ventilation is not effective.

In addition to carbon black, transfer papers are available in magenta, cyan, and yellow. With registration printing, these transfer papers produce full color similar to four-color printing processes and carbro printing.

**Carbro** (carbon plus bromide) uses a bromide print as the image source and transfers it chemically, irrespective of light, to a carbon sheet. To obtain the bromide print, a photo negative is used to expose commercially available bromide paper, which is developed in amidol, rinsed in water, and fixed in hypo. The carbon paper is sensitized with a solution of potassium ferricyanide, potassium dichromate, potassium bromide, succinic acid, and potassium alum. When the carbon sheet is in contact with the bromide print, the silver image of the print is bleached out and the gelatin on the carbon tissue becomes insoluble in proportion to the density of the silver of the original print. The bromide paper is peeled away, and the carbon paper rinsed and then transferred, as noted previously for carbon printing, to a prepared, receptive transfer paper.

In three-color carbro printing, bromide prints from three color-separated negatives are transferred to their corresponding magenta, yellow, and cyan carbon tissues. The tissues are applied to paper supports for developing and registration, and then printed onto the receiving transfer paper. The process is the same as carbon and carbro printing with the addition of pigments (unknown health effects) and of methanol in the magenta reducing solution.

For carbro printing, buy premixed solutions. Do not heat or add acid to potassium ferricyanide or sodium thiosulfate. Discard used hypo solutions. Wear gloves and goggles to apply dichromate sensitizer. For mixtures, use a fume hood or glove box, or wear a respirator.

In **gum dichromate** (bichromate) printing, the receiving surface, paper or canvas is sensitized with a solution containing gum arabic, ammonium dichromate, or potassium dichromate (suspected carcinogens), and a premixed watercolor pigment (color plate 9.12). Once dried, a contact negative is applied to the receiving surface, and the sandwich is exposed to strong light. The gum becomes insoluble in proportion to the amount of light that passes through the negative. The image is then rinsed in water to dissolve the gum from unexposed areas, carrying the pigment away. Exposed areas retain the pigmented image within the hardened gum coating.

Precautions for gum dichromate printing include use of premixed gum arabic, ammonium dichromate, and tube watercolors rather than powdered pigments. Wear gloves and goggles and use effective dilution ventilation. Handle gum arabic carefully; avoid inhalation of dusts. Mix dichromate powders in a fume hood or inside a glove box, or wear an approved respirator. Store away from sources of heat. In case of eye contact, rinse immediately with water for at least 20 minutes. Seek medical attention.

Avoid powdered materials. Replace carbon arc lamps with a sunlamp, photo floodlights, or sunlight. Avoid sizing paper with solutions containing formaldehyde. Because ammonium dichromate is a suspected carcinogen, this process should be used with extreme caution.

**Oil printing** combines the hazards of dichromate printing with those of lithography, described in chapter 7. Oil printing can also be combined with the **bromoil process** when an image is produced on bromide paper. Prints are developed in a nonhardening developer such as amidol, fixed with hypo and sodium sulfite, and bleached. Bleaching solutions contain copper sulfate, 28 percent acetic acid, potassium bromide, copper sulfate, and potassium dichromate or chromic acid. Bleaching causes the imaged gelatin remaining on the paper to swell in water, forming a matrix that will hold ink in the dark areas and reject it in light areas, relative to the gelatin deposits. A final hypo bath and a water rinse prepare the bromoil plate for inking.

Do not use chromic acid bleaches for bromoil printing. Ferricyanide bleach is a safer alternative. Use a water rinse between bleach and finishing bath to avoid contaminating potassium ferricyanide with acid. Do not heat ferricyanide or expose it to ultraviolet light. Wear gloves and goggles when handling acids and copper sulfate. Mix in a fume hood or in a glove box, or wear an approved toxic dust mask. Avoid inhalation of bromide powders. Use premixed inks and the least toxic solvents available.

The toxicity and health effects data on agents used in oil and bromoil processes (see Appendix I) suggest they should only be used with strict precautions against inhalation, skin contact, and ingestion.

## XEROGRAPHIC PROCESSES

Xerography combines light, optics, heat, static electric charge, and motion to transfer an image onto paper in a matter of seconds. Its development revolutionized print communications in the third quarter of this century. Experimentation with color soon followed, such that today, with digital imaging and laser printing technologies, we have reached noteworthy levels of visual expression and design applications (color plates 9.13–9.18).

Full-color copying machines are appearing in colleges, universities, art schools, and art supply stores as well as in high-technology graphic design studios and media centers, making them increasingly accessible to artists. The newest copiers permit considerable manipulation of both colors and image, accommodating a variety of surfaces such as mylar, cloth, and papers: 100 percent rag, transparent, and transfer. Running preliminary proofs on the least expensive paper will reduce costs.

Enterprising artists have been known to buy time on color copiers (as opposed to buying copies from a copy center) in the private sector as well as in art supply stores to actively participate in, and have control over, the image-making process. High-technology copiers range widely in their capabilities and price. Copiers, in general, are a relatively safe method of image making. Nevertheless, they:

- Use high voltages internally, which may cause ozone to be emitted

- Use powdered pigment toners or wet toners with alcohol or a paraffinlike aliphatic hydrocarbon solvent in the development or printing process (the Iris 3000 series color printer uses water-soluble dyes in a four-color ink-jet process)

- May employ lasers, whose beams can cause eye damage if not used in accordance with instructions

Ozone may be detected by its sharp, distinctive odor. It can impair lung function, aggravate respiratory conditions, and decrease resistance to infection. Its long-term health effects are not yet known but are under study. Copiers should be located and used in a well-ventilated environment.

Most toners come in sealed cartridges that prevent leakage of materials during copying. During placement and removal of the cartridge, however, exposure may occur, and replacement should be done only by persons trained to handle cartridges safely and dispose of them appropriately. Toner dust on a print or around the machine or solvent odors indicate that there is toner leakage that should be repaired by a trained professional before further use. Problems with the drum should also be serviced only by trained personnel.

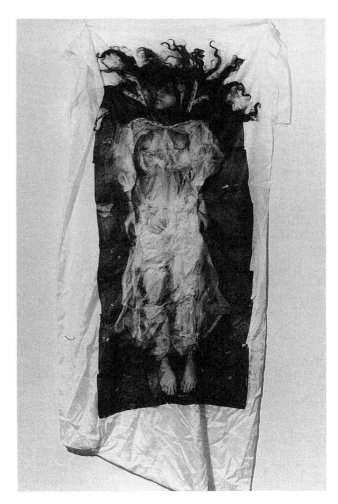

Figure 9.13.—**C**atherine Jansen: *Self-Portrait*, 1987, xerography on cloth, appliqué, and stitched, 8′ × 4′. The 3600 Xerox offers a safe, clean environment for photoimaging on cloth, with sharp, clear reproduction and a full range of colors.

Laser devices are regulated by the U.S. Department of Health and Human Services Radiation Performance Standard of the Radiation Control for Health and Safety Act of 1968 as amended to include lasers in 1976. Compliance must be stated on a label attached to the machine and include the manufacturer's name and address. A label stating "DANGER: AVOID DIRECT EXPOSURE TO BEAM" must also be applied to the copier in a visible location.

Laser beams must be completely confined within protective housings to prevent escape from the machine during any phase of user operation. Servicing and repair of laser copiers should be undertaken only by a properly trained and authorized service representative. Care must be taken when servicing or adjusting the optical system of the printer so that screwdrivers or other shiny objects such as wristwatches and rings do not enter the path of the laser beam and reflect it elsewhere. Though invisible, the reflected beam can cause permanent damage to eyes.

Copiers must be installed and used in accordance with the manual, and special installation specifications may be required to prevent radio interference. Read the operator's manual for the copier, and work only in compliance with it.

To summarize this chapter on photographic processes, we recommend the safe elimination and disposal of arsenic, lead, mercury, selenium, and uranium compounds from all photographic studios. The following chemicals are suspected carcinogens that we suggest be avoided and safely discarded: ammonium dichromate (bichromate), chromic acid, formaldehyde and formalin, 6-nitrobenzimidazole, potassium dichromate, sodium chloropalladite, thiourea, and trichloroethylene.

Whenever possible, minimize exposure to photochemicals classified as moderately or highly toxic. Use them only with rigorous precautions to protect the health and well-being of yourself and those close to you. Effective darkroom ventilation, intermediate water rinses, and tongs to transfer prints along the developing path are key commonsense elements of all safe photographic endeavors. Observation of appropriate precautions in photography does not compromise the quality of the work and may actually contribute to a longer, healthier, more productive and creative life.

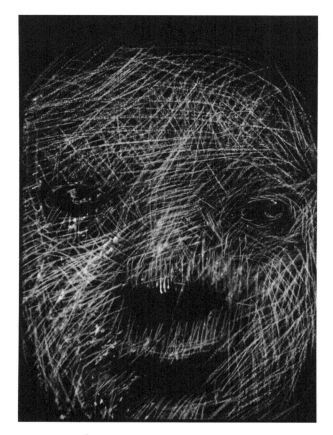

**Figure 9.14.—J. L. Steg:** *Face of Old Nick*, **1988, photocopy and spray paint, 17" × 11". I used a Panasonic copier in a well-ventilated room to lessen exposure to vapors. I use an exhaust fan and wear a respirator when spraying the final product with gloss acrylic medium.**

# 10

As the next millennium approaches, we are a little apprehensive. Every century seems to conclude with a tinge of craziness, and we cannot predict what surprises the last decade of this most developmental of all centuries will bring. We have already seen artists using molten lead

# Looking Ahead

in their artwork, despite clear indications that lead fumes present a significant health hazard. With the presumptive invincibility and omnipotence of youth, some artists have even declared that making art hazardously is an appropriate act to draw attention to the toxic world and society in which we live. Yes, the statement is political, polemical, and pungent. But we wonder how these individuals

◆ ◆ ◆

will feel when they discover themselves martyrs to the message. Will history see them as poignant or pathetic?

Equally disconcerting is the desire to make art with unstable materials that will decompose in a few years: throwaway art for a throwaway society. Such nihilism may wake up a few people, and the perpetrators may have their 15 minutes of fame, but the rest of us are left with the task of cleaning up the mess.

Art, which can pierce the opacity of the world and find truth beyond facts, has too great a responsibility to be so abused.

We think that the beginning of a new millennium, as an artifact of the decimal system, is not the real issue of the moment. Rather, it is the greater consciousness about the interrelatedness of all life that is growing among us. We now know that only when each individual takes responsibility for the personal implications of his or her actions can our healthy ecological future be sustained. No matter how difficult it may be to teach an old dog new tricks, making art safely, by reducing our use of toxic and potentially toxic materials and practices, has implications beyond our individual lives, families and studios.

This is the spirit with which we entrust the information contained herein to individual artists, educators, and administrators who will carry forth thoughtful, aware, and excellent art-making opportunities:

• To educate young people to make art safely, as well as knowledgeably and creatively, by making art safety a recurring theme in all art instruction

• To prepare courses of study that include instruction for safety in the studio and basic principles of chemistry and art product composition

• To establish studios in educational institutions and the workplace that are exemplary for their precautions and practices

As indicated in the introduction, we were inspired to create this book to share with artists the liabilities of working casually and freely with many art materials today. Typically, if we knew what we were getting into at the outset, we probably would never have had the temerity to even broach the subject. Early on, Snyder opined to Curtiss and Spandorfer that no one should be admitted to major in art in college without having passed at least high school chemistry, a course neither of us had taken.

We now agree wholeheartedly. Better yet, we recommend that to graduate from art school or with an art major, all students should be required to study hazards in the arts, basic principles of chemistry, and the precautions and safeguards for using all the art materials and processes they may encounter. Such a course would include how to obtain, read, and use Material Safety Data Sheets, set up a safe studio, and acquire sound personal working habits.

Most artists, at some point in their careers, question the value of what they are doing, especially if the public and art establishment are slow to wake up to their significant discoveries and contributions. One such artist, agonizing over the purpose of his life and art, awoke suddenly one night to the realization that artists do something truly spectacular: We create jobs for hundreds of thousands of others. Art materials manufacturers and retailers, chemists, factory workers, paper producers, conservationists, museum workers and curators, art schools and educators, art movers and handlers, art galleries, photographic suppliers and processors, critics, scholars, magazine and book publishers . . . the list could go on and on. He realized that by creating so many jobs for others who are thereby gainfully employed, thanks to struggling artists, we deserve a lot more credit than we realize!

There is one vocation, however, already overtaxed by the problems of a growing and aging society, namely the medical profession. Let us not, therefore, create too many jobs for the medical establishment by compromising our health with ignorant use of dangerous art materials, or unwittingly commend ourselves as case studies for pathologists of the future. We do not wish to make artists excessively fearful, but the fact is that removal of all actual and potential dangers in art materials is probably an unattainable ideal. Thus education— for the safety and well being of ourselves, our families, our students and employees—is imperative. It is a laudable part of the creative process to attend to the well-being of ourselves and those close to us by facing facts, staying in the know, and making art as safely as we are informed to do. Knowledge is our best and, in many cases, only defense.

Therefore, in order to make art safely with some of the desirable but hazardous materials, abide by the precautions and safeguards we have recommended. The development of first-rate work habits and observation of all of the precautions for making art will significantly decrease the possibility of both acute and chronic long-term health problems. That is the goal of *Making Art Safely*.

Join us in a chorus of praise for long, creative, and healthy lives for all artists!

# Appendices

# *Appendix* 1

**W**e have reviewed many of the chemicals found in art materials for two-dimensional media. A definitive list is impossible: some materials may have escaped our search, and new materials are continuously developed. If you find a chemical on a product label or Material Safety Data Sheet that is not on our list, we urge you to call your local poison control center or the Center for Safety in the Arts (212-227-6220) for more information.

The agents identified, described, and classified in the following table have been assembled from a number of sources.[1] It is important to understand, however, that with continuing research classifications may change.

The toxicity of art materials are rated as follows:

**High** if they:

- May cause major damage or death from a single exposure
- Have been clearly linked with chronic health problems following long-term or repeated exposures
- Are reported to cause significant numbers of life-threatening allergic reactions

**Moderate** if they:

- May cause minor damage from a single exposure
- May cause an allergic reaction in a large percentage of people
- May have chronic health effects
- May be lethal if exposure involves a massive dose

**Slight** if:

- A single exposure may cause minor injury that is readily reversible
- They have no known long-term health repercussions
- A small percentage of people may develop an allergy
- Massive overdoses could cause a more serious injury

**NKT** (No Known Toxicity) if current knowledge indicates that they:

• Are relatively nontoxic

• Are not associated with any short- or long-term health hazard

• Have toxicity believed to be negligible

Materials may also be designated as:

**CARCINOGEN:** the chemical is known to induce cancer in humans.

**SUSPECTED CARCINOGEN:** a substance that has been demonstrated to cause cancer in laboratory animals, but is as yet unproven for humans.

**REPRODUCTIVE TOXIN:** the material is associated with:

• Adverse effects on reproduction, fertility, or pregnancy

• Spontaneous abortion, fetal death, retarded growth, or premature birth

• Low birth weight, birth defects, and other complications

**ANESTHETIC:** the substance causes dizziness, unconsciousness, and, with high exposures, death.

**ASPHYXIANT:** the substance deprives the body of oxygen to result in unconsciousness, coma, and death.

Some materials, as relevant, are also identified as:

**OXIDIZER** (Oxidant): the substance contains oxygen and gives it up readily to support combustion.

**COMBUSTIBLE:** the material has a flashpoint above 100° Fahrenheit (38° Celsius).

**FLAMMABLE:** the material has a flashpoint lower than 100° Fahrenheit (38° Celsius).

**EXPLOSIVE:** this material, under specified conditions, will explode.

In response to increased knowledge, we hope that art material manufacturers are producing safer products. Informed artists, however, incorporate three major principles in their work:

• **Avoid** materials that are highly toxic whenever possible, and do not use suspected and known carcinogens.
• **Adopt** safer substitutes, a safe studio, personal protection and safe handling procedures.
• **Be alert** to clues or symptoms that may indicate a health problem. Seek prompt attention from professionals trained to recognize occupation-related health problems.

| Chemical Trade Name | Toxicity | Health Effects | Uses | Precautions |
|---|---|---|---|---|
| Acetic acid | Moderate | Via all routes of entry. Vapors and mists irritate respiratory passages; associated with bronchitis and reversible airway obstruction. Chronic exposure to dilute solutions associated with bronchitis. | Glacial acetic acid concentrate, used in film processing, stop baths, color developer starters; photo fixer, toner, hardener, bleach, replenisher, retoucher, hypo test kits; dye mordant. Vinegar is about 5% acetic acid. | Avoid by substituting water rinse to stop photo-developing. Wear gloves (rubber, neoprene) goggles and protective apron when handling concentrated acids. Use with effective exhaust ventilation. |
| Acetone | Slight to Moderate

Extremely flammable | Associated with eye, mucous membrane, and skin irritation; central nervous system depressant that may cause headache, dizziness, and unconsciousness if used with inadequate ventilation. No known long term health effects. | A ketone solvent for plastics, lacquers, inks paints and varnishes. Also used to clean brushes rollers and tools. | Use local exhaust ventilation; wear neoprene gloves. |
| Acridine | Unknown | | A dye, coloring agent. | |
| Acrylic polymer painting mediums & binders | Slight to Moderate | Ammonia vapor may irritate eyes, skin and mucous membranes, cause headache, nausea. Preservatives may be an allergen for some individuals. | Acrylic emulsions may contain ammonia and formaldehyde preservatives. | Use effective ventilation. Avoid mixing with bleach (sodium hypochlorite). |
| Agfa TSS | Slight | May cause allergy. | A color photo-developer. | |
| Alcohols (ethyl, isopropyl, methyl) | Variable

Flammable

Anesthetic | Via all routes. Irritant to eyes and mucous membranes; drying to skin. Central nervous system depressant. | Solvents for lacquer, paint, rosin, shellac, varnish. | Use exhaust ventilation; wear nitrile, neoprene or natural rubber gloves. Ethyl and isopropyl among the safer solvents. |
| Aldehydes (formaldehyde, succinaldehyde, glutaraldehyde) | Moderate | Via all routes of entry; irritants and sensitizers. | Preservatives used in photochemical solutions and compounds, water-based paints and inks. | Use local exhaust ventilation. |
| Aliphatic hydrocarbons (petroleum distillates) | Variable

Anesthetic

All flammable some explosive | Via all routes of entry. Irritating to mucous membranes and skin. Central nervous system depressants, narcotic effects by absorption and inhalation. | Multi-use, fast-drying solvents (benzine, gasoline, hexane, kerosene, mineral spirits, paint thinner, petroleum ether and distillates, Stoddard solvent, VM&P naphtha). | Substitute isopropyl alcohol or acetone. Use only with effective ventilation away from flame; wear nitrile gloves. See also N-hexane. |
| Alizarin Crimson | NKT | | Pigment red 83 (rose madder), natural or synthetic anthraquinone. | |

| Chemical Trade Name | Toxicity | Health Effects | Uses | Precautions |
|---|---|---|---|---|
| **Alkalis** | Moderate to high | Corrode skin. Irritate eyes, nasal and mucous membranes. Inhalation of alkali dusts or vapors can cause pulmonary edema. Ingestion can cause painful damage to mouth and digestive system; can be fatal. | Neutralize acids; in general cleaning agents; in dye baths, photo-developer baths, photo-film cleaners and activators. | Wear gloves and goggles when handling alkaline powders and concentrates. In case of spills, wash with plenty of water; flush eyes 15-20 minutes, call doctor. |
| **Alum** (potassium aluminum sulfate) | Slight | Frequent use can damage skin. | A photo/film hardener that may be used in clearing and fixing baths; also as a dye mordant. | Wear gloves if handling frequently. |
| **Alumina** (aluminum hydrate, oxide) | Slight | Nuisance dust; physical irritant. | Ingredient in whiting, used in printmaking and as a pigment and extender. | Wear a dust mask. |
| **Aluminum palmitate** | NKT | Dust or powdered form may present an expolosion hazard. | A paint stabilizer. | |
| **Aluminum stearate** | NKT | | A paint stabilizer. | |
| **Amidol** (diaminophenol hydrochloride, hydroxy-para-phenylene diamine hydrochloride) | Moderate | Via all routes. Causes skin irritation, allergies. Inhalation of powder may cause bronchial asthma; inhalation or ingestion may cause blood pressure elevation, gastritis, tremors, vertigo, convulsions, and coma. | Ingredient in photo developers. | **Avoid** if possible. Use with local exhaust ventilation. Wear gloves, goggles, and dust mask or full-face respirator. |
| **Amines, aliphatic** | Moderate to High<br><br>Flammable | Form strong alkaline solutions; corrosive to skin; vapors irritating to upper respiratory system; mist irritates eyes, may cause corneal damage, allergies. | Photography: color processing. Also as a hardener for epoxy glue and paint. | Use only with effective exhaust ventilation; wear goggles and gloves. |
| **Aminophenol** (para-aminophenol hydrochloride, 2-aminoethanol) | Moderate | Can cause skin irritation and allergies; ingestion or inhalation of powder may cause methemoglobinemia (an acute blood disease), cyanosis (blue lips and nails), breathing difficulties, and bronchial asthma. | Used in photography developers. | Use exhaust ventilation. Buy pre-mixed developer solutions. Wear gloves & goggles when preparing solutions. Use tongs. Powder should be mixed in a fumehood, glove box, or while wearing a dust mask. |
| **Ammonia** (ammonium hydroxide) concentrate 28% | High | Via all routes of entry. Particularly dangerous to eyes where it can increase intraocular pressure and cause corneal ulceration and blindness. | An alkali found in washing aids, hypo eliminator, photo rinses, bleach fixes, replenishers; dye mordant. | **Avoid** concentrates. Household ammonia (5% solution) is safer. |
| **Ammonia gas** | High | Respiratory irritant known to cause chronic lung problems and death. | Gas arises from all ammonia solutions and, to varying degrees, from all ammonia compounds. | Prevent inhalation with effective ventilation and by wearing an ammonia gas cartridge respirator. |

| Chemical Trade Name | Toxicity | Health Effects | Uses | Precautions |
|---|---|---|---|---|
| **Ammoniated mercury** | High | Most compounds containing mercury present significant health hazards. | An intensifier and photosensitizer. | **Avoid.** |
| **Ammonium alum** (ammonium aluminum sulfate) | Slight | May be an irritant and may cause allergy. | A photo toner, dye mordant. | |
| **Ammonium bromide** | Slight Moderate  High when heated | Via skin. Via inhalation and ingestion. Bromides affect the nervous system. Releases highly toxic hydrogen bromide gas. Symptoms are depression, mental confusion, somnolence, vertigo; in extreme cases, psychosis, mental deterioration. | A photo bleach bath. | Use only with exhaust ventilation or while wearing an approved ammonia/gas respirator. Avoid heating. |
| **Ammonium chloride** | Moderate High | Via all routes, an irritant. When heated may form highly irritating ammonia and hydrochloric acid gases. | Used in flux, patina and metal colorant chemicals. | **Avoid.** Use effective ventilation or wear a respirator with ammonia gas cartridge. |
| **Ammonium dichromate** | Moderate  Suspected carcinogen  Flammable | Via all routes. An irritant, possible allergen, and in large doses a poison that causes digestive distress and kidney damage. | An intensifier, photo-sensitizing agent. | **Avoid.** Use glove box and toxic dust respirator; wear protective gloves when mixing and handling. |
| **Ammonium EDTA** (iron salt of ammonium ethylene diamine tetra-acetic acid) | Slight to Moderate | Can be an irritant. | A silver bleach bath for photo color processes, bleach fixes, color reversers, and replenishers. | See EDTA. |
| **Ammonium hydroxide** | High | An irritant to eyes, respiratory and digestive tracts; corrosive to skin. | An ingredient in casein paints. | Use with local exhaust ventilation. Wear nonvented goggles and rubber gloves. |
| **Ammonium nitrate** | High Slight Oxidizer, Explosive | Via ingestion. Via skin and respiration. | Lithography. | Avoid ingestion. |
| **Ammonium persulfate** | Slight High  Oxidizer | Irritant via all routes. When heated, forms sulfur dioxide and hydrogen sulfide gases, which are extremely irritating to respiratory passages, eyes and skin. | Removes hypo; an ingredient in reducing solutions. | Wear goggles and gloves when handling. Keep away from sources of heat. Use pre-mixed solutions rather than powders. |
| **Ammonium phosphate** | Slight | Irritant via all routes. | Lithography. | |

| Chemical Trade Name | Toxicity | Health Effects | Uses | Precautions |
|---|---|---|---|---|
| **Ammonium thiocyanate** | Moderate Slight to Moderate High | Via ingestion.<br><br>Via inhalation or skin. When heated or combined with acid; may form poisonous hydrogen cyanide gas. | A silver bleach in photography color processing. | Avoid heating or mixing with acids. |
| **Ammonium thiosulfate** | Slight Moderate | Via ingestion. May decompose to form sulfur dioxide, especially with heating or adding acid. Ammonia gas may also be released. | In photography fixers (Rapid Fix), bleach, color reversers, hardeners, and toners. | Avoid heating or mixing with acids. |
| **Aniline** | Moderate<br><br>Suspected carcinogen | May cause drowsiness, eye irritation, methemoglobinemia. | A class of dyes derived from petroleum, used to tint photographs; not colorfast. | Use with exhaust ventilation. Avoid if possible. |
| **Anthraquinone** | NKT | No acute effects; under study for long-term health hazards. | A dye used in paints: red 83: alizarin crimson, madder lake, rose madder; indanthrene blues. | |
| **Antimony** (antimony sulfate, barium sulfate) | Moderate<br><br>High<br><br>Reproductive toxin | Via all paths of entry. Dusts and fumes irritate eyes, upper respiratory tract. Known to affect enzymes, heart, lungs. Ingestion may cause acute digestive upset, liver and kidney damage; in extreme exposures, possible respiratory failure, coma, death. Can cause ulcers on skin and anemia. Symptoms with ingestion may include: metallic taste, vomiting, diarrhea, irritability, fatigue, muscular pain. | Pigment; in antimony white 11, Naples yellow 41. Read labels. May be contaminated with arsenic, a suspected carcinogen. | Avoid powder; prevent skin contact. |
| **Aromatic hydrocarbons** | Moderate High | Via all routes of entry. Especially irritating to skin: dissolve waxy layers to increase skin absorption and cause local irritation. | Solvents: benzene, styrene, toluene, xylene. | **Avoid.** Substitute isopropyl alcohol or acetone. Use with local exhaust ventilation. Wear gloves and protective attire. |
| **Arsenic** | High<br><br>Reproductive effects<br><br>Suspected carcinogen | Via all paths of entry. Corrosive to skin and mucous membranes. Peripheral nervous system, kidney damage. Possible skin cancer, bone marrow damage, lung cancer. Symptoms: numbness in hands and feet; chest pain, headache, vomiting, diarrhea, coma, death. | A metal in pigments such as green 21 & 22, emerald green (cupric acetoarsenite), Scheele's green English (copper arsenite), Paris Schweinfurt, Veronese greens; yellow 39; and other colors. | **Do not use.** |

| Chemical Trade Name | Toxicity | Health Effects | Uses | Precautions |
|---|---|---|---|---|
| **Arylamide, Arylide** | NKT | No acute effects, long term health effects unknown. | Azo organic pigments. | |
| **Asphaltum** | Slight to moderate<br><br>Suspected carcinogen | Skin, eye and mucous membrane irritant, may photosensitize skin, cause pigmentation, or dermatitis. If excessively heated. | Ingredient in etching ground resists, usually dissolved in paint thinner or benzine. | Prevent skin contact, and inhalation of fume and dust. Use with local exhaust ventilation. |
| **Azo** | Slight | Under study for long term health hazards. | An organic pigment/dye in orange, red, yellow and brown paints and inks; red 3, yellow 1, 3, 5, 10 (Hansa, Napthol, others). | |
| **Azurite** | Slight<br>Moderate | Via skin.<br>Via ingestion and inhalation. May discolor skin, cause skin and respiratory allergy, irritation, and metal fume fever; possible ulceration and perforation of nasal septum. Acute ingestion may cause gastrointestinal irritation, shock, liver and kidney damage. | A natural copper-containing mineral in pigments: malachite, copper carbonate. | |
| **Barium Compounds** (barium acetate, carbonate, chloride, chromium, hydroxide, nitrate, sulfide)<br>**Barium sulfate** | Moderate<br><br><br><br><br><br><br><br>NKT | Absorption via inhalation or ingestion can cause muscle spasms, weakness, paralysis, gastroenteritis, tinnitus, impaired vision, and irregular heartbeat. Powders are irritating to eyes, skin, upper respiratory tract; may cause benign pneumoconiosis. | A metal used in compound forms in pigments, extenders, whiting, and as a precipitant in dyes; White 10; White 21, 22, permanent white, yellow 31. | Avoid ingestion. Wear gloves, goggles and dust respirator when handling powder, or use a fumehood. |
| **Beeswax** | Combustible | Overheating emits irritating gases. | Ingredient in etching grounds and painting mediums. | Use exhaust ventilation when heating in a double boiler or with a slow cooker. Avoid waxes containing solvents. If necessary, dissolve with alcohol or acetone, observing precautions. |
| **Benzaldehyde** | | See Aldehydes | Used in photo-etching (KPR Blue Dye) | Substitute presensitized etching plates. |
| **Benzene** (benzol) | High<br><br>Extremely flammable<br><br>Suspected carcinogen | All paths of entry; associated with aplastic anemia.<br>May cause some forms of leukemia; known to damage bone marrow where blood cells are formed. | An aromatic hydrocarbon solvent. | **Do not use.**<br>Substitute isopropyl alcohol or acetone. |

| Chemical Trade Name | Toxicity | Health Effects | Uses | Precautions |
|---|---|---|---|---|
| **Benzine** (VM&P Naphtha) | Slight to Moderate<br><br>Flammable | Less toxic than benzene, but also depresses nervous system; may be an eye, skin and respiratory irritant. | An aliphatic hydrocarbon solvent, one of the least toxic in this class. | Use with local exhaust ventilation; wear nitrile gloves, protective attire. Preferable to turpentine. |
| **Benzotriazole** (1,2,3-Benzo-triazole) | Slight to Moderate<br>Suspected carcinogen | An allergen. | A photo restrainer, anti-fogger, and developer. | Avoid. |
| **Bismuth** | Slight | Nuisance dust, may be physical irritant; In extremely high quantities causes symptoms similar to lead poisoning. | A metal occasionally used in pigments. | |
| **Blockout** | Slight to Moderate | Refer to specific ingredients for health effects and precautions. | Printmaking resist, may contain water-soluble glues; or enamel, lacquer, varnish, methyl Cellosolve acetate; polyurethane; or asphaltum. | Use local exhaust ventilation; wear protective attire; use water-soluble glue version. |
| **Borax** (sodium borate, sodium tetraborate) | Slight to Moderate | Contact with dust is irritating to eyes, skin, respiratory tract. Contact with moist skin may cause thermal burns. | A preservative (odorless white or gray crystals). | Avoid all skin contact. |
| **Boric acid** | Slight Moderate | Via skin. Via inhalation and ingestion. Entry into the bloodstream may cause gastroenteritis, skin rash, appetite loss, nausea, agitation and possible kidney damage. | A buffering agent in photo developers, bleach, fixers, replenishers, and hardeners. | Use local exhaust ventilation, wear gloves and a respirator when using powder or concentrated solutions. Use dilute solutions whenever possible. |
| **Boron** | Moderate | Local irritant to eyes, skin, respiratory tract, digestive system; may cause kidney damage and, with extreme exposure, death. May be absorbed through skin. Symptoms: nausea, stomach pain, vomiting, diarrhea. | Metal in pigments. | Avoid powdered pigments whenever possible |
| **Bromine** | Moderate | All routes of entry. Volatile, an irritant to eyes, skin and respiratory system (measles-like skin eruptions, delayed pulmonary edema). | A daguerreotype accelerator. | **Do not use.** |
| **Burnt plate oil** | Slight | Can be an allergen. | Boiled linseed oil used in etching. | |
| **Burnt Sienna** | NKT | | Pigment of natural and synthetic iron oxides. May contain impurities such as manganese. | |

| Chemical Trade Name | Toxicity | Health Effects | Uses | Precautions |
|---|---|---|---|---|
| **Butyl Cellosolve** | Moderate | All paths of entry: irritating to eyes, skin, upper respiratory tract; absorbable through skin to cause blood, kidney, liver damage. | A solvent, less toxic than methyl Cellosolve. | Use less toxic solvents such as acetone. |
| **Cadmium** (cadmium sulfide, cadmium selenide) | Moderate to High<br><br>Reproductive toxin<br><br>Suspected carcinogen | Irritating to eyes, skin, respiratory tract: cough, chest pain, chills, breath shortness, weakness. Ingestion: nausea, vomiting, diarrhea, abdominal cramps; kidney and lung damage; anemia. Associated with lung and prostate cancer. | Ingredient in red, orange, yellow pigments: red 108, 113; orange 20, 23; yellow 37. | **Avoid** powdered pigments. Wear gloves; avoid ingestion and inhalation. |
| **Calcium carbonate** (lime) | Slight | Nuisance dust; may cause irritation in susceptible individuals. | Whiting. A filler in paints, pastels, gessos; and used in printmaking procedures (found in lime). | Wear dust respirator. |
| **Calcium hydroxide** (lime) | Moderate | Corrosive to eyes, skin, respiratory and digestive tracts. | Ingredient in solution for grinding pigments for fresco painting. (found in lime). | Wear goggles, gloves. |
| **Calcium sulfate** (plaster of Paris) | NKT | Plaster dust is slightly irritating to eyes and respiratory system. Produces heat upon mixing; use in body casts has resulted in skin burns. May contain preservatives with health effects. | Used in modeling, casting and carving. | Wear a dust respirator when mixing; wear insulated gloves or allow to cool before handling; do not sweep, but damp mop or vacuum to clean-up. |
| **Carbitol** (diethylene glycol monoethylether) | Slight to Moderate<br><br>Reproductive effects | Mild irritation on direct contact. Can be absorbed through skin. May reduce sperm counts in men. Under study for long term effects. | A solvent used in color photo processing; a Cellosolve replacement. | Not clearly established as a safe alternative to Cellosolve. |
| **Carbolic acid** (phenol) | Moderate to High | Dilute solutions can cause serious skin, or eye burns without warning pain. Vapors irritate eyes, skin, respiratory tract. Absorbable by all routes. Systemic effects include nausea, vomiting, kidney and liver damage. High doses can be fatal. Skin contact with concentrated phenol for even several minutes can be fatal. | A preservative, insecticide, and fungicide; used in printmaking, especially lithography. | **Avoid.** Use only with local exhaust ventilation or fumehood. Wear gloves and other protective attire. Store away from heat. |
| **Carbon** | NKT | May be contaminated by impurities associated with tumors. | A common pigment in inks and paints: carbon, channel, lamp black. | Use quality products. |

| Chemical Trade Name | Toxicity | Health Effects | Uses | Precautions |
|---|---|---|---|---|
| **Carbon disulfide/ bisulfide** | High<br><br>Flammable | By all routes. Acute exposure may cause narcosis, nerve damage, psychosis and death. Chronic exposure causes peripheral and central nervous system damage; adverse effects on blood, liver, heart and kidney; may be fatal. | A fumigant used in conservation. | Poor odor warning. **Do not use.** |
| **Carbon monoxide** | High<br><br>Asphyxiant<br><br>Reproductive toxin | Via inhalation. Combines with hemoglobin in the blood preventing it from carrying oxygen to the body, and at high concentrations leads to heart attack, coma and death. Symptoms include nausea, headache, dizziness, vomiting. | Released in the incomplete combustion of carbon arc lamps. May also be formed in the body from exposure to methylene chloride, a paint solvent. | Use carbon arc lamps only with local exhaust vented directly to the outside. Substitute quartz halogen lamps or sunlight. |
| **Carbon tetrachloride** | High<br><br>Anesthetic<br><br>Suspected carcinogen | Skin absorption can cause dermatitis; kidney and liver damage. Central nervous system depressant. | Chlorinated hydrocarbon solvent. | **Do not use.** |
| **Carborundum** (silicon carbide) | NKT | Nuisance dust. | Used to sharpen carving, engraving tools; grit used in printmaking. | Wet stone to reduce dust or wear dust mask. |
| **Catechin** (catechol, pyrocatechol, pyrocatechol, o-dihydroxybenzene, 1,2-benzenediol) | Moderate to High | Via all paths of entry. Skin irritant, potential allergen; well absorbed through skin. Inhalation of powder may lead to anemia, convulsions, cyanosis, vomiting, hypertension, increased respiration, liver and kidney damage. Ingestion of small amounts of powder can be fatal. | An ingredient in some photography developers. | **Avoid.** Use less toxic developers, buy pre-mixed developer solutions whenever possible. Prevent skin contact by using tongs to handle prints. |
| **Caustic soda** | High | Via all routes of entry. | See sodium hydroxide. | |
| **Cellosolve** (glycol ether, 2-ethoxyethanol) | Moderate<br><br><br><br>Reproductive toxin | Mildly irritating on direct skin contact. Inhalation, ingestion and skin absorption may cause kidney damage, anemia, narcosis and behavioral changes. Testicular atrophy and low sperm count observed in animals. | Once common in color processes such as Cibachrome. | **Avoid.** If used, wear goggles and gloves. Use exhaust ventilation or an organic vapor respirator. |
| **Cellosolve acetate** | Moderate<br><br>Combustible<br><br>Reproductive toxin | Via significant skin absorption. An eye, nose and throat irritant; associated with central nervous system depression; kidney and lung damage. | Solvent for resins, lacquers, inks, paints, varnishes and dyes. | **Avoid.** |

| Chemical Trade Name | Toxicity | Health Effects | Uses | Precautions |
|---|---|---|---|---|
| Cellulose gum etch | Slight | Contains small amounts of phosphoric acid. | Lithography etch. | Prevent eye contact. Safest litho etch. |
| Cerulean blue | Slight | May induce symptoms associated with cobalt exposure. | Pigment formed of cobalt, tin and calcium sulfate. | |
| Chalk | NKT | Nuisance dust. | Pigment additive. | Wear dust mask. |
| Chinese white | Slight | Ingestion or inhalation may cause irritation of respiratory or gastrointestinal tract. | Zinc oxide pigment. | |
| Chloramine-T | Slight to Moderate | Releases chlorine gas in acid solutions. | Color photo retouching agent for yellow dyes. | Use exhaust ventilation, wear gloves. |
| Chlorinated hydrocarbons | Moderate to High<br><br>Anesthetic<br><br>Suspected carcinogen | Depress central nervous system; may cause dizziness, loss of consciousness; with massive exposure, death. Liver and kidney damage. Defat skin. Some decompose in heat or ultraviolet light to emit highly toxic phosgene. | Solvents: carbon tetrachloride, ethylene dichloride, methyl chloroform, chloroform, trichloroethylene, methylene chloride and others. | Poor odor warning. **Do not use.** |
| Chlorine | High | By all paths of entry. Severely irritating gas may cause burning, coughing, wheezing, hoarseness, pulmonary edema. Ingested solution may cause burning, sore throat, drooling; throat, chest and abdominal pain. | Used in photo-processes, Daguerreotype accelerator. | **Do not use.** |
| Chlorine dioxide | High | Acute exposures can cause bronchitis, pulmonary edema, and pneumonia. Chronic exposure to small amounts may lead to eye irritation, chronic bronchitis and emphysema. | Bleach for rag papers used by conservators and paper makers. | Use only with an exhaust fume hood. |
| Chlorine gas | Moderate to High | See chlorine. | Printmaking and photo by-product. Forms from heating or adding acid to hypochlorite (bleach) or potassium chlorochromate (an intensifier); and in the preparation of Dutch mordant (hydrochloric acid and potassium chlorate). | **Avoid.** Substitute ferric chloride for Dutch mordant. |
| Chlorox (household bleach) | Moderate to High | See sodium hypochlorite. | | See sodium hypochlorite. |
| Chlorquinol (chlorhydro-quinone) | Moderate High | Via skin and inhalation. Via ingestion, causes systemic effects similar to catechin. | A photo developer. | Use exhaust ventilation and wear gloves. |

| Chemical Trade Name | Toxicity | Health Effects | Uses | Precautions |
|---|---|---|---|---|
| **Chrome alum** | Moderate<br><br><br><br>High | Via skin and inhalation, may cause ulceration and perforation of nasal passages, allergies, and eye burns. Heating or adding acid causes release of toxic chlorine gas. | Common photo-chemical, hardener. | Wear goggles and gloves when handling powder or solutions. Mix in fume hood or wear a dust respirator. Do not heat or add acid. |
| **Chromic acid** | High<br>Suspected carcinogen | Via all routes. Associated with lung tumors. | In photographic bleaching solutions. | **Avoid.** Substitute copper sulfate. |
| **Chromium compounds Chromium III** (chromic oxide, chromic sulfate) | Moderate to High | Via all routes. Corrosive to skin and mucous membranes, causing dermatitis, respiratory irritation, lung damage; severe enteritis, fluid loss, even shock if ingested. | Metal in orange, yellow, green pigments. | **Avoid** if possible. Use local exhaust ventilation and wear gloves. Avoid powdered pigments. |
| **Chromium VI** (barium, lead, strontium and zinc chromate) | Suspected carcinogen | Associated with cancer. | | **Avoid.** |
| **Cibachrome developers 1A and 1B** (ethoxydiglycol and hydroquinone) | Moderate | Irritating to eyes, and skin. Altered vision and eye changes with chronic exposure. Ingestion leads to headache, dizziness, tinnitus, agitation and gastrointestinal problems. | Color photo process. | Flush eyes and skin immediately with water. If ingested, get immediate medical attention. Use developing tanks to avoid contact with chemicals. |
| **Cibachrome P-30 2A** (p-toluene-sulfonic acid) | Moderate | Via all paths of entry. Corrosive to skin, mucous membranes, and clothing; may cause burns. | Color photo process, bleach. | Flush immediately; if swallowed, do **not** induce vomiting, get immediate medical attention. |
| **Cibachrome P-30 2B** (ethoxydiglycol) | Moderate | Irritating to eyes, skin and through ingestion. | Color photo process. | |
| **Cibachrome P-30 2C** | Moderate | Highly alkaline, corrosive. | Color photo process. | If ingested, do **not** induce vomiting; get immediate medical attention. |
| **Cibachrome P-30 fixer 3** (n-methyl pyrolidone, ammonium thiosulfate) | Moderate | Via skin and ingestion. | Color photo process. | |
| **Citric acid** (citrazinic acid) | Slight to Moderate | Via all routes of entry; an irritant and a sensitizer. | Photo washing, rinsing aids. | Use exhaust ventilation. |
| **Coal tar naphtha** | Variable | See benzene. | Contains benzene. | **Do not use.** |
| **Cobalt** | Slight to Moderate | Eye, skin irritant, allergen; chronic inhalation may cause asthma, pneumonia, fibrosis. Ingestion may cause vomiting, diarrhea; heart damage. | Ingredient in blue, green, yellow, and violet pigments: cobalt arsenate, oxide, phosphate; potassium cobaltinitrite. | **Avoid** powdered pigments or paint sprays. Wear gloves. |

| Chemical Trade Name | Toxicity | Health Effects | Uses | Precautions |
|---|---|---|---|---|
| **Cobalt chloride** | Slight Moderate | Via skin and ingestion. Via inhalation of dusts: is a sensitizer causing both skin and respiratory allergies. Large doses associated with heart damage and gastroenteritis. | A photo toner. | Use exhaust ventilation and wear gloves. |
| **Cobalt linoleate** | Slight to Moderate | See cobalt and cobalt chloride. | A drier in inks comprised of cobalt, linseed oil, turpentine, petroleum distillates. | Least toxic drier. |
| **Cobalt naphthenate** | Suspected carcinogen | | A drier in inks. | **Do not use.** Substitute cobalt linoleate. |
| **Copal varnish** | Slight  Moderate | Powdered resins are occasionally allergenic. Additional hazards depend on specific driers and solvents in the product. | A painting medium with a resin base that contains driers and solvents. | Check solvent and drier ingredients, and observe appropriate precautions. |
| **Copper** | Slight   Moderate | Via skin, may discolor it. Contact with eyes may cause conjunctivitis and corneal ulcers. Elemental copper is poorly absorbed orally. Heavy exposure through skin absorption or inhalation can cause nausea, abdominal pain; respiratory allergy and irritation; metal fume fever, possible ulceration or perforation of nasal septum. | A metal in pigments. | **Avoid** powdered pigments; wear gloves. |
| **Copper iodide** | Slight Moderate | Via skin; Via ingestion and inhalation. May cause skin allergies and irritation to eyes, mucous membranes and skin. Prolonged absorption can cause severe skin boils, blisters, weakness, anemia, weight loss and central nervous system depression. | A photo intensifier. | Wear goggles and gloves when handling. Mix powders in a glove box, under a fumehood, or wear an approved toxic dust respirator. See also copper sulfate. |
| **Copper sulfate** (cupric sulfate, blue vitriol) | Slight Moderate | Via skin. Via ingestion and inhalation. A sensitizer and irritant. Ingestion leads to intestinal, liver, and kidney damage. | A bromoil process bleach bath; dye mordant. | Wear goggles and gloves when handling; avoid inhalation of powder. |
| **Cream of Tartar** (potassium acid tartrate) | NKT | | Dye mordant, reducing agent. | |
| **Cyanide** | | See potassium cyanide. | | |

| Chemical Trade Name | Toxicity | Health Effects | Uses | Precautions |
|---|---|---|---|---|
| Cyanoacrylate | Slight | Skin and respiratory irritant and allergen. Bonds skin. | Fast bonding glue: "Super Glue," "Krazy Glue." | Use less hazardous glues such as "Elmer's" white glue; use acetone as solvent. |
| Damar varnish | Slight | Dust may be an allergen. Check solvent hazards. | Resin derived painting medium. | Wear dust mask if using powdered form. |
| Diarylide, diarylanilide | Carcinogen | | Organic pigment/dye: contains benzidine. | **Do not use.** |
| 2,4-Diaminophenol | Moderate | A sensitizer. | Ingredient in film processing. | Wear gloves and use local exhaust ventilation. |
| Diatomaceous Earth | Slight | Excessive inhalation may cause silicosis. | Colloidal silica in pigments. | Wear dust respirator. |
| Diazo | Slight | An eye irritant. | A photo-sensitizer for photostencils in screenprinting. | Avoid dust inhalation. Wear respirator, gloves and goggles when making solutions. Safer than ammonium dichromate. |
| Dichlorophene | Slight to Moderate | An allergen. | Prevents mildew. | |
| Diethyl-para-phenylenediamine | Moderate to High | Causes severe skin allergies; may penetrate skin. Inhalation of powder can cause asthma and upper respiratory irritation. Ingestion associated with blurred vision, gastritis, liver and nervous system damage, weakness and vertigo. | A photo developer. | **Avoid.** Use less toxic developers and pre-mixed solutions to avoid inhalation of powders. Prevent skin contact: wear gloves, use tongs. |
| Dimethyl-para-phenylenediamine | Moderate to High | Similar hazards as previous entry. | A photo developer. | **Avoid.** See previous entry for precautions. |
| Dimezone (1-phenyl-4, 4-dimethyl-3-pyrazolidinone) | Slight | Skin and eye irritant; potential allergen. | A photo developer, similar to phenidone. | Use with exhaust ventilation. |
| Dioxane (dimer of ethylene glycol) | Moderate to High; Suspected carcinogen | All routes; eye, nose, throat, skin and respiratory irritant. Acute exposure may cause narcosis, gastrointestinal upset, liver, kidney damage. | Film cleaner. | **Do not use.** Substitute mineral spirits. |
| Dioxazine | Unknown | | Purple pigment. | |
| "Dippit" fixer (stannous chloride) | Moderate | An irritant by all paths of entry. Prolonged inhalation may cause benign pneumoconiosis. | For black and white instant photo-processes. | Wear goggles and gloves when handling. |
| Disodium EDTA | Slight to Moderate | See EDTA. | Color photo retouching agent for magenta dyes. | Avoid ingestion. |

| Chemical Trade Name | Toxicity | Health Effects | Uses | Precautions |
|---|---|---|---|---|
| **D-limonene** | Slight | Long term effects under study. May cause eye, mucous membrane, skin or gastrointestinal irritation. | A paint solvent in waterless hand cleaners and paint thinners; a terpene extracted from citrus peels. | Do not use around children as citrus odor renders it attractive to ingestion. Reasonable alternative to turpentine. |
| **Driers** | High | May increase lead or manganese exposure. | Contained in printing inks to accelerate drying. | Use inks and paints containing no driers, or those with cobalt linoleate drier. |
| **Dutch Mordant** | High<br><br>Combustible | Skin irritant, solution causes severe burns. Emits chlorine gas which irritates eyes and mucous membranes of the nose and throat. See potassium chlorate. | Hydrochloric acid, potassium chlorate and water solution used in printmaking for etching. | Prepare and use acid baths in a contained exhaust fume-hood. **Avoid** skin contact by wearing goggles and gloves. Substitute ferric chloride, which is safer and less expensive. |
| **EDTA** (ethylene diamine tetra-acetic acid) | Slight to Moderate | Weak solutions can be an irritant and allergen. Concentrates are corrosive to skin; repeated exposure via skin and inhalation can cause skin and respiratory allergies, irritation, liver and kidney damage. If ingested in large quantities, may cause kidney damage, calcium depletion, muscle spasms. | A developing bath for photo processes. | Wear goggles and gloves; use with local exhaust ventilation or wear a respirator approved for ammonia and amines. |
| **Elon (metol)** (monomethyl-p-aminophenol sulfate) | Moderate | Powder inhalation can cause breathing difficulty, allergies. Ingestion may cause nausea, dizziness. Irritated, sensitized skin may show tiny white blisters. Aminophenols can induce methemoglobinemia (cyanosis, chocolate-colored blood) if ingested or inhaled in large quantities. | A photo developing agent. | Use pre-mixed solutions. Wear goggles and gloves; mix powders in glove box, with a fume hood, or while wearing an approved toxic dust respirator. |
| **Eosin** | Slight | May cause dermatitis when used in cosmetics. | Red dye. | Avoid skin contact, wear gloves. |
| **Epoxy resin** (uncured) | Slight to Moderate<br><br>Suspected carcinogen | Skin irritant, allergen; chemically related to known carcinogens; under study. | Ingredient in glues, paints; must be cured with an amine or other hardener. May contain a variety of solvents. | **Avoid** epoxies containing glycidol, dimethylamine (DMA) glycidyl and glycol ethers, and 2-nitropropane. Substitute other epoxy products. |
| **Esters** (amyl, ethyl, isoamyl, isopropyl, methyl acetates) | Slight to Moderate | Irritants of skin, eyes, upper respiratory system; adverse effects on central nervous system, liver and kidney at high doses. | Solvent. | Least toxic class of solvents. Good odor (fruity) warning. Ethyl acetate the least toxic. |

| Chemical Trade Name | Toxicity | Health Effects | Uses | Precautions |
|---|---|---|---|---|
| **Etching Grounds** | Variable | Refer to specific ingredients for adverse effects and precautions. | Liquid grounds typically contain asphaltum, beeswax and rosin dissolved in paint thinner. | Avoid inhalation when heated. Use local exhaust ventilation and wear gloves. Substitute solvent-free hard ball ground. |
| **Ether** (diethyl ether) | Moderate Anesthetic<br><br>Extremely flammable potentially explosive | Irritant to eyes, skin, respiratory tract. Overexposure produces nausea, headache, dizziness, respiratory arrest. | For thinning etching grounds. | **Avoid.** Use ethanol, isopropyl alcohol or acetone. |
| **Ethyl alcohol** (ethanol, grain alcohol, denatured alcohol) | Slight<br><br><br><br><br><br><br>Reproductive toxin | Irritant to skin, mucous membranes; central nervous system depressant via inhalation and ingestion. Acute ingestion causes intoxication; chronic ingestion may cause liver damage, or adverse effects on fetal development. | Solvent. | Safest solvent. Use with exhaust ventilation and gloves. Denatured alcohol may contain small amounts of methyl alcohol which must **not** be ingested. |
| **Ethoxy-diglycol** | Moderate | Irritant to eyes and skin; harmful through ingestion. | Ingredient in Cibachrome developers 1B and 2B. | Induce vomiting if ingested. |
| **2-Ethoxy ethanol** | Moderate<br><br><br><br><br><br>Reproductive toxin | Mildly irritating on direct skin contact. Inhalation, ingestion and skin absorption may cause kidney damage, anemia, narcosis and behavioral changes. Testicular atrophy and low sperm count observed in animals. | Cellosolve, once common in color processes such as Cibachrome. | **Avoid.** If used, wear goggles and gloves; use exhaust ventilation or an organic vapor respirator. |
| **Ethylene diamine** | Moderate | Via all routes of entry; an irritant and sensitizer to both skin and respiratory system, can be absorbed through skin; potential for lung and kidney injury at high doses. | Developing bath for color photography. | Wear goggles and gloves; use local exhaust ventilation or wear respirator approved for ammonia and amines. |
| **Ethylene diamine tetra-acetic acid** (EDTA) | Slight to Moderate | Weak solutions can be an irritant and allergen. Concentrates are corrosive to skin; repeated exposure via skin and inhalation can cause skin and respiratory allergies, irritation, liver and kidney damage. If ingested in large quantities, may cause kidney damage, calcium depletion, muscle spasms. | A developing bath and clearing agent for photo processes. | Wear goggles and gloves; use with local exhaust ventilation or wear a respirator approved for toxic dust, ammonia and amines. |
| **Ethylene glycol** | Slight High | Via inhalation, skin. Via ingestion: may cause intestinal, heart, kidney, and metabolic disturbances; less than three ounces can be fatal to adults. | A solvent used in photo developers. | **Avoid** accidental ingestion. |

| Chemical Trade Name | Toxicity | Health Effects | Uses | Precautions |
|---|---|---|---|---|
| **Ethylene glycol dimethacrylate** | Slight | Irritant and allergen to skin and respiratory system. Bonds skin. | Fast bonding glue ingredient. | Use a safer alternative such as "Elmer's" white glue. |
| **Ethylene oxide** | Moderate to High<br>Suspected carcinogen<br><br>Reproductive toxin | Potent eye, mucous membrane and lung irritant. Chronic over-exposure may cause nerve damage. | A fumigant for conservation. | **Avoid.** Substitute O-phenyl phenol. |
| **Ferric ammonium citrate** | Slight | May be an irritant and sensitizer. | A blue toner; cyanotype and Van Dyke process photo-sensitizer. | Prevent skin contact to avoid sensitization. Wear gloves. |
| **Ferric ammonium sulfate** | Slight | May be an irritant and sensitizer. | A photo toner. | Wear gloves. |
| **Ferric chloride** | Slight to Moderate | Forms hydrochloric acid in solution which is irritating to skin, eyes, mucous membranes. | In photogravure process; an etching acid (iron perchloride). | Use with exhaust ventilation, wear gloves and goggles. A preferred acid for etching. |
| **Ferric oxalate** | Moderate<br><br>High | Corrosive to skin, eyes, respiratory system; Via ingestion, inhalation; can be fatal by ingestion; may cause kidney damage. | Used in photography as a platinum sensitizer. | Wear goggles and gloves when handling. Mix in a glove box or fume hood, or wear an approved toxic dust respirator. |
| **Ferric sulfate** | Slight | A slight irritant to skin, eyes, nose, throat. | A photo toner; dye mordant. | Use exhaust ventilation. |
| **Film cleaners** | Variable | Most are irritants. | Contain various toxic solvents such as toluene. | See Material Safety Data Sheet. |
| **Fixative** | Variable | Check ingredients, see lacquer and appropriate solvent for effects, hazards. | Resins disolved in solvents for spray fixing drawings. | Use only with effective exhaust ventilation that prevents inhalation of mist and vapors. |
| **Fixing baths** | Slight to High | Refer to specific ingredients for health effects and precautions in this Appendix. | Photography: contain sodium thiosulfate (hypo); a weak acid: acetic or sodium bisulfite; sodium sulfite as a preservative; potassium alum as a hardener; and boric acid as a buffer and anti-sludging agent. | Label solutions with all ingredients. Use local exhaust ventilation; wear gloves. Heat, decomposition or contamination with stop bath may release irritating sulfur dioxide gas. Do not use old hypo solutions. Cover trays when not in use. |
| **Flake white** | High | See lead carbonate. | | |
| **Fluorescent dyes** | Slight to Moderate | May irritate skin, eyes, and the respiratory system. | Used as brighteners in color developing processes. | Use exhaust ventilation. |

| Chemical Trade Name | Toxicity | Health Effects | Uses | Precautions |
|---|---|---|---|---|
| **Formaldehyde** | Moderate to High<br><br>Suspected carcinogen | Highly irritating to eyes, skin, and respiratory tract; an occasional sensitizer; may aggravate bronchitis and asthma. Frequently overlooked symptoms include shortness of breath , eye and skin irritation, headaches, dizziness, nausea and nosebleeds. | A preservative and hardener found in stabilizers, wetting agents, final baths, Kodalith developer, prehardeners, replenishers, retouching dyes and aquaeous paints and inks. | **Avoid.** Otherwise wear gloves and goggles. Use local exhaust ventilation or wear an approved formaldehyde respirator. Substitute succinaldehyde if possible. |
| **Formalin** (solution of 40% formaldehyde and up to 5% methanol) | Moderate to High Suspected carcinogen | Via all routes. See formaldehyde. | A hardener for photography, paints, etc. | **Avoid.** Substitute succinaldehyde if possible; observe precautions for formaldehyde. |
| **Formic acid** (methanoic acid) | Moderate | Corrosive to eyes, mucous membranes; may cause ulcerations in mouth and damage to gastrointestinal tract if ingested. | Dye mordant. | Wear goggles and gloves. Avoid ingestion. |
| **Fountain solutions** | Moderate to High<br><br>Suspected carcinogens | Corrosive to skin, may cause dermatitis. See dangers and precautions for each ingredient. | Used to maintain proper plate chemistry in lithography. Contain phosphoric or tannic acids as well as ammonium or potassium dichromate and formaldehyde. | **Avoid.** See dangers and precautions for each ingredient. Substitute a small amount of gum arabic in water. |
| **Gallic acid** | Slight to Moderate | Concentrate may irritate skin, mucous membranes; dilute solutions are less irritating. | 3,4,5-trihydroxybenzoic acid used in cyanotype printing; printmaking. | Wear goggles and gloves when handling concentrates. Use local exhaust ventilation. |
| **Gasoline** | Moderate<br><br>Extremely flammable | May cause dermatitis; vapors irritate eyes, respiratory tract; high doses by inhalation may cause incoordination, dizziness, headaches, nausea. | A solvent. May contain highly toxic benzene and/or organic lead. | **Do not use.** Substitute deodorized mineral spirits, isopropyl alcohol, ethyl alcohol, or acetone. |
| **Gesso, acrylic** | Slight | May contain small amounts of ammonia and formaldehyde; the latter may cause allergic reactions in sensitized individuals; is also a suspected carcinogen. | A primer for canvas and other surfaces; comprised of titanium white pigment and calcium carbonate (whiting) in an acrylic base. | If gessoing large areas, provide effective general ventilation. |
| **Gesso, traditional** | NKT | See effects for each ingredient. | A canvas primer of calcium carbonate (whiting) and calcium sulphate (plaster of Paris) in rabbit skin glue. | Do not use any gesso material containing lead. |
| **Gilsonite** | Moderate | Known to cause photosensitization of skin. | Ingredient in etching resist grounds. | Avoid skin contact; wear gloves. |
| **Glauber's salt** | NKT | See sodium sulfate. | | |

| Chemical Trade Name | Toxicity | Health Effects | Uses | Precautions |
|---|---|---|---|---|
| **Glutaraldehyde** | Variable | Eye, skin and respiratory irritant and sensitizer. | A photo print hardener; present in some paints. | Purity varies widely; beware of allergic dermatitis. |
| **Glycidol** (2,3-epoxy-propanol) | Moderate<br><br>Suspected carcinogen | Highly irritating to eyes on contact; burns may result; skin and respiratory irritant. | A dilutant/solvent in epoxy resins. | **Avoid.** Obtain Material Safety Data Sheets on epoxy products to avoid those containing glycidol. |
| **Glycin** (para-hydroxyphenyl aminoacetic acid) | Slight to Moderate | May cause skin irritation or dermatitis; breathing difficulty, nausea, and dizziness. | Ingredient in photo developers. | Use pre-mixed developers; use tongs to agitate solutions and handle prints. Wear gloves and goggles when preparing solutions. |
| **Glycol ethers** (cellosolves) | Slight to Moderate<br><br>Combustible | Mild skin irritants; can be absorbed through skin; vapors may irritate mucous membranes, upper respiratory tract. May reduce sperm counts in men. Under study for reproductive effects. | Used to disperse pigments in solution. Solvents in photo resists, offset lithography and many paint products and varnishes. Contained in dyes, lacquers, resins, epoxy products, cleaners. | **Avoid** Cellosolves. Propylene glycol ethers are preferred over ethylene glycol ethers. Use with local exhaust ventilation, and wear neoprene or natural rubber gloves. |
| **Gold chloride** | Slight to Moderate | A sensitizer. Chronic inhalation or ingestion may cause anemia, and kidney, liver and nervous system damage. | A photo toner. | Avoid repeated skin contact and inhalation of powder. |
| **Graphite** | NKT | | A common drawing medium. | |
| **Gum arabic** (gum acacia) | Slight to Moderate | Inhalation of powder may cause respiratory allergy or "printer's asthma." | Ingredient in photo-retouching mediums, lithography, gum printing; as a sealer; binder of pastels, water-soluble inks and paints. | Use premixed gum arabic to avoid inhalation of powder. |
| **Gum damar** | Slight | May cause allergies. | Ingredient in photo retouching mediums. | Avoid inhaling resin dusts. |
| **Gum tragacanth** | Slight | May cause allergies. | Ingredient in photo retouching mediums. | Avoid inhaling powder. |
| **Hansa** | | See Azo. | Organic pigment. | |
| **Hard Ball Ground** | Slight | Refer to specific ingredients for hazards and precautions. | Commercially available etching ground made without solvents. | A safer substitute for solvent-containing liquid grounds. |
| **Haze Remover** | Variable | Many are highly corrosive. Obtain Material Safety Data Sheets to learn ingredients, hazards and precautions. | Removes ink stains and ghosts from printing screens. | **Avoid.** Replace screen mesh. Use only with local exhaust ventilation and wear gloves. |
| **Hematoxylin** | Slight | Ingestion of large quantities may cause gastrointestinal upset. | A pigment extracted from wood. | |

| Chemical Trade Name | Toxicity | Health Effects | Uses | Precautions |
|---|---|---|---|---|
| **Heptane** | Slight<br><br>Flammable | Mild skin, eye and respiratory irritant; may predispose heart to abnormal rhythms. | A rubber cement solvent. | One of the least toxic aliphatic hydrocarbons. Preferred over hexane. |
| **Hexane**<br>(N-hexane) | High<br><br><br><br><br><br><br><br>Extremely flammable | Mild skin, eye and respiratory irritant. High dose exposure may produce headache, dizziness, gastrointestinal upset. Major concern is onset of peripheral nerve damage. Early symptoms include cramping, numbness or weakness in hands, legs and feet; loss of appetite and weight; fatigue. | An aliphatic hydrocarbon solvent of rubber cement, contact glues; an ingredient in multi-purpose solvents. | **Do not use** except with effective local exhaust ventilation and protective attire. Substitute safer adhesives such as those containing heptane. Methyl ethyl ketone enhances neurotoxicity of n-hexane. |
| **Hydrazine sulfate** | Moderate<br>Suspected Carcinogen | A sensitizer. | A photo developer. | **Avoid.** |
| **Hydrochloric acid**<br>(muriatic acid) | High | Direct contact with concentrated solutions may cause acid burns. Vapors are highly irritating to eyes, respiratory tract. Also associated with lung edema (excess fluid), chronic bronchitis, and tooth erosion. | A photo intensifier. Ingredient in Dutch Mordant used in printmaking. | Wear goggles and gloves when handling concentrates. Slowly add acid to solution to decrease generation of heat (exothermic reaction). Wear rubber apron, goggles; use with exhaust ventilation. |
| **Hydrofluoric acid** | High | Direct contact with solutions causes severe penetrating burns to eyes and skin. Pain and redness may be delayed for 24 hours. Vapors irritate mucous membranes, eyes, respiratory tract. Anemia, weight loss, dental problems associated with chronic overexposure. | Printmaking acid. | **Avoid**. Wear goggles, rubber apron and gloves when handling concentrates. Use in a fume hood. Substitute bifluoride pastes. |
| **Hydrogen peroxide**<br>(usually 3% solution) | Slight<br>to<br>Moderate | By skin and ingestion; may cause irritation of eyes, skin, mucous membranes. Concentrates (>3%) can cause more extensive eye, skin and mucous membrane damage. | A hypo eliminator; used in photo-sensitizing printing plates and screens. | Wear gloves and goggles; use local exhaust ventilation. |
| **Hydrogen selenide** | Moderate to High<br><br><br><br>Explosive in high concentration | Vapors extremely irritating to eyes and respiratory tract; low level exposures are associated with nausea, diarrhea, vomiting, metallic taste, odor of garlic on the breath, dizziness, fatigue and liver injury. | Selenium toner (sepia). In photography, those containing sulfides may release hydrogen selenide and hydrogen sulfide gas, especially if contaminated with acids carried over from the fixer into the toning bath. | Causes olfactory fatigue, reducing odor warning. Wash prints thoroughly to remove residual acid. Use local exhaust ventilation. Use pre-mixed solutions to avoid hazards of mixing. Do not use toners that require heating. |

| Chemical Trade Name | Toxicity | Health Effects | Uses | Precautions |
|---|---|---|---|---|
| **Hydroquinone** (1,4-dihydroxy benzene) | Moderate<br><br>High | Via skin and inhalation; is an allergen and irritant. Chronic occupational exposure may cause discoloration of cornea and conjunctiva, with structural changes in cornea that may alter vision. Ingestion causes tinnitus, nausea, headaches, dizziness, increased respiration, delirium, muscle twitches, anemia and, in severe cases, unconsciousness and coma. | A common ingredient in photographic developers, and replenishers. | **Avoid.** Use pre-mixed solutions or less toxic developer. |
| **Hydroxylamine hydrochloride** | Moderate<br><br>Suspected reproductive toxin | A sensitizer and irritant to skin, eyes, and mucous membranes by inhalation or ingestion. May cause methemoglobinemia, cyanosis, convulsions and coma. | A color fog restrainer, neutralizer and replenisher in photography. | Wear goggles and gloves; use with local exhaust ventilation. **Avoid** during pregnancy and nursing. |
| **Hydroxylamine sulfate** | Moderate<br><br>Suspected reproductive toxin | A sensitizer and irritant to skin, eyes, mucous membranes. Inhalation or ingestion may cause methemoglobinemia, cyanosis, convulsions and coma. | A color neutralizer, photography. | Wear goggles and gloves; use with local exhaust ventilation. **Avoid** during pregnancy and nursing. |
| **Hypo** | Variable | Refer to specific ingredients for health effects and precautions. Check Material Safety Data Sheets. | Common photo fixing bath. Contains: sodium or ammonium thiosulfate; a weak acid, usually acetic to further neutralize the alkaline developer; and sodium sulfite as a preservative. Acid-hardening hypos usually contain potassium aluminum sulfate (alum) and boric acid as a buffer and anti-sludging agent. | Label solutions with all ingredients. Use local exhaust ventilation; wear gloves and use tongs. Heat, decomposition or contamination with stop bath may release irritating sulfur dioxide gas. Cover trays between printing sessions. Do not use old hypo solutions. Use a water rinse step between developer and stop bath to reduce formation of sulfur dioxide gas. |
| **Indian red** | Slight | Fumes and dusts can produce a benign pneumoconiosis (siderosis) with shadows on chest radiographs. | Iron oxide pigment, red 101. | Avoid powdered pigments. |
| **Iodine** | Moderate to High | Dermal exposure or inhalation of vapors can cause respiratory irritation, chronic lung problems, insomnia, irregular heartbeat, and circulatory disorders. | A reducer and daguerreotype sensitizer. | Wear gloves when handling. If heating, use local exhaust ventilation or an air-supplied full-face respirator or an approved respirator with an acid gas cartridge to prevent inhalation of vapors. |

| Chemical Trade Name | Toxicity | Health Effects | Uses | Precautions |
|---|---|---|---|---|
| **Iron oxide** | Slight | Fumes and dusts can produce a benign pneumoconiosis (siderosis) with shadows on chest radiographs. | Pigment: Indian red, red iron oxide, Tuscan red, Mars red, Mars orange, Red 101. | Avoid powdered pigments. |
| **Isophorone** (trimethyl-cyclohexanone) | Moderate to high | Central nervous system depressant associated with kidney, liver damage. Irritating to eyes, skin and upper respiratory tract. Chronic inhalation may lead to fatigue, nausea, headache and dizziness. | In paint and varnish removers, lacquer and plastic solvent. | **Avoid.** Very toxic ketone. Use only with local exhaust ventilation. |
| **Isopropyl alcohol** (household rubbing alcohol, isopropanol) | Slight to Moderate<br><br>Flammable | In high concentrations, vapors produce mild eye and respiratory irritation. Ingestion of 50-250 ml of rubbing alcohol may cause drunkenness, abdominal pains, vomiting, low blood sugar and central nervous system depression. | A solvent. | Good odor warning; one of least toxic solvents. No evidence of long-term hazards. |
| **Ivory black** | NKT | | Carbon from charred animal bones, black 9. | |
| **Kerosene** | Slight to Moderate<br><br>Flammable | A central nervous system depressant. Can be absorbed through skin; skin and mucous membrane irritant. | Aliphatic hydrocarbon, petroleum distillate, relatively inexpensive solvent. | **Avoid.** Use local exhaust ventilation; wear nitrile or neoprene gloves. If ingested, do **not** induce vomiting, which could cause fatal pulmonary edema. |
| **Ketones** (acetone, cyclohexanone, isophorone ketone, methyl butyl ketone, methyl ethyl ketone, methyl isobutyl ketone) | Variable<br><br>Flammable<br><br>Anesthetic | Methyl butyl ketone associated with neuropathy. All cause mucousal irritation, dermatitis. High exposures associated with central nervous system depression, narcosis. | Organic solvents for lacquers, oils, plastics, vinyl inks and waxes; paint removers and photo-supplies. | Acetone less toxic, but extremely flammable. Wear gloves and use local exhaust ventilation. |
| **Kodak color developers CD-2, CD-3, CD-4, and CD-6** | Variable | Check Material Safety Data Sheets for hazards. | Photography | Use gloves and local exhaust ventilation. Follow Material Safety Data Sheet precautions for ingredients and product. |
| **Kodalith Developer** | Moderate<br><br>Slight | Part A: contains hydroquinone, sodium formaldehyde bisulfite.<br><br>Part B: Contains sodium carbonate. May cause skin and eye irritation and allergic skin reaction. | For photography and printmaking. | Avoid skin, eye and clothing contact; inhalation of dust. Use gloves and local exhaust ventilation. Wash thoroughly after handling. In case of contact, immediately flush with water; for eyes get medical attention. |

| Chemical Trade Name | Toxicity | Health Effects | Uses | Precautions |
|---|---|---|---|---|
| **KPR Photo Resist** | Variable | See ingredients for hazards and precautions. | For photo-etching. Contains ethylene glycol monomethyl ether acetate and 5 to 10% polymers. | **Avoid** this process if possible. Buy presensitized zinc plates to eliminate using photo resist and KPR blue dye. |
| **KPR Developer** | Variable | See ingredients for hazards and precautions. Check Material Safety Data Sheets. | For photo-etching. Contains 75% xylene, 20-25% ethylene glycol monomethyl ether acetate. | **Avoid.** Use gloves and local exhaust ventilation. |
| **KPR Blue Dye** | Variable | See ingredients for hazards and precautions. Check Material Safety Data Sheets. | For photo-etching. Contains ethylene glycol monoethyl ether (Cellosolve), xylene and benzaldehyde. | **Avoid.** Use gloves and local exhaust ventilation. Buy presensitized zinc plates to avoid using this material. |
| **KPR thinner** | Variable | See methyl Cellosolve acetate for hazards and precautions. | For photo-processes. Contains (methyl Cellosolve acetate) ethylene glycol monomethyl ether acetate or 2-methoxy ethyl acetate. | **Avoid.** Use gloves and local exhaust ventilation. Buy presensitized zinc plates to avoid using this product. |
| **Lacquers** | Variable Anesthetic Flammable | See ingredients for hazards and precautions. Check Material Safety Data Sheet. | A mixture of resins and solvents for printmaking, drawing, and painting; fixative sprays. | **Avoid.** Use only with local exhaust ventilation; wear gloves and protective attire. |
| **Lacquer C** (vinyl base) | Variable Anesthetic Combustible | See ingredients for hazards and precautions. Check MSDS. | Used in printmaking. | **Avoid.** See previous entry. |
| **Lacquer thinner/solvent** (lacquer film adherents) | Variable Anesthetic Combustible | See ingredients for hazards and precautions. Check Material Safety Data Sheets. | A solvent mixture that may contain various alcohols, aliphatic, aromatic or chlorinated hydrocarbons, ketones and acetates; used as lacquer film adherent. | **Avoid.** Substitute acetone, ethyl or isopropyl alcohol if effective. Use only with local exhaust ventilation or an approved organic vapor respirator. Observe fire precautions. |
| **Lake** | Variable | See Barium Compounds. Check label for barium. By ingestion when precipitated by barium carbonate. | A barium precipitate of an organic chemical dye in paints and inks: red lake, red 53. | Avoid powdered pigments. Observe barium precautions. Some grades contain soluble barium. |
| **Lead acetate** | High | See lead compounds. | A cyanotype toner. | **Avoid.** Substitute tannic acid or gold chloride. |

| Chemical Trade Name | Toxicity | Health Effects | Uses | Precautions |
|---|---|---|---|---|
| **Lead compounds** (lead antimoniate, lead carbonate, lead chromate, lead molybdate, and others) | High<br><br>Reproductive toxin | Inhalation and ingestion are major routes that cause lead poisoning, symptoms of which are: anemia, gastroenteritis, weakness, malaise, headaches, irritability, joint and muscle pain, kidney, liver and nervous system damage. May affect neurological development in fetuses and children. Accumulates in bones and tissues; children and fetuses are more susceptible to lower doses. | Pigment, paint ingredient: flake white, Naples yellow, chrome yellow; white 2, 4; red 103, 104, 105; orange 21, 45; yellow 34, 46; green 15. Ingredient in ink driers. | **Avoid.** Substitute non-lead gesso, paints and inks. |
| **Lead carbonate** (flake white lead white) | High | See lead compounds. | A pigment used in painting and primers: white 1, flake white. lead white. | **Avoid.** Substitute non-lead gesso and paints. |
| **Lead nitrate** | High<br>Oxydizer | See lead compounds. | A salted paper toner used in photography. | **Avoid.** Substitute gold chloride and platinum chloride. |
| **Lead oxalate** | High | See lead compounds. | A platinum toner used in photography. | **Avoid.** Substitute potassium phosphate. |
| **Lepages liquid glue** | Slight | Contains animal hide glues and preservatives that may be allergenic. | Common adhesive, can be used in screenprinting. | |
| **Lime** (calcium oxide, calcium hydroxide, calcium carbonate) | Moderate | Corrosive to eyes, skin, respiratory and digestive tracts. | Calcium hydroxide in solution to grind pigments for fresco painting. | Use effective ventilation. Wear goggles and gloves. |
| **Limonene** (d-limonene) | Slight | Long term effects under study. May cause eye, mucous membrane, skin or gastrointestinal irritation. | A paint solvent in waterless hand cleaners and paint thinners; a terpene extracted from citrus peels. | Do not use around children: citrus odor renders it attractive for ingestion by children. Reasonable alternative to turpentine. |
| **Linseed oil** | NKT<br><br>Combustible | An allergen for some individuals when heated on etching plates. When heated. | Common binder in oil paints, printing inks. Pressed and distilled from flax seed. | May result in spontaneous combustion in rags. Dispose used rags in metal safety containers. |
| **Lithium** | Slight to Moderate | Mild skin, eye, mucous membrane irritant. If ingested, may cause fatigue, dizziness, gastrointestinal upset. Systemic absorption can cause kidney damage, tremors, muscular weakness, seizures, coma, death. | Found in some pigments. | Avoid inhalation of powdered pigments. |

| Chemical Trade Name | Toxicity | Health Effects | Uses | Precautions |
|---|---|---|---|---|
| **Lithographic crayons** | Slight<br><br>Suspected carcinogen | Repeated exposures associated with skin cancer attributed to impurities in some carbon pigments. | Lithography crayons containing lamp black or carbon black. | Wear gloves. |
| **Lithotine** | Slight to Moderate | Eye, mucous membrane and skin irritant. | A solvent containing mineral spirits and pine oil. | Use local exhaust ventilation and nitrile gloves. If ingested, do not induce vomiting as accidental entry of lithotine into lungs can cause fatal pulmonary edema. |
| **Magnesium** | Slight | Large exposure to magnesium oxide fumes could result in metal fume fever, or eye and upper respiratory irritation. | A metal in pigment. | Avoid powdered pigments. |
| **Magnesium carbonate** | NKT | Nuisance dust. | Ingredient in whiting, used in printmaking. | Wear dust mask. |
| **Manganese compounds** (manganese ammonium phosphate, dioxide, silicates; barium manganate) | Moderate to High | Can irritate eyes, mucous membranes and respiratory tract. Chronic inhalation can produce behavioral disturbances and a degenerative nervous system disorder resembling Parkinsonism. Reproductive effects. | Drier in inks. Also a pigment ingredient in Mars brown, raw and burnt umber, manganese blue and violet, red 48, blue 33, violet 16, black 14, 26. | **Avoid.** Wear gloves, do not use powdered pigments. |
| **Mars colors** | Slight | See iron oxide for hazard. Fumes and dusts can produce a benign pneumoconiosis (siderosis) with shadows on chest radiographs. | Includes black, orange, red 101, violet, yellow 42, 43, brown. | Avoid powdered pigments. |
| **Mastic varnish** | Moderate to High | Obtain Material Safety Data Sheet for hazards. | A painting medium that will not mix with mineral spirits. Contains alcohols, turpentine, other solvents. | Observe appropriate ingredient precautions on MSDS. Use exhaust ventilation and gloves. |
| **Mercuric chloride** and **mercuric iodide** | High | (See mercury) Mercury salts are corrosive to skin, eyes and intestines. | Salts used as Daguerre-otype intensifiers and preservatives. | **Avoid.** With proper precautions, chromium intensifiers are safer alternatives. |

| Chemical Trade Name | Toxicity | Health Effects | Uses | Precautions |
|---|---|---|---|---|
| **Mercury** | High | Acute inhalation of large amounts of elemental mercury vapor may cause respiratory irritation and pulmonary edema. Skin contact leads to irritation and/or sensitization. Mercury salts, if acutely ingested, can cause intestinal, liver, kidney, and nervous system damage. Chronic inhalation of elemental vapor, or chronic ingestion of mercury salts, can cause gum disorders, kidney damage and permanent impairment of nervous system. | Daguerreotype process (mercuric sulfide). Pigment ingredient: vermillions, cinnabar, mercadium colors; red 106. A former preservative in water-based paints, may still be in latex house paints. | **Avoid** all mercury compounds. Vapor has no odor warning. Use the Becquerel method in Daguerreotype. |
| **Methanoic acid** (formic acid) | Slight to Moderate | Corrosive to eyes, mucous membranes; may cause ulceration. | Dye mordant. | Wear goggles, gloves. If ingested, call poison control center. |
| **Methyl alcohol** (methanol, denatured, wood alcohol) | High | By ingestion. As little as three tablespoons has led to blindness; larger doses have resulted in death. Inhalation, skin absorption can contribute to injury of internal organs and the nervous system. In high concentrations, vapors are irritating to eyes and skin. | A solvent. | **Avoid.** Use isopropyl alcohol. |
| **Methyl bromide** | High | Vapors irritate eyes, skin and respiratory tract. Inhalation of high concentrations may lead to lung edema or nervous system impairment. | A preservative and fungicide. | **Avoid.** Substitute O-phenyl phenol. |
| **Methyl butyl ketone** (MBK, methyl n-butyl ketone, 2-hexanone) | High  Flammable | All paths of entry. Skin, eye, mucous membrane irritant; inhalation: headache, lightheadedness, nausea, vomiting, dizziness, unconsciousness; peripheral nerve damage occurs with repeated exposures. | A ketone solvent. | **Do not use.** Substitute acetone. |
| **Methyl Cellosolve** (ethylene glycol monomethyl ether, or 2-methoxy ethanol) | Moderate  Reproductive toxin | Via all paths of entry. Demonstrated in animal tests to cause testicular atrophy, miscarriages, and birth defects. May cause bone marrow and central nervous system depression, heart, lung and kidney problems. | A glycol ether in some photo-chemicals; in KPR developer, KPR photo resist, KPR blue dyes and thinner used in photo etching. | **Avoid.** Permeates most glove materials in minutes; can be absorbed through skin. Use presensitized zinc plates to eliminate using KPR photo resist, developer and blue dye. |

| Chemical Trade Name | Toxicity | Health Effects | Uses | Precautions |
|---|---|---|---|---|
| **Methyl Cellosolve Acetate** (2-methoxy-ethyl acetate, ethylene glycol monomethyl ether, or 2-methoxy ethanol) | Moderate<br><br>Reproductive toxin | Absorbed through skin. Eye, skin, lung irritant. Based on animal studies, may cause bone marrow and nervous system depression, kidney damage, decreased sperm counts, and birth defects. | A glycol ether in some photo-chemicals; in KPR developer, KPR photo resist, KPR blue dyes and thinner used in photo etching. | **Avoid.** Permeates most glove materials in minutes (see above). |
| **Methyl cellulose** | NKT | Powder may be an allergen for some individuals. | An archival adhesive with many applications. | |
| **Methyl Chloroform** (1,1,1-trichloroethane) | Slight to Moderate<br><br><br><br><br><br><br><br><br><br><br>Flammable<br>High | Absorbed through skin. Vapors irritate eyes, skin, respiratory tract. Chronic overexposure may cause dermatitis. Acute inhalation or ingestion of large amounts may lead to nausea, euphoria, incoordination, dizziness, drowsiness, lung, liver and kidney injury, abnormal heart rhythms, and nervous system impairment. If heated. Decomposes in heat to produce phosgene gas and hydrochloric acid. | Chlorinated hydrocarbon in photo color retouchers and correction fluids; a printmaking solvent; component in aerosol sprays and film cleaners. | **Do not use.** Substitute water-soluble inks and correction tapes or film. If necessary to use, wear gloves; use local exhaust ventilation. |
| **Methyl ethyl ketone** (MEK, 2-butanone) | Anesthetic<br><br><br><br><br>Flammable<br><br>High | May cause dizziness, headache, lightheadedness, nausea, vomiting, and unconsciousness if inhaled at high concentrations. Vapors irritate eyes, skin, mucous membranes and respiratory tract. Enhances nervous system damage caused by methyl butyl ketone (MBK) or by n-hexane. | Organic solvent for lacquers, oils, plastics, vinyl inks and waxes. | Use with local exhaust ventilation; natural rubber gloves best, but will permeate in 10 minutes. Substitute acetone. Do not use with n-hexane. |
| **Methylene chloride** (dichloromethane, methylene dichloride) | Moderate<br><br><br>Asphyxiant<br><br><br><br>Suspected carcinogen | Eye, skin and upper respiratory tract irritant. Metabolized (transformed) to carbon monoxide. Inhalation of large amounts has been associated with heart attacks. Causes lung, liver, breast cancer in laboratory animals. | Chlorinated hydrocarbon used in printmaking, lithography; spray paints. | **Avoid.** |
| **Metol (elon)** (monomethyl para-aminophenol sulfate) | Moderate | Powder inhalation can cause breathing difficulty. Ingestion may cause nausea, dizziness. Irritated, sensitized skin may show tiny white blisters. | A photo developing agent. | Use pre-mixed solution, avoid direct skin contact. Wear gloves; mix powders in glove box, with a fume hood, or while wearing an approved toxic dust mask. Use tongs, local exhaust ventilation. |

| Chemical Trade Name | Toxicity | Health Effects | Uses | Precautions |
|---|---|---|---|---|
| **Mica** | Slight to Moderate | Chronic inhalation of respirable dusts can cause pneumoconiosis with cough, shortness of breath, weakness and weight loss. | A mineral in iridescent paint colors. | Avoid inhalation of fine dusts. Wear an approved toxic dust mask. |
| **Mineral spirits** (naphtha, Stoddard solvent, Varsol) | Slight to Moderate <br><br> Combustible | Absorbed through skin, cuts or abrasions. Vapors irritate eyes, skin and respiratory tract. Chronic over-exposure associated with fatigue, headache, bone marrow depression. | Petroleum distillate solvents. | Use with local exhaust ventilation; wear nitrile gloves. Do **not** induce vomiting if ingested. |
| **Molybdate** | High | If it contains lead. | Organic orange and red pigments, red 104. | Avoid. |
| **Monastral blue** | Slight to Moderate Possible carcinogen | Stains skin. Inhalation of powdered pigment may cause breathing difficulty. If contaminated with PCBs. | Phthalocyanine blue pigment, blue 15, brand-name blue. | Avoid powdered pigment. See phthalocyanine blue. |
| **Naphtha** | Variable | See mineral spirits. | A solvent that can be a petroleum distillate (aliphatic) or coal tar derivative (aromatic hydrocarbon). | Use with local exhaust ventilation; wear nitrile gloves. Different from VM&P Naphtha. |
| **Naphthol** | Slight | Skin irritant. If absorbed through the skin in large amounts, may cause gastrointestinal and systemic symptoms. Long-term effects under study. | Organic pigment: reds, Acid green, green 12. | Avoid powdered pigments and spray paints. |
| **N-Hexane** (Normal hexane) | High <br><br><br><br><br><br><br> Extremely flammable | Mild skin, eye and respiratory irritant. High dose exposure may produce headache, dizziness, gastrointestinal upset. Major concern is onset of peripheral nerve damage. Early symptoms include cramping, numbness or weakness in hands, legs and feet; loss of appetite and weight; fatigue. | An aliphatic hydrocarbon solvent of rubber cement, contact glues; ingredient in multi-purpose solvents. | **Do not use** except with effective local exhaust ventilation. Substitute safer adhesives such as waxers or those containing heptane. Methyl ethyl ketone enhances neurotoxicity of n-hexane. |
| **Nickel** | Moderate <br><br><br><br><br><br><br> Suspected carcinogen | Common cause of severe skin allergy, chronic eczema. Ingestion of salts can cause giddiness, nausea. Fumes irritate respiratory tract, may cause pulmonary edema. Inhalation of some nickel compounds associated with nasal and lung tumors. | Metal used in photo processes; pigments: yellow 53, 57; green 10. | Avoid powdered pigments, spray paints. Use with local exhaust ventilation or air-supplied respirator. |
| **Nigrosine** | Slight | May be a skin and respiratory allergen. | Organic pigment. | |

| Chemical Trade Name | Toxicity | Health Effects | Uses | Precautions |
|---|---|---|---|---|
| **Nitric acid** (aqua fortis, engraver's acid) | High<br><br>Oxidizer | In concentrated form via all routes. Repeated inhalation may cause chronic bronchitis, pulmonary edema, emphysema, and erosion of dental enamel. Corrosive to skin, (penetrating burns) irritating to eyes. Vapors and gases have poor olfactory warning. | For etching, lithography; photography tray cleaners, and ambrotype. | Keep acid bath covered as much as possible; use with local exhaust ventilation. Wear goggles and gloves when handling; there are no approved respirator cartridges. Substitute phosphoric acid. When mixing, always add acid slowly to water, never the reverse. |
| **Nitrobenzene** | Moderate to High<br><br>Reproductive toxin | Well-absorbed via all routes. Irritating to eyes, skin, mucous membranes. May cause headache, cyanosis, weakness, intestinal upset, liver injury, and methemoglobinemia. Chronic exposure associated with anemia, bladder irritation; adverse reproductive effects. | A solvent. | **Avoid.** Use only with local exhaust ventilation. Wear supported polyvinyl alcohol (PVA) gloves. Odor of shoe polish is a good warning property. |
| **6-nitrobenzimidazole** | High<br>Suspected carcinogen | | In phenidone developers. | **Avoid.** |
| **Nitrogen oxides** | Moderate to High | Gases/vapors irritate eyes and respiratory tract; can cause pulmonary edema, chemical pneumonia, and permanent lung damage. | Emitted from carbon arc lamps and from etching zinc plates with nitric acid. | Use exhaust ventilation. Do not use carbon arc lamps unless they are vented directly to the outside. Keep acid trays covered when not in use. |
| **Oleum** (foaming sulfuric acid) | High | Vapors and concentrated solutions cause severe burns, irritations to skin, eyes, mucous membranes, respiratory tract. May release irritating sulfur oxides. Dilute solutions dry and may ulcerate skin. | Printmaking; dye mordant. | Wear goggles and gloves. Keep away from heat to prevent release of irritating sulfur dioxide gas. |
| **Ortho-phenyl phenol** (o-phenyl phenol) | Slight | Eye and upper respiratory irritant; chronic exposure associated with kidney damage. | A preservative for photography and conservation. | Avoid inhalation of powder. Wear gloves and an approved respirator with organic vapor cartridge. |
| **Ortho-phenylenediamine** | Moderate | See Para-phenylenediamine. | A photo developer. | |
| **Overprint Clear** | Variable | Check Material Safety Data Sheets for hazards and precautions. | Used in screenprinting. Contains hydrocarbon solvents. | **Avoid.** Substitute water-based equivalent. |

| Chemical Trade Name | Toxicity | Health Effects | Uses | Precautions |
|---|---|---|---|---|
| Oxalic acid | High | Via all paths of entry, both powder and solution. Corrodes skin, causes skin ulcers and with repeated skin contact may cause gangrene in extremities. Inhalation may cause severe irritation and possible ulceration in the respiratory tract; ingestion causes kidney damage, potentially fatal. | Used as a photo toner; dye mordant. | Wear goggles and gloves; use local exhaust ventilation. Add acid slowly to solution. Cover acid bath when not in use. |
| Ozone | Moderate | Strong lung irritant, can cause cell damage that increases lung permeability. Irritating to skin, eyes, and mucous membranes. Acute exposure can result in pulmonary edema and hemorrhage; may be fatal. Symptoms of low exposure include headache, malaise, dizziness. | Released from some photocopiers, carbon arc lamps and color laser printers. Also produced by decomposition of nitrogen-containing compounds. | Use carbon arc lamps only with exhaust ventilation that vents directly to the outside. Substitute halogen quartz lamps or sunlight. Place color copiers in a well-ventilated location that exhausts to the outside. |
| Paint Thinner | Variable <br> Combustible to Flammable | Check label and Material Safety Data Sheets for ingredient health effects and precautions. | Most are mineral spirits, but some may contain aromatic and chlorinated hydrocarbons or ketones. | "Odorless" paint thinners have most aromatic hydrocarbons removed. Avoid any product containing benzene. |
| Palladium chloride | Slight Suspected carcinogen | A sensitizer and skin irritant. | A photo toner and sensitizer. | Use with exhaust ventilation and gloves. |
| Para-Aminophenol | Moderate | A sensitizer. See aminophenol. | Photo developer. | Avoid skin contact and inhalation of powder. Wear approved respirator and gloves when preparing solutions. Use exhaust ventilation. Buy pre-mixed developer solutions. |
| Para-Phenylenediamine | Moderate | Via all routes. Causes severe skin allergies. Inhalation of powder can cause asthma and upper respiratory irritation. Ingestion has been associated with blurred vision, gastritis, liver and spleen damage, nervous system damage, weakness and vertigo. | A photo developer for both black and white and color. | Use pre-mixed developer solutions and tongs. Wear gloves and goggles while mixing solutions. Mix powders only in a glove box or fume hood, or while wearing an approved toxic dust mask. |
| Para-toluene sulfonic acid | Moderate | Via all routes of entry. Respiratory irritant, corrosive to skin, eyes, mucous membranes, digestive system. | Cibachrome (color) processing chemical. | Wear goggles and gloves; use local exhaust ventilation. If swallowed, do **not** induce vomiting. Seek medical attention. |

| Chemical Trade Name | Toxicity | Health Effects | Uses | Precautions |
|---|---|---|---|---|
| **Paris blue** | Slight High | If heated. See potassium cyanide. | Contains ferric ferrocyanide, pigment blue 27, Prussian blue. | Avoid heating. If exposed to heat, acid, or ultraviolet light, will emit highly toxic hydrogen cyanide gas. |
| **Petroleum distillates** (gasoline, n-hexane, heptane, Stoddard solvent, kerosene, benzine, mineral spirits, VM&P naphtha) | Variable Combustible to Flammable | Vapors irritating to eyes, skin, respiratory tract. Central nervous system depressants. Narcotic effects by absorption and inhalation. See individual listings. | Common solvents, paint thinners, cleaning fluids. Check Material Safety Data Sheets on specific agents. | **Avoid.** Use only with effective local exhaust ventilation and nitrile gloves. If accidentally ingested, do **not** induce vomiting. |
| **Petroleum ether** (petroleum benzine, lactol spirits) | Variable Flammable | Vapors irritating to eyes, skin, respiratory tract. Central nervous system depressant. | Common solvent in paint thinners, cleaning fluids. Check Material Safety Data Sheet on specific agents. | **Avoid.** Use only with effective local exhaust ventilation; nitrile gloves. |
| **Phenidone** (1-phenyl-3-pyrazolidone) | Slight | Ingestion of large amounts may lead to liver damage. Prolonged contact with powder can cause skin allergies. | A photo developer. | Use pre-mixed solutions, tongs, local exhaust ventilation; keep tray covered whenever possible. |
| **Phenol** (carbolic acid) | Moderate to High | Dilute solutions can cause serious skin or eye burns without warning pain. Vapors irritate eyes, skin, respiratory tract. Absorbed by all routes. Systemic effects include nausea, vomiting, kidney and liver damage; high doses can be fatal. | A preservative, insecticide, and fungicide. Used in lithography. | **Avoid.** Substitute thymol or o-phenyl phenol. Use only with local exhaust ventilation or fume hood. Wear gloves and other protective attire. |
| **Phenyl mercuric chloride** or **acetate** | Moderate High | To skin. Via inhalation and ingestion. See mercury effects. | A preservative. | Use pre-mixed solution rather than powder; wear gloves and use exhaust ventilation or an approved respirator when handling concentrates or powders. |
| **Phosphoric acid** (orthophosphoric acid) | High | By all routes; a corrosive irritant throughout the body. Less toxic than nitric and sulfuric acids. | In photography and printmaking. | Wear safety goggles and gloves; use local exhaust ventilation. Add acid slowly to water. |
| **Photo-etching** | Moderate Combustible | See methyl Cellosolve acetate and xylene. See KPR listings. | Photo-resist contains methyl Cellosolve acetate and xylene. | Use with fume hood, wear butyl gloves. Expose plate with photo-floods, not carbon arc lamps. |
| **Photo fixers** | Slight | See sodium thiosulfate. | | |
| **Phthalocyanine** | Slight to Moderate Possible carcinogen Possible reproductive toxin | If contaminated by polychlorinated biphenyls (PCBs). | Blue and green pigment ingredient, green 7, blue 15, Monastral blue, manufacturer's name blue or green. | Use recently prepared, labeled products. Avoid powdered pigments, paint sprays. Discard products purchased prior to 1982. |

| Chemical Trade Name | Toxicity | Health Effects | Uses | Precautions |
|---|---|---|---|---|
| **Plaster of Paris** | NKT | If it contains no preservatives. | Calcium sulfate. | See calcium sulfate. |
| **Platinum chloride** | Moderate<br><br><br><br><br><br>Unknown | May cause skin irritation or allergy, worsen pre-existing dermatitis, or cause respiratory allergies, severe asthma (platinosis), lung scarring and emphysema. Via ingestion. | A photo toner and sensitizer. | Wear goggles and gloves. Mix salts under a fume hood or in a glove box; or wear an approved toxic dust mask. |
| **Polyacrylamide** (cationic) | Slight to Moderate Explosive Flammable | Eye, skin irritant.<br><br>Dust.<br>Solution is a slip hazard. | A retention aid for papermaking. | Use glove box, or dust mask, goggles and gloves when mixing powders into solution; gloves and goggles when handling solution. Keep powder away from oxidants, sources of ignition. |
| **Polychlorinated biphenyls** | Slight to Moderate<br><br>Suspected carcinogen<br><br>Possible reproductive toxin | High concentrations irritate eyes and mucous membranes. Known animal carcinogen; massive exposures may also cause the skin disorder, chloracne; liver and kidney damage. | PCBs. Prior to 1982, a possible contaminant of phthalocyanine colors. | **Avoid.** Discard phthalocyanine colors purchased prior to 1982. |
| **Polyethyl alcohol** | Slight | Via all routes; irritating to eyes. | P-tertiary-octylphenoxy-polyethal alcohol, a wetting agent used in photography; has low volatility. | Use exhaust ventilation. |
| **Poly-methyl-methacrylate** | None to Slight | Liquid is an irritant and sensitizer. Once poly-merized, it is generally without hazard. | Fast bonding glue ingredient. | Obtain Material Safety Data Sheet and observe precautions. |
| **Polyurethane** | Variable | Can cause lung irritation. Check Material Safety Data Sheet or product label for contents. See mineral spirits. | Sealer for screen printing frame and base. | Use exhaust ventilation or wear an approved respirator. Do not spray. Substitute duct tape. |
| **Polyvinyl acetate** (PVA) | Slight | Dusts and preservatives may be allergens. | Common white glue; pH neutral, but non-reversible, so archivists dislike it. May be ingredient in screenprinting direct photo-sensitive emulsions. | Relatively safe; use with gloves, ventilation. |
| **Potassium acid tartrate** | NKT | | Reducing agent. | |
| **Potassium aluminum sulfate** (alum) | Slight | Via all routes; may be an irritant and sensitizer. | Alum; a photo/film hardener. | Avoid inhalation of powders and skin contact. |

| Chemical Trade Name | Toxicity | Health Effects | Uses | Precautions |
|---|---|---|---|---|
| **Potassium bisulfite** (potassium hydrogen sulfite) | Slight to Moderate | Irritating to eyes, skin and respiratory tract. Adverse reactions more frequent in asthmatics. | Photography for a hypo eliminator or reducer. | Avoid heating or contamination with acids. Wear goggles and gloves when handling powders or concentrates, mix in a glove box or under a fume hood, or wear an approved toxic dust mask. |
| **Potassium bromide** | Slight to Moderate | Through inhalation, skin. Ingestion induces vomiting; chronic ingestion may cause depression, weight loss, drowsiness, mental confusion, deterioration, vertigo, and even psychosis. | In photography developers, bleach baths; as a restrainer, neutralizer, hardener, and toner. | Wear goggles and gloves when mixing solutions or powders; use local exhaust ventilation, glove box, fume hood or wear an approved toxic dust mask when handling powders. |
| **Potassium carbonate** (potash) | Moderate to High | Irritates skin, eyes, mucous membranes and respiratory tract. Ingestion of large amounts can be fatal. | A photo developer. | Wear goggles and gloves when mixing solutions or powders; use local exhaust ventilation, glove box, fume hood or wear an approved toxic dust mask when handling powders. |
| **Potassium chlorate** | Moderate High <br><br><br><br><br><br><br><br>Combustible <br><br>Explosive Oxidizer | Via skin and ingestion; via inhalation. Its combination with hydrochloric acid to prepare Dutch mordant releases highly toxic chlorine gas. Ingestion may cause abdominal pain, vomiting, diarrhea, damage to red blood cells, liver, kidneys, and methemoglobinemia. Fire hazard in contact with flammable materials. With salts, resins, dusts, heat or friction. | A photo sensitizer; ingredient in Dutch mordant. | Use in a fume hood or wear an approved respirator with an acid gas cartridge. Store away from heat and combustible materials. Use ferric chloride as a safer alternative. |
| **Potassium chlorochromate** | Slight to Moderate | Via skin, ingestion. Via inhalation. May cause skin, respiratory irritations and allergies. Ingestion leads to gastroenteritis, muscle cramps, vertigo and kidney damage. | A photo intensifier. | Do not heat or add acids which will release highly toxic chlorine gas. |
| **Potassium chloroplatinate** (potassium tetrachloroplatinate) | Moderate | Via skin, ingestion or inhalation. A sensitizer that causes skin and respiratory allergies, lung scarring and emphysema. | A photo toner. | Wear gloves or a barrier hand cream and goggles. Mix in a fume hood or glove box, or wear an approved toxic dust mask. |
| **Potassium chrome alum** (potassium chromium sulfate) | Moderate | Via skin, ingestion or inhalation. An irritant and sensitizer. Ingestion of large amounts can cause gastroenteritis, kidney damage, and shock. | A photo/film hardener. | Wear goggles and gloves when handling solutions or powders. Mix powders in a fume hood, glove box, or while wearing an approved toxic dust mask. |

| Chemical Trade Name | Toxicity | Health Effects | Uses | Precautions |
|---|---|---|---|---|
| **Potassium cyanide** | Moderate to High | Absorbed through skin. By ingestion and inhalation can be rapidly fatal from arrest of cellular respiration. | A photo reducer and intensifier. | **Do not use.** Substitute potassium ferricyanide. |
| **Potassium dichromate** | Moderate to High Flammable<br><br>Suspected carcinogen<br><br>Oxidizer | Via skin, ingestion. Via inhalation. Dusts and mists are irritating and sensitizing to eyes, skin, and respiratory tract. Kidney damage associated with high doses. | A photography intensifier and sensitizer. A dye mordant. | Do not add hydrochloric acid during intensification process as it will produce highly toxic chromic acid. Mix powders into solution only in a glove box or while wearing an approved toxic dust respirator and gloves. |
| **Potassium ferricyanide** or **ferrocyanide** (Farmer's reducer) | Variable | Excessive heat or mixing with strong acid may release toxic hydrogen cyanide. | In photo bleaches as a reducer, and in palladium, platinum and carbro printing processes. | Potassium ferricyanide can be recovered. Discharge into sewer systems may be regulated. Keep away from heat and acid; do not use with carbon arc lamps or strong sunlight. |
| **Potassium hydroxide** (caustic potash) | High | Corrosive to skin; vapors and mists irritating to eyes, mucous membranes, skin; may erode nasal septum; dilute solution exposure may cause dermatitis. | Ingredient in etching resist removers. | One of the most hazardous alkalis. **Avoid.** Use only with local exhaust ventilation; wear nitrile, neoprene, or rubber gloves. |
| **Potassium iodide** | Slight Moderate | Via skin. Via ingestion and inhalation. Ingestion in large amounts may cause depression, weight loss, gastroenteritis, drowsiness, mental confusion and deterioration, vertigo, and psychosis. | An ingredient in photo developers, bleach fix and reducers. | Heating may release highly toxic iodine vapor. Avoid inhalation of powder; mix in a fume hood, glove box, or wear an approved toxic dust mask. |
| **Potassium metabisulfite** | Slight Moderate | Via skin. Via ingestion and inhalation. Irritant to respiratory and digestive system, central nervous system depressant. | An ingredient in photo developers, stop baths, fixers and neutralizer. | Heating or acids cause formation of irritating sulfur dioxide gas. Use exhaust ventilation and gloves. |
| **Potassium oxalate** | Moderate High | Via skin. Via inhalation and ingestion. Corrosive to eyes, skin, respiratory and digestive systems; can cause kidney damage. Ingestion can be fatal. | A platinum developer and toning solution. | Wear goggles and gloves when handling. Mix solution in fume hood, glove box or wear an approved toxic dust mask. |
| **Potassium perborate** | Slight Moderate<br><br>Explosive Oxidizer | Via skin. Via ingestion and inhalation. A respiratory and skin irritant; ingestion causes diarrhea. Upon contact with combustible materials. | A hypo eliminator. | Wear gloves; avoid inhalation of powder. |

| Chemical Trade Name | Toxicity | Health Effects | Uses | Precautions |
|---|---|---|---|---|
| **Potassium permanganate** | High<br><br>Oxidizer<br><br>Slight | Via all routes. Corrodes skin, eyes, mouth, respiratory tract, lungs, and stomach.<br>From dilute solutions. | A photo reducer, hypo eliminator; in Kodak stain remover. | Wear rubber gloves. Mix powders in fume hood, glove box or wear an approved toxic dust mask. |
| **Potassium persulfate** | Moderate<br><br><br><br><br>Explosive Oxidizer | Via all paths of entry. Skin, eye, and respiratory irritant, allergen. Ingestion causes diarrhea.<br>Upon contact with combustible materials. | A photo reducer and hypo eliminator. | Do not heat as it will form irritating sulfur dioxide gas. Wear goggles and gloves. |
| **Potassium phosphate** | Slight | Via all routes. | In bleach-fix, stop-fix and developers. | Wear gloves when handling powders and solutions. |
| **Potassium sulfide** (liver of sulphur, polysulfide) | Moderate | Via all routes of exposure. Causes skin softening, irritation; may corrode like alkalis. May release hydrogen sulfide gas through inadvertent mixture with acids in the studio or in the stomach. | A photo toner. | **Do not use.** Emits rotten egg odor, but induces olfactory fatigue (loss of ability to detect its presence by smell). If used, wear goggles and gloves and use exhaust ventilation. **Do not** contaminate with acids. |
| **Potassium sulfite** | Slight<br>Moderate | Via skin.<br>Via ingestion and inhalation. An irritant to respiratory and digestive systems. | An ingredient in photo developers, stop baths, hypo clearing agent, fixers, neutralizer. | Wear goggles and gloves. Avoid heating as it causes formation of irritating sulfur dioxide gas. Use exhaust ventilation. |
| **Potassium thiocyanate** | Slight<br>Moderate<br><br><br><br><br>High | Via inhalation, skin.<br>Massive ingestion leads to nausea, muscle spasms, tinnitus, weakness, disorientation.<br>If heated or mixed with acid. | In photography developer and bleach-fixer; a silver bleach in color processing. | Heating or combination with acid forms highly toxic hydrogen cyanide gas. |
| **Propionic acid** (methylacetic acid) | Slight to Moderate | Vapors and concentrated solutions are irritating to eyes, skin and respiratory tract. A food additive with little toxicity by ingestion. | For photography reversal baths. | Wear goggles and gloves when handling concentrates. Always add acids to solution, never the reverse. Use exhaust ventilation. |
| **Prussian blue** (ferric ferrocyanide) | Slight<br>High | Via all routes.<br>If heated or mixed with acid. | Pigment: Paris, Berlin, Iron blue, blue 27. | Will form highly toxic hydrogen cyanide gas if heated, mixed with acids, allowed to decompose or exposed to strong sunlight. |
| **Pumice** (aluminum silicate) | Slight | Dust may cause eye, respiratory irritation. Long-term inhalation of fine aluminum powder may contribute to lung scarring. | Used to sharpen carving, engraving tools. | Wet stone to reduce dust or wear an approved dust mask. |
| **Pyrazolone** | NKT | | Azo organic pigment. | |

| Chemical Trade Name | Toxicity | Health Effects | Uses | Precautions |
|---|---|---|---|---|
| **Pyrogallic acid** (pyrogallol, 1,2,3-trihydroxybenzene) | High | Via all routes. An allergen and irritant. Ingestion, inhalation and skin exposure associated with anemia, cyanosis, methemoglobinemia, convulsions, vomiting, kidney, liver, heart, lung, muscle damage. Ingestion and skin exposure are potentially fatal. | A tanning developer. | **Avoid.** If used, wear goggles and gloves, and use a fume hood or wear an approved respirator when mixing. |
| **Quinacridine** | NKT | | A dye; coloring agent in yellow, orange and red paints and inks. | |
| **Rabbit skin glue** | NKT | | Sealer, canvas primer. | |
| **Rapid fix** | Slight to Moderate | Respiratory irritant. Chronic inhalation of large doses may contribute to lung problems. See ammonium thiosulfate. | Photographic fixer containing 60% ammonium thiosulfate solution. | Avoid heating. Use local exhaust ventilation and tongs. |
| **Raw sienna** | NKT | May be contaminated with impurities. | Pigment brown 6; clay with iron and aluminum oxides. | Avoid powdered pigments. |
| **Resins** (natural, synthetic) | Slight | Natural resin dusts provoke allergies in some individuals. | Hardened natural tree secretions, or synthetic. Must be dissolved or suspended in solvents such as alcohol or turpentine. | Wear dust mask. Safest if dissolved in ethyl or isopropyl alcohol. |
| **Resist removers** | Moderate to High | Via all routes. Many contain potassium hydroxide which is corrosive to skin; vapors and mists irritate eyes, mucous membranes and skin; may erode nasal septum. Exposure to dilute solution may cause dermatitis. | Highly alkaline for removing etching resists. | Use local exhaust ventilation; wear nitrile, neoprene, rubber gloves. Use less toxic alkalis. |
| **Resorcinol** | Moderate | Sensitizes, irritates, and may burn skin and eyes. Absorbed through skin, may cause methemoglobinemia. | A photo developer. | Wear gloves and protective attire. Provide effective local exhaust ventilation or use a fume hood. |
| **Retarders** | Variable | Obtain Material Safety Data Sheet for ingredients, effects and precautions. | Chemical additives to paints and inks to slow drying time. | |
| **Retention aids** (cationic polyacrylamide) | Slight Explosive | Eye and skin irritant. Dust and powders. | An agent used in papermaking to hold colors. | Use glove box, or dust mask, goggles, gloves when mixing powders into solution. Keep powder away from oxidants, sources of ignition. Avoid spilling. |
| **Rhodamine** | Suspected carcinogen | No known short term hazards. | A dye in Day-Glow colors. | Wear gloves. Avoid powdered pigments, spray paints. |

| Chemical Trade Name | Toxicity | Health Effects | Uses | Precautions |
|---|---|---|---|---|
| **Rhoplex** | Slight to Moderate | Contains ammonia and preservatives. Vapors may irritate eyes, skin, mucous membranes, cause headache, nausea. | An acrylic emulsion. | Use exhaust ventilation and wear gloves. Avoid mixing with acid cleaners or those that contain bleach (sodium hypochlorite). |
| **Rosin** | Slight to Moderate<br><br><br><br><br>Explosive | Its dust is an allergen for some people; chronic inhalation may cause pneumoconiosis and pulmonary fibrosis.<br>Rosin dust. | Resinous residue left after turpentine extraction. For etching, aquatint; ingredient in painting mediums. | Use in a glove box or wear a respirator with a toxic dust cartridge. |
| **Rubber cement** | Variable | May cause central nervous system effects. See n-hexane, naphtha, benzol. | Multi-purpose adhesive. | **Do not use.** Substitute non-toxic glues and adhesives, or a waxer. |
| **Rubber cement thinner** | Variable | May cause central nervous system effects. See n-hexane, naphtha, benzol. | Solvent mixture. | **Do not use.** If unavoidable, use sparingly with exhaust ventilation. Wear nitrile or PVA gloves; store in leak-proof containers. |
| **Screen Degreaser** | Variable | Check Material Safety Data Sheet for ingredient health effects and precautions. | To clean and prepare screens for screenprinting. | Some may contain caustic cleaning agents. Wear gloves; use exhaust ventilation. |
| **Screen Haze Remover** | Moderate to High Combustible | Very caustic to skin, eyes, mucous membranes. Can cause nausea and vomiting. | Removes ink stains, photo ghosts from silkscreens. | **Avoid.** Replace screen mesh. Industrial product: use only with effective ventilation, protective attire. |
| **Screen photo emulsion** | Variable | Check Material Safety Data Sheet for ingredient health effects and precautions. | Photo-sensitizing silkscreens. | Use exhaust ventilation; avoid skin contact, wear nitrile, neoprene, or rubber gloves; follow manufacturer's directions. Diazo is safer than ammonium dichromate. |
| **Screen stencil Remover/ Reclaimer** | Moderate | A caustic oxidant, skin irritant. | Screenprinting liquid or paste cleaner that usually requires high pressure water spray. | Wear gloves, protect eyes, skin. Read manufacturer's directions and Material Safety Data Sheet before using. Avoid contact with water mist that contains reclaiming solution. |
| **Screen Wash** | Moderate to High | Usually contains aromatic hydrocarbons, ketones, alcohols. Check Material Safety Data Sheets for ingredients, health effects and precautions. | Silkscreen cleaner. | **Avoid.** Substitute water-soluble inks that may be cleaned with water. |

| Chemical Trade Name | Toxicity | Health Effects | Uses | Precautions |
|---|---|---|---|---|
| **Selenium** (sodium selenide, cadmium selenide) | Moderate<br><br><br><br><br><br>Suspected carcinogen<br><br>Reproductive toxin | Dusts, fumes, vapors irritate eyes, skin, respiratory tract. Many selenium compounds are absorbed through skin. Chronic, high-level exposure causes anxiety, depression, dermatitis, gastrointestinal upset, metallic taste, garlicky breath, loss of fingernails and hair, kidney, liver, bone marrow and thyroid injury. | Pigment ingredient in red 108, yellow 35. A photo toner. | Avoid powdered pigment, spraying or grinding. Mix toner in a fume hood, glove box, or wear an approved toxic dust mask. Wear goggles and gloves. |
| **Sepia** | NKT | | Natural ink derived from the cuttle fish; brown 9. | |
| **Shellac** | Variable | Depends on type of alcohol solvent. Check Material Safety Data Sheet. | A common sealer, varnish; a resin suspended in methyl or ethyl alcohol. | Use effective ventilation; avoid shellac containing methyl alcohol. |
| **Sienna** | NKT | Except for possible contamination from impurities | Natural and synthetic iron oxides used as pigment; brown 6. | Avoid use of powdered pigments. |
| **Silica** | Variable<br><br>Suspected carcinogen | Amorphous silica (pre-cipitated diatomaceous earth and gel silica) is unlikely to cause silicosis. If comprised of greater than 1% of quartz (crystalline form of silica), chronic inhalation of dust may cause silicosis, a progressive scarring of the lungs. | Basic ingredient of glass; used in paints, pigments, lubricating powders. | Wear dust mask if using in powdered form. Avoid crystalline forms: quartz, cristobolite, tridymite, Tripoli. |
| **Silver** | Slight to Moderate | Chronic ingestion leads to silver deposits in skin, other tissues (argyria). Can stain (especially silver nitrate) skin and irritate mucous membranes. | The most common photosensitive material used in photo emulsions. | Wear gloves, use tongs. |
| **Silver iodide** | Slight to Moderate | See silver. | Forms on sensitized plates in daguerreotype, other historic processes. | Wear gloves, use tongs. |
| **Silver nitrate** | Slight to Moderate | Via all routes. Allergen and irritant. Ingestion may cause severe gastroenteritis leading to vertigo, shock, convulsions and coma. Can cause skin burns and has caused blindness when splashed in eyes. | For hypo test kit, as a toner, to photosensitize printmaking plates and screens. | Avoid skin contact. Wear goggles and gloves. |
| **Sodium acetate** | Slight | Via all routes. Concentrated solution may irritate eyes, skin, respiratory tract. | An ingredient in stop bath, stabilizers, neutralizers. | Wear goggles and gloves when handling. |

| Chemical Trade Name | Toxicity | Health Effects | Uses | Precautions |
|---|---|---|---|---|
| **Sodium bisulfite** (sodium hydrogen sulfite) | Slight to Moderate | Irritating to eyes, skin, and respiratory tract. Adverse reactions more frequent in asthmatics. | An ingredient in photo washing aids, pre-hardeners, fixers and replenishers. | Excess heat or acid may release irritating sulfur dioxide gas. Wear goggles and gloves; use local exhaust ventilation. |
| **Sodium bromide** | Slight Moderate | Through inhalation. By ingestion. Acute ingestion causes nausea and vomiting. Chronic ingestion causes "bromism:" drowsiness, loss of appetite, rashes, confusion, constipation, weakness, vertigo, and psychosis. May be a mild skin irritant. | A color developer starter. | Avoid inhalation of powder. Mix in glove box, fume hood or wear an approved toxic dust mask. |
| **Sodium carbonate** (soda ash) | Moderate to High | Irritates skin, eyes, respiratory tract, mucous membranes. Ingestion of large amounts can be fatal. | An ingredient in photo developers and bleach-fixes; dye mordant. | Wear goggles and gloves. Use with local exhaust ventilation. |
| **Sodium chloride** | Slight | Common table salt. Excessive ingestion can be harmful, especially to young children who may develop drowsiness, weakness, abnormal behavior. | Contained in some dyes as a mordant. | |
| **Sodium chloropalladite** (sodium tetrachloro-palladate) | Slight  Unknown  Suspected carcinogen | Via skin contact and inhalation. Via ingestion. | A photo toner. | Avoid repeated skin contact and inhalation; Use local exhaust ventilation. |
| **Sodium chloroplatinate** (sodium tetrachloro-platinate) | Slight Moderate | Via skin. Via inhalation. A sensitizer that causes skin and respiratory allergies, lung scarring and emphysema. | A photo toner. | Use with effective local exhaust ventilation and gloves. |
| **Sodium cyanide** | High  Moderate  Metabolic asphyxiant | Via ingestion, inhalation: can be fatal. Caustic action can promote skin absorption. Symptoms include headache, nausea, confusion, seizures. | A photo reducer and intensifier. | **Do not use**. Potentially lethal hydrogen cyanide gas can form in solutions. Substitute potassium ferricyanide. |
| **Sodium dichromate** | Moderate  Oxidizer  Suspected carcinogen | See chromium compounds. | Oxidizing agent used in lithography. | **Do not use.** |
| **Sodium fluoride** | Moderate High | Via skin. Via inhalation and ingestion; ingestion can be fatal. Symptoms include salty, soapy taste, gastrointestinal upset, weakness, depression and shock. | A preservative. | **Do not use.** Substitute zinc chloride or thymol. Use pre-mixed solution; wear gloves and a respirator. |

| Chemical Trade Name | Toxicity | Health Effects | Uses | Precautions |
|---|---|---|---|---|
| **Sodium hydrosulfite** (sodium dithionite) | Slight<br><br>Flammable | Corrosive to respiratory system, eyes and skin. | Dye mordant; color photo retouching agent, used in a cyan bleach. | Decomposition, heat or mixture with acid may release irritating sulfur dioxide gas. |
| **Sodium hydroxide** (caustic soda) | High | Via all routes of entry. Corrosive alkali will burn eyes, skin. Inhalation of large amounts can cause pulmonary edema; irritating even in dilute solutions. Ingestion can be fatal. | Caustic soda used in photo developers, toners; dye mordant; linocut etch. | One of the most toxic alkalis. Wear gloves and goggles when handling concentrates. |
| **Sodium hypochlorite** (household bleach, a 5% solution) | Slight<br>Moderate | Via ingestion.<br>Via skin and inhalation. An irritant, repeated contact can cause dermatitis. Inhalation may cause respiratory problems, pulmonary edema. | A hypo eliminator; used in photo-printmaking processes. | Heating or addition of acid forms highly toxic chlorine gas; mixture with ammonia forms highly toxic chloramine gas. Wear goggles and gloves; use local exhaust ventilation. |
| **Sodium metabisulfite** | Moderate | A respiratory irritant. Ingestion may cause diarrhea, circulatory disorders and depression. Adverse reactions more frequent in asthmatics. | Used in photo bleach-fix, stop bath, intensifier. | Heating or addition of acid forms irritating sulfur dioxide gas. Wear goggles and gloves; use local exhaust ventilation. |
| **Sodium metaborate** (Kodalk, combination of sodium carbonate and sodium borate) | Moderate | Via all routes. Corrosive to skin, eyes. Irritating to respiratory tract, mucous membranes. An irritant even in dilute solutions. Acute ingestion has caused death. | Used in photo developers, bleach-fixes. | Wear goggles and gloves; use local exhaust ventilation. |
| **Sodium nitrate** | Slight<br><br>Fire hazard<br>Oxidizer | Via skin, inhalation, or ingestion.<br>Will accelerate fire in the presence of organic materials. | A silver bleach in color processing. | Keep away from heat, spark or flame. |
| **Sodium phosphate** (trisodium phosphate) | Moderate<br><br>Slight | Via skin and inhalation; an irritant.<br>Via ingestion. | Ingredient in photo bleach, developer, stabilizer, and reversal bath solutions. | Wear goggles and gloves; use local exhaust ventilation. |
| **Sodium potassium tartrate** (Rochelle salt) | NKT | | An accelerator, buffer and re-agent in photography. | |
| **Sodium selenite** | Moderate | Via all routes. Can cause respiratory irritation, pulmonary edema and bronchitis. Symptoms include garlic odor on breath, nausea, vomiting, nervous disorders and possible kidney and liver damage. | A photo toner. | Avoid mixture with acid as it forms highly toxic hydrogen selenide gas. Wear goggles and gloves; use local exhaust ventilation. |
| **Sodium sulfate** (salt cake, Glauber's salt) | Slight | Irritant to skin, eyes, nose and throat. | An ingredient in photo bleaches, developers and pre-hardeners. A dye mordant. | Avoid heating as it forms irritating sulfur dioxide gas. |

| Chemical Trade Name | Toxicity | Health Effects | Uses | Precautions |
|---|---|---|---|---|
| **Sodium sulfide** | Moderate | Via all routes. Causes skin softening and irritation; may corrode like alkalis. May release hydrogen sulfide gas through inadvertent mixture with acids in the studio or in the stomach. | In photo toner and testing solutions. | **Do not use.** Avoid mixture with acids. If used, wear goggles, gloves with local exhaust ventilation. |
| **Sodium sulfite** | Slight to Moderate | Ingestion may cause circulatory disorders, diarrhea, depression. Is a respiratory and skin irritant. | In photo developers, fixers, bleach-fixes, fixing baths, photo papers, toners, hypo clearing agents. | Avoid heating as it forms irritating sulfur dioxide gas. Wear goggles and gloves; use local exhaust ventilation. |
| **Sodium thiocyanate** | Slight Moderate | Via skin and inhalation. Massive ingestion leads to nausea, muscle spasm, tinnitus, weakness, and disorientation. | Ingredient in photo developer, and as a silver bleach in color processing. | Do not heat or mix with acid as it forms highly toxic hydrogen cyanide gas. |
| | High | With heat or acid. | | |
| **Sodium thiosulfate** | Slight | Via skin, ingestion or inhalation. | Ingredient in hypo, fixing baths, as a bleach. | Decomposition, heating, or addition of acids may release irritating sulfur dioxide gas. Do not use old solutions; cover when not in use; use with local exhaust ventilation. |
| **Soft Ground** | Variable | Review solvent content and observe appropriate precautions. | For etching. Contains petroleum jelly, and hard ball ground softened with tallow or lard. | Wear gloves; use local exhaust ventilation. If solvents are needed, use alcohol or acetone. |
| **Spinel** | NKT | No known adverse health effects as used by artists. | Natural or synthetic oxides in pigments, known for their stability. | |
| **Stand oil** | NKT | | A painting medium of linseed oil heated with a special process. | |
| **Stannic chloride** (tin chloride, tin tetrachloride) | Moderate | An irritant via all paths of entry. May react with water to form acid. Prolonged inhalation of dust may cause benign pneumoconiosis. | "Dippit" fixer for black and white instant processes. | Wear gloves; use tongs and local exhaust ventilation. Wear goggles when handling large amounts of solution. |
| **Stannous chloride** | Moderate | Via all routes of entry. Irritant to skin, eyes, respiratory and gastro-intestinal systems. May react with water to form acid. Prolonged inhalation of dust may cause benign pneumoconiosis. | Color photo retouching agent for magenta dyes. | Wear gloves and goggles; use with local exhaust ventilation. |

| Chemical Trade Name | Toxicity | Health Effects | Uses | Precautions |
|---|---|---|---|---|
| **Stoddard solvent** | Variable<br><br>Combustible | May be absorbed through skin. Vapors irritate eyes, skin and respiratory tract. Chronic over-exposure associated with fatigue, headache, bone-marrow depression. | Petroleum distillate solvent. | Use with general or local exhaust ventilation; wear nitrile gloves. Kerosene-like odor and irritation are good warning properties. |
| **Stop bath** | Variable | Continued inhalation of acid vapors can cause chronic bronchitis. Some ingredients may cause skin and respiratory allergies. | For photo-processes, contains acetic acid, hardening agents such as potassium chrome alum, and an anti-swelling agent such as sodium sulfate. | Substitute water rinse between developer and stop bath to reduce formation of irritating sulfur dioxide gas. Keep bath covered between printing sessions. Use local exhaust ventilation and tongs; wear gloves. |
| **Stop-Out** | Variable<br><br>Suspected carcinogen<br>Explosive | See ingredient health effects and precautions. May contain asphaltum. Rosin dust. | For printmaking: etching. Rosin and alcohol, or asphaltum dissolved in mineral spirits. | Wear gloves; use exhaust ventilation. Avoid brand name stop-outs that contain unspecified solvents. |
| **Strontium** | NKT | May be combined with other hazardous ingredients. | Metal, ingredient in pigments, yellow 32. | |
| **Styrene**<br>(vinyl benzene) | Moderate<br>Anesthetic<br><br>Suspected carcinogen | Via all routes. Can cause neurological impairment (narcosis). Eye, skin, and respiratory system irritant. Associated with liver and nerve damage. | Aromatic hydrocarbon solvent. | **Do not use.** More toxic than toluene and xylene. Attractive, sweet odor at low concentrations; penetrating in larger doses. Substitute isopropyl alcohol or acetone (flammable). |
| **Succinaldehyde** | Slight to Moderate | Via all routes of exposure. Eye, skin, respiratory irritant. | A pre-hardener and hardening agent used in photography. | Avoid skin contact; wear gloves and goggles; use local exhaust ventilation. |
| **Succinic acid**<br>(ethylene dicarboxylic acid) | Slight to Moderate | Via all routes. Concentrate is an irritant to skin, eyes, and respiratory system. | Bromide printing, photography. | Wear goggles and gloves when handling acids. Use local exhaust ventilation. |
| **Sulfamic acid**<br>(amidosulfonic acid) | Moderate | Via all routes. Corrosive to eyes, skin, respiratory and gastrointestinal systems. | Ingredient in photo bleaches. | Do not heat as it releases irritating sulfur dioxide gas. Use exhaust ventilation; wear gloves and goggles. |
| **Sulfur dioxide gas** | Moderate | Highly irritating to respiratory tissues, can cause chronic lung problems. Fatal pulmonary edema has been reported. | Released when photo fixer solution becomes contaminated from stop bath, or mixes with any acid. | A common hazard in photoprocessing. Prevent darkroom build-up with local exhaust and dilution ventilation. Cover trays as much as possible. Do not heat solutions except under a properly functioning fume hood. |

| Chemical Trade Name | Toxicity | Health Effects | Uses | Precautions |
|---|---|---|---|---|
| **Sulfuric acid** | High | Vapors and mists irritate respiratory passages, skin, eyes, stomach. Associated with reversible airway obstruction, and chronic bronchitis. Ingestion of small amount can be fatal. Dilute solutions will dry skin and cause ulceration. | In photo developers, bleaches, retouchers, stain removers; dye mordant; used in printmaking to etch. | **Avoid.** Do not heat as it forms toxic sulfur oxides. Substitute phosphoric acid or ferric chloride. Wear goggles and gloves; use tongs and local exhaust ventilation. |
| **Surfactants** | Slight to Moderate | Some are irritants. | In photo resists. | |
| **Super Blox** | Variable | See ingredients for health effects and precautions. Check Material Safety Data Sheet. | Etching resist, contains methanol methylene chloride, and methyl cellulose. | **Avoid.** Use only with local exhaust ventilation; wear PVA gloves. |
| **Talc** (hydrated magnesium silicate) | Slight Moderate | If certified as asbestos-free. Long-term inhalation of dust has been associated with pneumoconiosis and pulmonary fibrosis (lung scarring). Talc with asbestos has been linked to lung and peritoneal cancer. | Used in printmaking; also in pigment white 26. | Use baby powder certified to be free of asbestos. Avoid talc and pigments containing asbestos or silica. |
| **Tannic acid** | Slight to Moderate Suspected carcinogen | Via all routes. Skin irritant. Ingestion causes digestive upset. Associated with liver injury and cancer. | A toner in cyanotype; printmaking; dye mordant. | Wear goggles and gloves when handling acids. Use local exhaust ventilation; cover acid baths when not in use. Concentrated solutions should be handled with caution. |
| **Tartaric acid** | Moderate | Via all routes. An irritant. | Used in Van Dyke printing. | Wear goggles and gloves when handling acids, especially concentrated solutions. Use local exhaust ventilation. |
| **Terpenes** | Slight | May cause eye, mucous membrane, skin or gastrointestinal irritation. Undergoing further study for health effects. | Hydrocarbons extracted from plants, limonene a common example. Used in waterless hand cleaners, paint thinners, cleaning products. | Reasonable alternative to turpentine. Keep away from children. |
| **Tertiary-Butylamine Borane** (TBAB) | Moderate | Via all routes. Extremely irritating to eyes, respiratory tract; associated with nervous system damage. | Used in color photography developing bath and fogging agent. | **Avoid.** Use both lateral slot and dilution ventilation. Wear an approved respirator, goggles and rubber gloves when mixing chemicals. When possible, use light instead of TBAB between first and second developer in color processes requiring re-exposure. |
| **Thiourea** (thiocarbamide) | Slight Suspected carcinogen | Via all routes. Associated with cancer in laboratory animals. | A photo toning solution and stain remover. | Avoid repeated contact with powders or solutions. |

| Chemical Trade Name | Toxicity | Health Effects | Uses | Precautions |
|---|---|---|---|---|
| **Thymol** (methyl isopropyl phenol) | Slight<br><br>Moderate | To skin: absorbable, mild skin irritant. Via inhalation, ingestion. Overexposure and ingestion may lead to gastric pain, nausea, vomiting, shock, convulsions and coma. | A preservative with pungent, irritating odor good for warning against overexposure. | Substitute less toxic o-phenyl phenol. |
| **Tin chloride** | Slight to Moderate | Irritating to skin, eyes, mucous membranes and gastrointestinal system. Dust or fume deposits in lungs may cause benign pneumoconiosis. | Pigment ingredient; a dye mordant. | Use exhaust ventilation when spraying; avoid repeated skin contact; wear gloves. |
| **Titanium oxide** (titanium dioxide) | Slight | Mild lung irritant; heavy exposure associated with benign pneumoconiosis. | Titanium dioxide pigment, white 6, usually derived from the mineral rutile. | Avoid powdered pigment; use in pre-mixed formulations. |
| **Toluene** (methyl benzene, toluol) | Moderate<br><br>Flammable<br><br>Anesthetic | Absorbed through skin, may cause dermatitis. Vapors irritate eyes, skin and respiratory tract. Chronic over-exposure associated with headache, dizziness, nausea, drowsiness. Acute inhalation of large amounts may cause lung, liver, and kidney injury; abnormal heart rhythms, nervous system impairment, and muscle damage. | A common aromatic hydrocarbon solvent used as solvent for lacquers, varnishes, paints. Ingredient in photo-resist materials and retouching fluids. | **Avoid.** Substitute isopropyl alcohol or acetone. |
| **Toluidine** (methyl aniline) | Slight to Moderate<br><br>Suspected carcinogen | A corrosive alkali. In concentrated form can cause severe burns. Absorbed through skin, may cause methemoglobinemia. | Azo organic pigment, Hansa red, red 3. | Use only in premixed formulations. Avoid skin contact. |
| **Transparent base acrylic** | Slight | May be an allergen. | Screenprinting. | Use effective general ventilation; wear gloves. |
| **Transparent base oil/alkyd** | Moderate | Contains resins plasticizers, various solvents. Check Material Safety Data Sheet for ingredients, health effects and precautions. | Screenprinting. | **Avoid.** Use acrylic transparent base and inks. |
| **1,1,1-Trichloroethane** (methyl chloroform) | Slight to Moderate<br><br><br><br><br>Flammable<br><br>Suspected carcinogen | Absorbed through skin. Vapors irritate eyes, skin, respiratory tract. Chronic overexposure may cause dermatitis. Acute inhalation or ingestion of large amounts may lead to nausea, euphoria, incoordination, dizziness, drowsiness, lung, liver and kidney injury, abnormal heart rhythms, and nervous system impairment. If heated. Decomposes in heat to produce phosgene gas and hydrochloric acid. | A chlorinated hydrocarbon in photo color retouchers and correction fluids; a printmaking solvent; component in aerosol sprays and film cleaners. | **Do not use.** Substitute water-soluble inks, paints, and correction fluids or correction film. If necessary to use, wear gloves; use local exhaust ventilation. |

| Chemical Trade Name | Toxicity | Health Effects | Uses | Precautions |
|---|---|---|---|---|
| **1,1,2-Trichloroethane** | | See health effects for 1,1,1-trichloroethane. | | Observe precautions for 1,1,1-trichloroethane. |
| **Trichloroethylene** | Suspected carcinogen | See health effects and precautions for trichloroethane. | Chlorinated hydrocarbon solvent in varnishes, paints, correction fluids. | **Do not use.** Substitute less toxic solvent such as acetone or alcohol. |
| **Tungsten** | Slight Flammable | May irritate respiratory tract. Some salts may release acid on contact with moisture. Chronic exposure to tungsten carbide has been associated with lung scarring. | Metal used in pigments. | Avoid powdered pigments and spray paints. |
| **Turpentine** | Moderate Flammable | Via all routes. Vapors irritate eyes, skin, mucous membranes. Repeated exposure associated with chronic bronchitis, pulmonary edema, and kidney disease. Absorbed through skin, and may cause dermatitis. Ingestion of one-half ounce can be fatal to children. | Oleoresin solvent distilled from coniferous trees. | **Avoid.** Substitute deodorized mineral spirits such as Turpenoid. Use with local exhaust ventilation, and hand and skin protection: nitrile gloves. Gum turpentine is less irritating than wood or steam-distilled. |
| **Tusche** | None Slight | In its water-soluble form. Some ingredients are irritants or toxins. Check Material Safety Data Sheet. | Contains lamp or carbon black pigment, small amount of solvent such as lithotine, mineral spirits, turpentine, alcohol or water. | Use water-based liquid tusche or use water with solid tusche. |
| **Ultramarine** | NKT | | Pigment ingredient formed from sodium, and aluminium with sulfur. In blue, green, red, violet. | May contain impurities or manganese. |
| **Umber** | NKT | Unless it contains manganese or impurities. | Natural and synthetic iron oxides as brown pigments. | Avoid powdered pigments. |
| **Underpainting white** | NKT High | If it does not contain lead. If it contains lead. | Painting primer of (whiting) calcium carbonate, titanium or zinc white. | **Avoid** lead carbonate (flake white). |
| **Uranium nitrate** | Moderate | Via all paths of entry; can cause severe kidney, liver damage. | A photo intensifier, toner. | **Do not use.** Substitute silver nitrate or gold chloride. |
| **Urea** | NKT | | A dye mordant. | |
| **Vanadium chloride** (vanadium tetrachloride) | Moderate | Corrosive to eyes, skin, respiratory and digestive tracts. May cause skin allergy. Chronic exposure may lead to green tongue, metallic taste, intestinal disorders, heart problems. | A photo toner; pigment ingredient. | Wear goggles and gloves when handling. Use with local exhaust ventilation. |

| Chemical Trade Name | Toxicity | Health Effects | Uses | Precautions |
|---|---|---|---|---|
| **Varnish** | Variable | Depending upon ingredients. Check Material Safety Data Sheet for health effects and precautions. | A mixture of resins, plasticizers, solvents such as alcohols, mineral spirits, aromatic hydrocarbons, terpenes. | Use only with care appropriate to ingredients: local exhaust ventilation; eye, hand, skin protection. |
| **Varsol** | Slight to Moderate Combustible | See Mineral spirits. | | |
| **Vinyl Plate Lacquers** | Variable <br><br> Flammable | Check ingredients and Material Safety Data Sheet for health effects and precautions. | For lithography, a mixture of ketones, aromatic hydrocarbons, resins. | **Avoid.** Use effective local exhaust ventilation; wear nitrile gloves. |
| **VM&P naphtha** (varnish makers' and painters' naphtha) | Slight to Moderate <br><br> Flammable | Irritating to eyes, skin, and respiratory tract. Depresses central nervous system at high doses. | An aliphatic hydrocarbon solvent, also known as benzine, ligroin, high-boiling petroleum ether and paint thinner. | One of the least toxic solvents in its class. Good substitute for turpentine. Use with local exhaust ventilation; eye, skin, and hand protection; nitrile gloves. Avoid any product containing benzene. |
| **Wax** | NKT | Check ingredients for presence of organic solvents. | Petroleum wax for designers' waxers is non-toxic. Painting mediums containing wax will include solvents. | Avoid chlorinated synthetic waxes. Substitute beeswax or non-toxic petroleum wax. If necessary, use alcohol or acetone as solvents. |
| **Whiting** | NKT to Slight | Nuisance dust; may cause irritation in susceptible individuals. | Used in printmaking, contains calcium carbonate, French chalk, alumina hydrate, barytes, magnesium carbonate. | Wear dust mask. |
| **Xylene** (dimethyl benzene, xylol) | Moderate <br><br> Combustible <br><br> Anesthetic | Absorbed through skin, may cause dermatitis. Vapors irritate eyes, skin and respiratory tract. Chronic over-exposure associated with headache, nausea, dizziness, drowsiness. Acute inhalation of large amounts associated with lung, liver and kidney injury, abnormal heart rhythms, nervous system impairment and muscle damage. | An aromatic hydrocarbon solvent of lacquer, paint, and varnish; metal cleaner. | **Avoid.** Substitute isopropyl alcohol, acetone, or deodorized paint thinner. |
| **Yellow ochre** (yellow oxide) | NKT | | Pigment. | |
| **Zinc** | Slight | Significant inhalation of fumes of zinc oxide, formed when elemental zinc is heated, may cause metal fume fever: headache, fever, chills, muscle aches, vomiting. | Metal in pigments: Chinese white, white 4, yellow 36. | Avoid heating. Avoid powdered pigments. |
| **Zinc chromate** | Moderate to High Suspected carcinogen | May cause lung cancer; see chromium. | Ingredient in zinc yellow pigment, yellow 36. | **Avoid.** Avoid powdered pigments and spraying. |

| Chemical Trade Name | Toxicity | Health Effects | Uses | Precautions |
|---|---|---|---|---|
| **Zinc chloride** | Moderate | Caustic and highly irritating to skin; burns and ulcers may occur. Irritating to mucous membranes. Inhalation of large amounts can cause chronic bronchitis, severe respiratory irritation and possibly pulmonary edema. | Photography: salted paper sensitizer. Used as a preservative. | Wear goggles and gloves; mix powders in fume hood, glove box or wear an approved toxic dust mask. Avoid heating. |
| **Zinc palmitate** | NKT | In quantities used. | A paint stabilizer. | |
| **Zinc stearate** | NKT | In quantities used. | A paint stabilizer. | |
| **Zinc white** (zinc oxide) | Slight to Moderate | By skin, may cause dermatitis. Fumes from heating may irritate upper respiratory tract. | Ingredient in zinc white, white 4, Chinese white pigments. | May contain small amounts of impurities. Do not heat. **Do not use those that contain lead.** |
| **Zirconium** | Slight | Some compounds are irritating; zirconium tetrachloride may release hydrochloric acid on contact with moisture. | Metal used in pigments. | Avoid powdered pigments and spraying. |

**AMERICAN INDUSTRIAL HYGIENE ASSOCIATION**
National Office
45 White Pond Drive
Akron, OH 44320
216-873-2442

**ART AND CRAFT MATERIALS INSTITUTE**
100 Boylston Street
Suite 1050
Boston, MA 02116
617-426-6400

**ARTS, CRAFTS AND THEATER SAFETY**
181 Thompson Street
#23
New York, NY 10012
212-777-0062

**CANADIAN CENTRE FOR OCCUPATIONAL HEALTH AND SAFETY**
250 Main Street East
Hamilton, Ontario L8N 1H6
Canada
416-572-2981
CCINFOdisc

**CENTER FOR HAZARDOUS MATERIALS RESEARCH**
University of Pittsburgh
Applied Research Center
320 William Pitt Way
Pittsburgh, PA 15238
412-826-5320 Fax 412-826-5552
Hotline: 800-334-CHMR (2467)

**CENTER FOR SAFETY IN THE ARTS**
5 Beekman Street,
Suite 1030
New York, NY 10038
212-227-6220 (10–5, M–F)

**CHEMICAL REFERRAL CENTER**
2501 M. St., N.W.
Washington, DC 20037
Hotline: 800-262-8200

**EASTMAN KODAK CO.**
800 Lee Rd.
Rochester, NY 14650
24-Hour Hotline:
716-722-5151
Material Safety Data
Sheet and technical
information
8 A.M. to 6:30 P.M.
800-242-2424

**ENVIRONMENTAL RESEARCH FOUNDATION**
P.O. Box 73700
Washington, DC
20056-3700
202-328-1119

**GRAPHIC ARTS INTERNATIONAL UNION**
1900 L Street NW
Washington, DC 20036
202-833-3970

**HAZARDOUS SUBSTANCE FACT SHEETS**
New Jersey Department
of Health
CN 364 Trenton, NJ
08625-0368
609-984-2202

**ILFORD PHOTO**
W. 70 Century Rd.
Paramus, NJ 07653
For technical information
call: 201-265-6000
24-Hour Hotline
For medical emergency:
800-842-9660

***MSDS POCKET DICTIONARY***
Genium Publishing
Corporation
1145 Catalyn Street
Schenectady, NY
12303-1836
518-377-8854

**NATIONAL ART MATERIALS TRADE ASSOCIATION**
178 Lakeview Ave.
Clifton, NJ 07011
201-546-6400

**NATIONAL FIRE PROTECTION ASSOCIATION**
1 Batterymarch Park
P.O. Box 9101
Quincy, MA 02269-9101
800-344-3555

**NATIONAL INSTITUTE FOR OCCUPATIONAL SAFETY AND HEALTH**
4676 Columbia Parkway
Cincinnati, OH 45226
800-356-4674 (9–4:30, M–F)

**NATIONAL SAFETY COUNCIL**
444 North Michigan Avenue
Chicago, IL 60611
312-527-4800

**OCCUPATIONAL SAFETY AND HEALTH ADMINISTRATION**
Local offices are listed in telephone directories under U.S. Government, Department of Labor.

National Office
Information and Consumer Affairs
200 Constitution Avenue NW
Washington, DC 20210
202-523-8151

**ONTARIO COLLEGE OF ART**
Ted Rickard, Director of Health and Safety
100 McCaul Street
Toronto, Ontario M5T1W1 Canada
416-977-5311

**ONTARIO CRAFTS COUNCIL**
35 McCaul Street
Toronto, Ontario M5T 1V7 Canada
416-977-3551

**POISON INFORMATION CENTERS AND POISON CONTROL CENTERS**
Listed in telephone directories, often with emergency listings, 24 hours per day, year round

**POLAROID RESOURCE CENTER**
Customer service:
800-225-1384

**RACHEL (Remote Access Chemical Hazards Electronic Library)**
Environmental Research Foundation
P.O. Box 73700
Washington, DC 20056-3700
202-328-1119

**RCRA/SUPERFUND/OUST ASSISTANCE HOTLINE INFORMATION**
Source on E.P.A. solid and hazardous waste regulations
Hotline: 800-424-9346 (8:30 A.M.–7:30 P.M. EST, M–F)

**SOCIETY OF TOXICOLOGY**
1101 14th Street N.W., Suite 1100
Washington, DC 20005
(202) 371-1393

**WATER POLLUTION CONTROL FEDERATION**
601 Wythe Street
Alexandria, VA 22314-1994
703-684-2400

# SAFETY EQUIPMENT SUPPLIERS

**ANSELL EDMONT PRODUCTS**
Box 6000, 1300 Walnut St.
Coshocton, OH 43812
614-622-4311
Provides job-specific gloves and
protective clothing

**BILSOM SAFETY PRODUCTS**
Bilsom International Inc.
109 Carpenter Drive
Sterling, VA 22170
703-834-1070

**CABOT SAFETY CORP.**
5457 West 79th St.
Indianapolis, IN 46268
508-764-5500

**GRAINGER**
333 Knightsbridge Parkway
Lincolnshire, IL 60069-3639
800-772-7868

**HAZARD TECHNOLOGY COMPANY, INC.**
1419 Forest Drive
Annapolis, MD 21403
410-268-6560
Provides vacuum systems for toxic
materials and hazardous dusts.

**INDUSTRIAL PRODUCTS COMPANY**
21 Cabot Boulevard
Langhorne, PA 19047
215-547-1900 (locally)
800-827-2338

**LAB SAFETY SUPPLY, INC.**
P.O. Box 1368
Janesville, WI 53547-1368
800-356-0783; FAX 608-754-1806;
FAX 800-543-9910
TWX 910-288-2921

**LENPAR CORP.**
10 Leek Crescent
Richmond Hill,
Ontario, Canada
416-731-7850

**3M OCCUPATIONAL HEALTH AND
ENVIRONMENTAL SAFETY DIVISION**
Building 275-6W-01, 3M Center
St. Paul, MN 55144-1000
800-328-1667

**MINE SAFETY APPLIANCE**
121 Gamma Drive
Pittsburgh, PA 15235
412-967-3000; 800-672-2222

**SAFETY SUPPLY CANADA**
90 W. Beaver Creek Rd.
Toronto, Canada
416-222-4111

| ACGIH | American Conference of Governmental Industrial Hygienists |
|---|---|
| ACMI | Art and Craft Materials Institute |
| ACTS | Arts, Crafts, and Theater Safety |
| AEA | Artists' Equity Association |
| ANSI | American National Standards Institute |
| AP | ACMI designation as Approved Product; with the word NONTOXIC, one that conforms to ASTM D-4236 as having no acute or chronic health hazard |
| ASTM | American Society for Testing Materials |
| CAG | Carcinogens Assessment Group of the Federal EPA |
| CAS | Chemical Abstracts Service: assigns identification numbers to specific chemicals |
| CD-ROM | Compact Disk-Read Only Memory |
| CHMR | Center for Hazardous Materials Research |
| COSH | Committee on Occupational Safety and Health |
| CP | ACMI designation as Certified Product, one that meets quality performance standards ASTM D-4302 or American National Standards Institute ANSI Z356.1-5; with the word NONTOXIC, additionally conforms to ASTM D-4236 |
| CPSC | Consumer Product Safety Commission |
| CSA | Center for Safety in the Arts |
| DEP | State Departments of Environmental Protection |
| DOT | U.S. Department of Transportation: regulates chemical transport |
| EPA | U.S. Environmental Protection Agency |
| FHSA | Federal Hazardous Substances Act of 1968 |
| GAIU | Graphic Arts International Union |
| HAMA | Hazardous Art Materials Act of 1988 |
| HAZCOM | Hazards Communications Act of 1983 |
| IARC | International Agency for Research on Cancer |
| MSDS | Material Safety Data Sheet |

**MSHA**   Mine Safety and Health Administration: federal agency that regulates mining and evaluates and approves respirators

**NAMTA**   National Art Materials Trade Association

**NCI**   National Cancer Institute

**NFPA**   National Fire Protection Association

**NIOSH**   National Institute for Occupational Safety and Health

**NSC**   National Safety Council

**NTF**   National Toxicology Program

**OEL**   Occupational Exposure Limits: the Canadian equivalent of PELs set for airborne chemicals in the workplace

**OSHA**   Occupational Safety and Health Administration

**OSHRC**   Occupational Safety and Health Review Commission

**PEL**   Permissible Exposure Limits: regulates exposure to airborne chemicals in the workplace as set by OSHA

**RACHEL**   Remote Access Chemical Hazards Electronic Library

**STEL**   Short-Term Exposure Limits: exposure periods limited to 15 minutes, with a maximum of four STELs permitted during a shift and at least one hour separating each exposure

**TLV**   Threshold Limit Values: airborne substance standards set by the American Conference of Governmental Industrial Hygienists (ACGIH)

**TLV-C**   Threshold Limit Values-Ceiling limits: the maximum exposure time that should not be exceeded during any part of a work day

**TWA**   Time-Weighted Average: airborne concentrations of substances averaged over eight hours; used to prevent adverse effects upon workers exposed to substances at this concentration over the normal 8-hour day and 40-hour work week

## TREATMENT PRIORITY

- Stay calm.
- Flush acids or alkalis from eyes or skin with cool water.
- Call for medical assistance.
- Restore breathing with artificial respiration.
- Stop bleeding by applying pressure.
- Treat for shock by laying patient down with head low to assist blood flow to the brain.
- If victim has head injury or broken bones, do not move unless threatened by fire or noxious vapors.
- Remove victim from hazardous conditions.
- Provide fresh air.

## ACCIDENTAL INGESTION OF ACIDS, ALKALIS OR SOLVENTS

- Drink two glasses of water or milk.
- Do not induce vomiting.
- Call physician or poison control center.

## INHALATION OF NOXIOUS GAS OR VAPOR

- Leave area, seek fresh air, breathe deeply, and expectorate (do not swallow) phlegm that is coughed up.

## BONE FRACTURES

- Seek immediate medical attention.
- Do not move victim unless threat of more serious injury is present; support fracture with splint before moving.

## BURNS

- Cool first- and second-degree burns by immersing affected area in cold water.
- With second-degree burns over a large area or third-degree burns, seek immediate medical attention. Cover burned area with sterile dressings soaked in a solution of five teaspoons bicarbonate of soda per quart of water.

## ELECTRIC SHOCK

- Turn off electric current.
- If the victim is unconscious, apply artificial respiration or cardiopulminary resuscitation (CPR); continue resuscitation efforts until medical help arrives.

Unless you know pertinent procedures, it can be better to do nothing than to treat yourself or others inappropriately.

### Emergency Telephone Numbers

Primary care physician _____

Ophthalmologist _____

Fire _____

Poison Control Center _____

Center for Safety in the Arts     212-227-6220

# Notes

## INTRODUCTION

**1** Curtiss, Deborah, 1987: *Introduction to Visual Literacy, A Guide to the Visual Arts and Communication*, Englewood Cliffs, NJ, Prentice-Hall.

## CHAPTER 1

**1** There has been no manufacture of benzidine-based dyes since 1979. No new art products containing benzene or benzidine should be on the market since these agents are on the federal list of recognized carcinogens.

**2** American Conference of Governmental Industrial Hygienists, 1988: *Threshold Limit Values for Chemical Substances and Physical Agents in the Work Environment Adopted by the ACGIH, with Intended Changes for 1988–89*, Cincinnati, ACGIH; California Department of Health Services, 1988: *Art & Craft Materials Acceptable for Kindergarten and Grades 1–6*, Sacramento, CDHS; McCann, Michael, 1979: *Artist Beware: The Hazards and Precautions in Working with Art and Craft Materials*, New York, Watson-Guptill; Moses, Cherie, et al., 1978, *Health and Safety in Printmaking: A Manual for Printmakers*. Edmonton, Alberta, Occupational Hygiene Branch, Alberta Labor; National Fire Protection Association, 1987: *Flammable and Combustible Liquids*, Quincy, Mass., NFPA; Occupational Safety and Health Administration, 1989: *Air Contaminants: Permissible Exposure Limits*, (OSHA 3112), Washington, D.C., OSHA; Seeger, Nancy, 1982–84: *Guides to Safe Use of Materials*, Chicago, School of the Art Institute of Chicago; Shaw, Susan, 1983: *Overexposure: Health Hazards in Photography*, Carmel, Calif., Friends of Photography; and others.

## CHAPTER 2

**1** Alaska, Arizona, California, Connecticut, Hawaii, Indiana, Iowa, Kentucky, Maryland, Michigan, Minnesota, Nevada, New Mexico, New York, North Carolina, Oregon, Puerto Rico, South Carolina, Tennessee, Utah, Vermont, Virginia, Virgin Islands, Washington, Wyoming. New Jersey has its own right-to-know laws.

**2** Artists' Equity Association; Center for Safety in the Arts; Arts, Crafts, and Theater Safety; American Society for Testing Materials; Art and Craft Materials Institute; and National Art Materials Trade Association.

**3** Berkeley, California, Art and Craft Toxicology Unit, Hazard Evaluation Section, Department of Health Services, California State Department of Education, 721 Capitol Mall, Sacramento, CA 95814.

**4** Chapter 29, Code of Federal Regulations 1910.1200, 1983.

**5** Public Law No. 100-695, 102 Stat. 4568 (1988).

**6** Graphic Arts International Union, 1989: *Health & Safety Guide for the Graphic Arts Industry*. Chapters 17 and 18 provide guidelines and checklists for evaluating and establishing safety in the workplace.

## CHAPTER 3

**1** Clark, Nancy, Thomas Cutter, and Jean-Anne Mc-Grane, 1984: *Ventilation: A Practical Guide*, New York, Nick Lyons, is a good resource for this topic.

**2** Chapter 29, Code of Federal Regulations 1910.1200 (1983), as amended by Public Law No. 100-695, 102 Stat. 4568 (1988), Labeling of Hazardous Materials Act.

## CHAPTER 4

**1** Listed in California Department of Health Services, *Art & Craft Materials Acceptable for Kindergarten and Grades 1–6*, are B&J—The Masters Artist Hand Soap, by General Pencil Co.; and Art Gel, Art Wipes, and Artguard by Winsor & Newton.

**2** Sacks, Oliver, 1991: *Awakenings*, New York, Harper Collins/Viking, p. 2.

## CHAPTER 5

**1** *Acceptable Art and Craft Materials for Grades K–6* is available from Center for Safety in the Arts and from Arts, Craft and Theater Safety, listed in Appendix II. Berol Charcoal Pencil and Dick Blick charcoal are listed as approved, 1988.

**2** op. cit., Rembrandt pastels made by Royal Talens and Alphacolor pastels by Weber Costello have been approved as safe for use in Grades K–6 in California schools.

**3** op. cit., The California list includes, with no pigment restrictions, several lines of oil pastels made by Pentel of America, Sakura Color Products, Sanford, and Winsor & Newton.

**4** op. cit., Redimark 9500 and 9600 permanent markers by Dixon Ticonderoga have been approved as safe for use in grades K–6 in California schools, as have a number of brands of watercolor markers, some with pigment restrictions.

## CHAPTER 6

**1** Rossol, Monona, 1990: *The Artist's Complete Health and Safety Guide,* New York, Allworth.

**2** "Acceptable Art and Craft Materials for Grades K-6," California State Department of Education, is available from Center for Safety in the Arts, and from Arts, Craft and Theater Safety, listed in Appendix II.

## CHAPTER 7

**1** James E. Renson, Executive Director, National Association of Printing Ink Manufacturing, letter to *The New York Times,* June 10, 1990.

**2** Karin Victoria, Crown Point Press, Woodblock Program.

**3** The shift away from silk in screens has generated a shift in terminology as reflected in the magazine title, *Screen Printing.*

## CHAPTER 8

**1** Pinsky, Mark, 1987: *The VDT Book,* New York: New York Committee For Occupational Safety and Health (NYCOSH).

**2** Lewis, Peter H., 1990: "Worries about Radiation Continue, as Do Studies," *The New York Times,* F8, July 8, 1990.

**3** Broudeur, Paul, 1990: "The Magnetic Field Menace," *Macworld* 7(7):136–45.

**4** "Report Ties Electrical Fields to Cancer." *The New York Times,* A12, December 15, 1990.

**5** "Rules on VDTs Are Passed in San Francisco," *The New York Times* A22, December 18, 1990.

## CHAPTER 9

**1** It is a dangerous practice to mix developing solutions together to neutralize them before disposing, a misconception that some photographers hold. While the acids and alkalis may neutralize one another, the presence of chlorine or sulfur compounds, upon contact with acids, will form potentially lethal chlorine and sulfur dioxide gases. Where no chlorine or sulfur is involved, a manufacturer may suggest mixing used developer, bleach, and fixer before discarding, but *only* in that order.

**2** Personal communication from Bruce Katsiff on this section.

---

\* As we go to press, we have learned that the California State Department of Education's "Acceptable Arts and Crafts Materials for Grades K–6" is no longer available.

# Bibliography

Accrocco, Joseph O., ed., 1988: *The MSDS Pocket Dictionary*, Schenectady, N.Y., Genium.

ACGIH, 1988: American Conference of Governmental Industrial Hygienists: *Threshold Limit Values for Chemical Substances and Physical Agents in the Work Environment Adopted by the ACGIH, with Intended Changes for 1988–89*, Published annually by ACGIH, Building D7, 6500 Glenway Avenue, Cincinnati, OH 45211.

ACTS: see Rossol, *ACTS Facts.*

AHN: see McCann, *Art Hazards News.*

Antreasian, Garo, and Clinton Adams, 1971: *The Tamarind Book of Lithography: Art and Techniques*, New York, Abrams.

Arnold, Wilfred Niels, 1988: "Vincent Van Gogh and the Thujone Connection." *Journal of the American Medical Association*, 260:3042.

Babin, Angela, and Monona Rossol, 1988: *Glove Selection*, New York: Center for Safety in the Arts.

Barazani, Gail, 1978: *Safe Practices in the Arts and Crafts: A Studio Guide.* New York, College Art Association.

Batchelor, Anthony J., 1983: *Water-Based Screen Printing, Materials and Methods*, Cincinnati, Art Academy.

———, 1988: *Health Hazards in Water-Based Screen Printing*, Cincinnati, Art Academy.

———, and Lawrence D. Reed, 1984: "Waterbased Ink in Education," *Screen Printing Magazine*, 74:11 (Cincinnati, Signs of the Time).

Bocour, Leonard, and Kenneth Nelson, 1974: "How Acrylics Are Made," *American Artist Product Directory*, New York.

Brodeur, Paul, 1989: *Currents of Death: Power Lines, Computer Terminals and the Attempt to Cover Up Their Threat to Your Health.* New York, Simon and Schuster.

———, 1990: "The Magnetic-Field Menace," *Macworld*, July (Los Angeles, PCW Publications).

California Department of Health Services, 1988: *Art & Craft Materials Acceptable for Kindergarten and Grades 1–6*, Sacramento, CDHS.

California Public Interest Research Group, 1984: *Not a Pretty Picture: Art Hazards in California Public Schools*, Los Angeles, CalPIRG.

Caplan, Ruth, 1990: *Our Earth, Ourselves*, New York, Bantam.

Cardamone, Tom, 1981: *Advertising Agency and Studio Skills*, (3d edition), New York, Watson-Guptill.

Carnow, Bertram W., 1976: "Health Hazards in the Arts." *American Lung Association Bulletin*, January–February.

Carson, Rachel, 1962: *Silent Spring*, Boston, Houghton Mifflin.

Center for Hazardous Materials Research, 1989: *Hazardous Waste Minimization Manual for Small Quantity Generators*, Pittsburgh, Applied Research Center, University of Pittsburgh.

Center for Women in Government, 1989: *Ergonomics: Implementing Workplace Change*, Albany, State University of New York.

Clark, Nancy, Thomas Cutter, and Jean-Anne McGrane, 1984: *Ventilation, A Practical Guide*, New York, Nick Lyons.

Craig, James, 1980: *Designing with Type*, New York, Watson Guptill.

Cullen, Mark R., Thomas Rado, James A. Waldron, Judith Sparer, and Laura S. Welch, 1983: "Bone Marrow Injury in Lithographers Exposed to Glycol Ethers and Organic Solvents Used in Multicolor Offset and Ultraviolet Curing Printing Processes," *Archives of Environmental Health*, 38:6.

Cunning, Sheril, 1982: *Handmade Paper: A Practical Guide to Oriental and Western Techniques*, Escondido, Calif., Hatrack.

Curtiss, Deborah, 1987: *Introduction to Visual Literacy: A Guide to Visual Arts and Communication*, Englewood Cliffs, N.J., Prentice-Hall.

Dadd, Debra Lynn, 1984: *Nontoxic and Natural*, Los Angeles, Jeremy P. Tarcher/St. Martins.

Danneberg, John E., 1989: *Green Circle Newsletter*, Pine Brook, N.J., Green Circle.

Doull, John, Curtis D. Klassen, and Mary O. Amdur, eds., 1980: *Toxicology*, New York, Macmillan.

Dreger, Marianne, ed., 1989: *ACOM Report* (monthly newsletter of the American College of Occupational Medicine, Arlington Heights, Ill.), March.

Duccilli, Steve, 1989: "Making a Splash (again): Water-based Textile Inks Stage a Comeback," *Screen Printing Magazine*, 79:9.

Earth Works Group, 1989: *50 Simple Things You Can Do to Save the Earth*, Berkeley, Calif., Earth Works.

Eastman Kodak Co., 1979: *Safe Handling of Photographic Chemicals*, Rochester, N.Y., Eastman.

———, 1987: *General Guidelines for Ventilating*, Rochester, N.Y., Eastman.

Eaton, George T., 1980: *Photographic Chemicals in Black and White and Color*, Dobbs Ferry, N.Y., Morgan and Morgan.

Fanning, Deborah M., 1989: "Read the Label!" Boston, ACMI (unpublished paper).

Fitzgerald, Karen, Indira Nair, and M. Granger Morgan, "Electromagnetic Fields: The Jury's Still Out," *IEEE Spectrum*, August 1990.

Foote, Richard T., 1977: "Health Hazards to Commercial Artists," *Job Safety and Health* (Washington, D.C., OSHA), November.

Francekevich, Al, 1964: "That Duplex Super Safelight: It's Like Working Outdoors!" *Popular Photography*.

GAIU: see Graphic Arts International Union.

Gerber, Frederick H., 1975: "The Investigative Method of Natural Dyeing," published by the author.

———, 1977: "Indigo and the Antiquity of Dyeing," published by the author.

———, 1978: "Cochineal and the Insect Dyes," published by the author.

Goldfrank, L. R., Flomenbaum, N. E., Lewin, N. A., Weisman, R. S., and Howland, M. A., 1990: *Goldfrank's Toxicologic Emergencies*, (4th edition), Norwalk, Conn., Appleton & Lang.

Goldhaber, M. K., M. R. Polen, R. A. Hiatt, 1988: "The Risk of Miscarriage and Birth Defects among Women Who Use Visual Display Terminals during Pregnancy," *American Journal of Industrial Medicine*, June.

Graphic Arts International Union, 1989: *Health & Safety Guide for the Graphic Arts Industry*, Washington, D.C., GAIU.

Greenpeace, 1990: *Stepping Lightly on the Earth: Everyone's Guide to Toxics in the Home*, Washington, D.C., Greenpeace.

Haddad, Lester M., and James F. Winchester, 1990: *Clinical Management of Poisoning and Drug Overdose*, (2nd edition), Philadelphia, Penn., W. B. Saunders.

*Industrial Ventilation: A Manual of Recommended Practice*, 19th edition, 1986: The Committee on Industrial Ventilation, ACGIH.

Jenkins, Cate, 1978: *Organic Pigments*, New York, Center for Safety in the Arts.

Jenkins, Catherine, 1978: "Textile Dyes are Potential Hazards," *Journal of Environmental Health*, 40:256.

Johnson, Lois, and Hester Stinnett, 1987: *Waterbased Inks: A Screenprinting Manual for Studio and Classroom*, Philadelphia, University of the Arts.

Johnson, Margaret K., and Dale K. Hilton, 1980: *Japanese Prints Today, Tradition with Innovation*, Tokyo, Shufunotomo.

Karlen, Arno, 1984: *Napoleon's Glands and Other Ventures in Biohistory*, New York, Little Brown/ Quantum Research Association.

Krenneck, Lynwood, 1988: *Shop Notes on Water Based Screenprinting*, Lubbock, Tex., Hogarth Mesa.

Loeffler, J. J., ed., 1984: *Industrial Ventilation*, Lansing, Mich., American Conference of Governmental Industrial Hygienists.

Magnus, Günter Hugo, 1986: *Graphic Techniques for Designers and Illustrators*, Woodbury, N.Y., Barron's.

Massachusetts Public Interest Research Group, 1983: *Poison Palettes: A Study of Toxic Art Supplies in Massachusetts Schools*, Boston, Mass., MassPIRG.

Mayer, Ralph, 1957: *The Artists' Handbook of Materials and Techniques*, New York, Viking.

McCann, Michael, ed., *Art Hazards News*, since 1978, New York, Center for Safety in the Arts.

———, 1978: "Impact of Hazards in Art on Female Workers," *Preventive Medicine*, 7:338.

———, 1979: *Artist Beware: The Hazards and Precautions in Working with Art and Craft Materials*, New York, Watson-Guptill.

———, 1985: *Health Hazards Manual for Artists* (3d ed.), New York, Nick Lyons.

———, 1989: Center for Safety in the Arts annual

"Health Hazards in the Arts," a course of study, June 5–9.

———, and Gail Barazani, 1980: *Health Hazards in the Arts and Crafts,* Proceedings of the Society for Occupational Safety and Health Conference on Health Hazards in the Arts and Crafts, Washington, D.C.

———, and Monona Rossol, 1981: *Health Hazards in the Arts and Crafts,* New York, Center for Safety in the Arts.

McGrane, Jean-Ann, 1987: *Reproductive Hazards in the Arts and Crafts,* New York, Center for Safety in the Arts.

Moret, Christina, 1989: "Waiting in the Wings: A Look at the State of Water-Based Graphics Inks," *Screen Printing Magazine,* 79:9.

Moses, Cherie, et al., 1978: *Health and Safety in Printmaking: A Manual for Printmakers,* Edmonton, Alberta, Occupational Hygiene Branch, Alberta Labor.

Nadel, Brian, 1988: *Understanding and Using Material Safety Data Sheets,* New York, Center for Safety in the Arts.

National Art Materials Trade Association: *The Safe and Successful Use of Art Materials,* videotape.

National Fire Protection Association, 1987: *Flammable and Combustible Liquids,* Quincy, Mass., NFPA.

Needleman, H. L., A. Schell, D. Bellinger, A. Leviton, and E. N. Allred, 1990: "The Long Term Effects of Exposure to Low Doses of Lead in Childhood: An Eleven Year Follow-up Report," *New England Journal of Medicine,* 322:83.

Neumark, Devora, 1985: "Water-Base Screen Printing," *Art Hazards News,* 8:5, 8:6.

New Jersey Department of Health, ongoing: *Hazardous Substance Fact Sheets,* Trenton, N.J., Right to Know Project.

New York Committee for Occupational Safety and Health, 1987: *The VDT Book,* New York.

Occupational Safety and Health Administration, 1989: *Air Contaminants: Permissible Exposure Limits* (OSHA 3112), Washington, D.C., OSHA, U.S. Department of Labor.

———, 1977: "Essentials of Machine Guarding," *Safe Work Practices* (OSHA 2227), Washington, D.C., OSHA, U.S. Department of Labor.

Olson, Kent R., ed., 1990: *Poisoning and Drug Overdose,* Norwalk, Conn., Appleton & Lang.

Ontario Lung Association, 1982: *Health Hazards in Arts and Crafts,* New York, American Lung Association.

Ottoboni, Alice, 1984: *The Dose Makes the Poison: A Plain-Language Guide to Toxicology,* Berkeley, Calif., Vincente. (This is an excellent, readable introduction to toxicology for the nontoxicologist.)

Peterdi, Gabor, 1980: *Printmaking,* rev. ed., New York, Macmillan.

Pinsky, Mark, 1987: *The VDT Book: A Computer User's Guide to Health and Safety,* New York, NYCOSH.

Proctor, Christine, 1987: *OSHA Hazard Communication Standard,* New York, Center for Safety in the Arts.

Qualley, Charles A., 1986: *Safety in the Artroom,* Worcester, Mass., Davis. (A guide for art teachers that includes many helpful suggestions.)

Rickard, Ted, and Ronald Angus, 1986: *A Personal Risk Assessment for Craftsmen and Artists,* Ontario Crafts Council and College, University and School Safety Council of Ontario. New York, Center for Safety in the Arts.

Rifkin, Jeremy, ed., 1990: *Green Lifestyle Handbook, 1001 Ways You Can Heal the Earth,* New York, Henry Holt.

Romano, Clare, and John Ross, 1980: *The Complete Collagraph,* New York, Macmillan/Free Press.

Ross, John, and Clare Romano, 1972: *The Complete Printmaker.* New York, Macmillan/Free Press.

———, 1990: *The Complete Printmaker,* rev. ed., New York, Macmillan/Free Press.

Rossol, Monona, 1982: *Air-Purifying Respirators,* New York, Center for Occupational Hazards.

———, 1984: *Dye Hazards and Precautions,* New York, Center for Occupational Hazards.

———, 1985: *Ceramics and Health,* New York, Arts, Crafts, and Theater Safety.

———, ed., *ACTS Facts,* since 1986, New York, Arts, Crafts, and Theater Safety.

———, 1987: "Art Painting," *Just Paint* (New Berlin, N.Y., Golden Artist Colors Newsletter) 1:1.

———, 1988: *Pigments Used in Paints and Inks,* N.Y., Arts, Crafts, and Theater Safety.

———, 1990: *The Artist's Complete Health and Safety Guide,* New York, Allworth.

Saff, Donald, and Deli Sacilotto, 1978: *Printmaking History and Process,* New York, Holt Rinehart and Winston.

Seeger, Nancy, 1982: *A Painter's Guide to the Safe Use of Materials,* Chicago, Art Institute of Chicago.

———, 1983: *A Photographer's Guide to the Safe Use of Materials,* Chicago, Art Institute of Chicago.

———, 1984: *A Printmaker's Guide to the Safe Use of Materials,* Chicago, Art Institute of Chicago.

Shaw, Susan, and Monona Rossol, 1991: *Overexposure: Health Hazards in Photography* (2nd edition), New York, Allworth.

———and Rossol, Monona, 1989: "Warning: Photography May Be Hazardous to Your Health," American Society of Magazine Photographers ASMP Bulletin, 8:6.

Stellman, Jeanne, and Mary Sue Henifin, 1989: *Office Work Can Be Dangerous to Your Health,* New York: Ballantine.

Sturge, John M., ed., 1977: *Neblette's Handbook of Photography and Reprography, Materials, Processes and Systems,* 7th edition, New York, Van Nostrand Reinhold.

Tell, Judy, ed., 1988: *Making Darkrooms Saferooms, a National Report on Occupational Health and Safety.* National Press Photographers Association, Durham, N.C.

Waller, Julian: *Safer Practices in the Arts and Crafts,* New York, College Art Association.